Wireless Telecom FAQs

McGraw-Hill Telecommunications

- Ali: *Digital Switching Systems*
- Ash: *Dynamic Routing in Telecommunications Networks*
- Azzam: *High-Speed Cable Modems*
- Azzam/Ransom: *Broadband Access Technologies*
- Bartlett: *Cable Communications*
- Bates: *Broadband Telecommunications Handbook*
- Bayer: *Computer Telephony Demystified*
- Bedell: *Cellular/PCS Management*
- Clayton: *McGraw-Hill Illustrated Telecom Dictionary*
- Davis: *ATM for Public Networks*
- Gallagher: *Mobile Telecommunications Networking with IS-41*
- Harte: *CDMA IS-95*
- Harte: *Cellular and PCS: The Big Picture*
- Harte: *GMS Superphones*
- Heldman: *Competitive Telecommunications*
- Lachs: *Fiber Optics Communications*
- Lee: *Mobile Cellular Telecommunications*, 2/e
- Lee: *Mobile Communications Engineering*, 2/e
- Macario: *Cellular Radio*, 2/e
- Muller: *Desktop Encyclopedia of Telecommunications*
- Muller: *Desktop Encyclopedia of Voice and Data Networking*
- Muller: *Mobile Telecommunications Factbook*
- Pattan: *Satellite-Based Cellular Communications*
- Pecar: *Telecommunications Factbook*, 2/e
- Richharia: *Satellite Communications Systems*, 2/e
- Roddy: *Satellite Communications*, 2/e
- Rohde/Whitaker: *Communications Receivers*, 2/e
- Russell: *Signaling System #7,* 3/e
- Russell: *Telecommunications Pocket Reference*
- Russell: *Telecommunications Protocols*, 2/e
- Shepard: *Telecommunications Convergence*
- Simon: *Spread Spectrum Communications Handbook*
- Smith: *LMDS*
- Smith: *Practical Cellular and PCS Design*
- Smith/Gervelis: *Cellular System Design and Optimization*
- Turin: *Digital Transmission Systems*
- Winch: *Telecommunication Transmission Systems*, 2/e

Wireless Telecom FAQs

Clint Smith, P.E.

McGraw-Hill

New York • San Francisco • Washington, D.C. • Auckland • Bogotá
Caracas • Lisbon • London • Madrid • Mexico City • Milan
Montreal • New Delhi • San Juan • Singapore
Sydney • Tokyo • Toronto

Library of Congress Cataloging-in-Publication Data

Smith, Clint.
Wireless telecom FAQs / Clint Smith.
p. cm.
ISBN 0-07-134102-1
1. Wireless communication systems. I. Title.

TK5103.2.S6526 2000
621.3845—dc21 00-041820

McGraw-Hill

A Division of The ***McGraw-Hill*** *Companies*

1 2 3 4 5 6 7 8 9 0 DOC/DOC 0 6 5 4 3 2 1 0

ISBN 0-07-134102-1

The sponsoring editor for this book was Stephen S. Chapman, the editing supervisor was Stephen M. Smith, and the production supervisor was Sherri Souffrance. It was set in Sabon per the NM2 design by Joanne Morbit and Michele Pridmore of McGraw-Hill's Hightstown, N.J., Professional Book Group composition unit.

Printed and bound by R. R. Donnelley & Sons Company.

This book is printed on recycled, acid-free paper containing a minimum of 50% recycled, de-inked fiber.

To Sam, Rose, and Mary, who once again provided the support and patience that made this effort possible; and to all my many colleagues, who once again supplied much of the inspiration regarding the contents of this book.

Contents

Preface

The wireless industry continues to be an exciting place. The field supplies many opportunities on both the business and technical sides. I have had the fortune of working with very talented people who have made and continue to make this fast-paced industry enjoyable.

A sentiment I often hear expressed is, "I am never bored—just at times frustrated." The frustration that many people experience is obtaining the information needed to either make a decision or understand why a decision was made.

The knowledge needed by someone in marketing, technical, sales, real estate, or construction often spans multiple departments and therefore disciplines, making the acquisition of this material difficult due to logistics alone. The information needed by, say, real estate relative to a wireless system is not necessarily technology-dependent but often deals with implementation and time. Technical needs to have information related to the various aspects of construction and real estate to better understand these groups, which directly helps implement technical designs. Marketing and sales, of course, need to understand or have the ability to find out what all the jargon that is used so fluently means.

Therefore, when I was approached by McGraw-Hill to write a book that comprises frequently asked questions for wireless, the need was obvious. I believe it will provide many people, such as middle and upper management, sales and marketing, engineering, and new entrants into the wireless field, with a quick source of information on the wide variety of disciplines and of descriptions of the many important topics and issues encountered in the wireless industry.

The frequently asked questions are those that either colleagues or myself have been asked on numerous occasions, and each is followed by a brief and pointed response. I hope that this book will assist you in your job and career development.

Clint Smith, P.E.

Acronyms and Abbreviations

1FB	one-flat-rate business service
1FR	one flat rate
1MR	one measured rate
3/1/0	DS-3– or DS-1– or DS-0–compatible equipment
A link	access link
ac	alternating current
ACELP	algebraic code excited linear predictive [vocoder]
ADPCM	adaptive differential pulse code modulation
ADSL	asymmetric digital subscriber line
A&E	architectural and engineering
AE	architectural engineer
AGT	air-to-ground transmitter [system]
AIF	antenna interface frame
AIN	advanced intelligent network
AM	amplitude modulated; amplitude modulation
AMPS	Advanced Mobile Phone System
ANI	automatic number identification
ASK	amplitude shift keying
AUC	authentication center
AWG	American Wire Gauge
AWS	aerial wire service
B link	bridge link
B8ZS	binary 8 zero substitution
B911	basic 911 service
BER	bit error rate
BERT	bit-error-rate test
BH	busy hours
BPSK	bipolar phase shift keying
BSC	base site controller; base station controller
BSS	base station system
BTA	basic trading area
BTS	base transceiver station
BWM	broadcast warning message
BWS	buried-wire service
C link	control link
C7	common channel signaling 7
CCC	clear channel coding

CCITT	Consultative Committee on International Telegraphy and Telephony
CCS	centa call seconds
CCS7	common channel signaling 7
CDMA	code division multiple access [system]
CDPD	cellular data packet data
CDR	call detailed record
CGSA	cellular geographic service area; cellular geographic statistical area
C/I	carrier-to-interference [ratio]
CIC	caller identification code
CID	circuit identification [code]
CLEC	competitive local exchange network
CLLI	common language location identifier
CM	circuit merit
CN	change notice [Lucent]
C/N; CNR	carrier-to-noise [ratio]
CO	certificate of occupancy
COS	class of service
COW	cell on wheels
CP	circular polarization
CPE	customer premise equipment
CPFSK	continuous phase frequency shift keying
CPI	Consumer Price Index
CSU	channel circuit unit
CSU/DSU	channel service unit/data service unit
D link	diagonal (or quad) link
D-AMPS	Digital Advanced Mobile Phone System
DACS	digital-access cross-connect system
dc	direct current
DCA	dynamic channel allocation
DCC	digital color code
DCCH	digital control channel; dedicated control channel
DID	direct inward dialing
DMS	digital multiplex switch [DMS-100, Nortel]
DMX	digital mobile exchange
DOD	direct outward dialing [circuit]
DQPSK	differential quadrature phase shift keying
D/R	distance-to-radius [ratio]
DS	direct sequence
DS-0	digital signal, level zero

DS-1	digital signal, level 1
DS-1c	digital signal, level 1c
DSB	double sideband
DTC	digital control channel
DTMF	dual-tone multifrequency [signaling]
DXX	digital-access cross-connect system
E	electric [field; vertical field]
E link	extended link
E_b/E_o	energy per bit per noise energy
E-TACS	Extended Total Access Communication System [Extended TACS]
E911	enhanced 911 service
EIA	Electronic Industries Association
EIR	equipment identity register
EIRP	effective isotropic radiated power
E&M	ear and mouth
EMF	electromotive force
EMS	emergency medical services
E/N	energy-to-noise [ratio]
EPA	Environmental Protection Agency
ER	engineering recommendation
ERP	effective radiated power
ESF	extended superframe
ESMR	enhanced specialized mobile radio
ESN	electronic serial number
FAA	Federal Aviation Administration
FDMA	frequency division multiple access [system]
FER	frame erasure rate; frame error rate
FES	fixed end system
FFSK	fast frequency shift keying
FG	feature group
FM	frequency modulated; frequency modulation
FNE	fixed network equipment
FOA	first office application
FOM	figures of merit
FSK	frequency shift keying
FTP	file transfer protocol
GC	general contractor
GMSK	gaussian minimum shift keying
GOS	grade of service
GPS	Global Positioning Satellite [system]

GSM	Global System for Mobile Communications
H	magnetic [field; horizontal field]
HDSL	high-bit-rate subscriber line
HF	high-frequency
HH	hold harmless
HI	horizontal isolation
HLR	home location register
HO	handoff
IBP	in-building portable [unit]
ICP	in-car portable [unit]
iDEN	Integrated Dispatch Enhanced Network
IEC	interexchange carrier
IF	intermediate frequency
ILEC	local exchange network
IM; IMD	intermodulation distortion
IMP	intermodulation product
IMTSs	improved mobile telephone systems
IN	intelligent network
IP2	second-order intercept point
IP3	third-order intercept point
I&Q	intensity and quadrature
IS-3-D	interim standard for the AMPS
ISDN	integrated services digital network
ISUP	integrated services user part
ISWR	current standing-wave ratio
ITU	International Telecommunications Union
IXC	interexchange carrier
JTACS	Japan Total Access Communications System
LAC	linear amplifier
LAN	local-area network
LAPDm	link access procedure—D mobile
LATA	local-access transport area
LC	lost call
LD	long distance
LEC	local exchange network
LMDS	local multipoint distribution service
LNA	low-noise amplifier
LOS	line of sight
LPDA	log periodic dipole array
MAHO	mobile-assisted handoff
MDBS	mobile database system

MDIS	mobile database intermediate system
MDS	minimal discernible signal
MES	mobile end system
MF	multifrequency
MFC	multifrequency compelled
MFJ	modified final judgment
MIN	mobile identification number; model identification number
MMI	man-machine interface [commands]
MOP	method of procedure
MOS	mean opinion score
MSAs	metropolitan statistical areas; metropolitan service areas
MSC	mobile switching center
MSK	minimum shift keying
MTA	metropolitan trading area
MTBF	mean time between failures
MTS	mobile telephone system; mobile telephone service
MTSO	mobile telephone switching office
NACN	North American Cellular Network
NADC	North American Digital Cellular [system]
NAMPS	Narrow-Band Advanced Mobile Phone System
NMT	Nordic Mobile Telephone [standard]
NOC	network operation center
NPA	numbering plan area [code]
OC-1	optical carrier, level 1
OQPSK	offset quadrature phase shift keying
OSI	open-systems interconnection
OSS	operations and support system
O&Ts	originations and terminations
PA	power amplifier
PACS	personal access communication system
PBX	private branch exchange
PCCH	primary control channel [iDEN]
PCM	pulse code modulation
PCSs	personal communication services
PDC	Personal Digital Cellular [standard]
PF	power factor
Pi/4 DQPSK	Pi/4 differential quadrature phase shift keying
PIC	primary interexchange carrier
PIM	passive intermodulation
PL	private line

PM	phase modulation
PN	pseudo-random number
POI	point of interconnect
POP	point of presence
POTS	plain old telephone service
PRI	primary rate interface
PRS	primary reference source
PSAP	public safety answering point
PSK	phase shift keying
PSTN	Public Service Telephone Network; Public Switched Telephone Network
QAM	quadrature amplitude modulation
QOS	quality of service
QPSK	quadrature phase shift keying
RBOC	regional Bell operating company
RCC	recent change command
RELP	residual excited linear predictive [vocoder]
RF	radio frequency
RFI	radio frequency interference
RFP	request for proposal
RSAs	rural statistical areas; rural service areas
RSSI	received signal strength indication
Rx	receiver; receive path
SA	site acceptance
SAF	site acceptance form
SAT	supervisory audio tone
SCP	signaling control point
SF	superframe
SFDR	spurious free dynamic range
SI	slant-angle isolation
SID	system identification
SIM	subscriber identification module
SIT	special information tone
SMS	short-message service
S/N; SNR	signal-to-noise [ratio]
SONET	synchronous optical network
SOW	scope of work; switch on wheels
SP	signaling point
SQE	signal quality estimate
SQT	site qualification test

SS	switching system
SS7	switching system signaling 7
SSB	single sideband
SSP	signaling service points
STP	signaling transfer point
STS	synchronous transport signaling
SWR	standing-wave ratio
TACS	Total Access Communications System
TCAP	transactions capabilities application part
TCP/IP	Transmission Control Protocol/Internet Protocol
TDD	time division duplexing
TDMA	time division multiple access
TDR	time domain reflectometer
TI	tenant improvement
TIA	Telecommunication Industry Association
TMSI	temporary mobile station identifier
TN	true north
TTA	tower-top amplifier
Tx	transmitter; transmit path
UPS	uninterruptible power supply
VCELP	vector-sum excited linear predictive [vocoder]
VI	vertical isolation
VLR	visitor location register
VPN	virtual private network
VSB	vestibule sideband
WAN	wide-area network
WATS	wide-area telephone service
WLL	wireless local loop
WSP	wireless service provider

Introduction

The rapid expansion of wireless technology in the marketplace has resulted in a plethora of radio systems being released for use over the last few years. Consequently, many people with little or no practical experience with this type of telecommunications have been thrust into the wireless arena. The amount of information that needs to be assimilated by the current wireless workforce is daunting to say the least. Indeed, even within the wireless industry, the documentation needs are tremendous and often depend upon data distributed across multiple references and loose pieces of paper upon which are jotted the key facts pertaining to particular jobs.

This book attempts to bring into one place many of the issues that span cross functions between departments. It accomplishes this task by answering many of the more commonly asked questions pertaining to wireless communication.

The concept of cellular radio was initially developed by AT&T at its Bell Laboratories in an effort to provide additional radio capacity within geographic customer service areas. The initial mobile systems from which cellular evolved were called the *mobile telephone systems* (MTSs). Later, these systems were improved, and were then referred to as *improved mobile telephone systems* (IMTSs). One of the main problems with these systems was that a mobile call could not be transferred from one radio station to another without loss of communication. This problem was resolved with the reuse of the allocated frequencies of the system. Reusing the frequencies in cellular systems also increased a particular area's capacity for radio traffic, thus allowing more users to operate in that service area than the MTSs or IMTSs allowed.

The cellular systems are broken into *metropolitan statistical areas* (MSAs) and *rural statistical areas* (RSAs). Each MSA and RSA has two different cellular operators that offer service. The two cellular operators are referred to as *A-band* and *B-band* systems. The A-band system is the non-wire-line system, and the B-band is the wire-line system.

Personal communication services (PCSs) is the next generation of wireless communications. PCS is a general name given to wireless systems that have recently been developed out of the need for more capacity and design flexibility than that provided by the initial cellular systems. The similarities between PCS and cellular lie in the mobility of the user of the service. The differences between PCS and cellular fall into the applications and spectrum available for PCS operators to provide to the subscribers.

In the United States PCS operators obtained their spectrum through an action process set up by the Federal Communications Commission (FCC). The PCS band was broken into A, B, C, D, E, and F blocks. The A, B, and C blocks involve a total of 30 MHz while the D, E, and F blocks are allocated 10 MHz.

The spectrum allocation for both cellular and PCS in the United States is shown in Fig. 1.1. It should be noted that the geographic boundaries for PCS licenses are different from those imposed on cellular operators in the United States. Specifically, PCS licenses are defined as MTA (metropolitan trading area) and BTA (basic trading area). The MTA has several BTAs within its geographic region. In all, there are 93 MTAs and 487 BTAs defined in the United States. Therefore, there are 186 MTA licens-

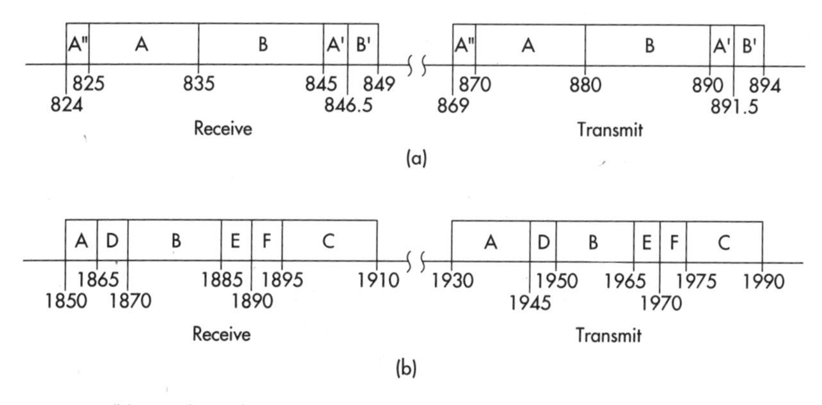

Figure 1.1 Cellular (*a*) and personal communication service (PCS) (*b*) allocation in the United States.

es awarded for the construction of a PCS network, each with 30 MHz of spectrum to utilize. In addition, 1948 BTA licenses were awarded in the United States. Of the BTA licenses, the C band will have 30 MHz of spectrum while the D, E, and F blocks will have only 10 MHz available.

Numerous types of cellular systems are used in the United States and elsewhere. The following is a brief list of some of the more common cellular systems:

1. The *Advanced Mobile Phone System* (AMPS) is the cellular standard that was developed for use in North America. This type of system operates in the 800-MHz frequency band. AMPSs have also been deployed in South America, Asia, and Russia.

2. The *Code Division Multiple Access* (CDMA) standard is an alternative digital cellular standard that was developed in the United States. The CDMA system utilizes the IS-95 standard and has been implemented as the next generation for cellular systems. The CDMA system coexists with the current analog system.

3. The *Digital Advanced Mobile Phone System* (D-AMPS), also called the North American Digital Cellular (NADC) system, is the digital standard for cellular systems developed for use in the United States. Rather than develop a completely new standard, the AMPS standard was developed into the D-AMPS digital standard. This was done to quickly provide a means to expand the existing analog systems that

were growing at a rapid pace. The NADC system has been designed to coexist with current cellular systems and relies on both the IS-54 and IS-136 standards.

4. The *Global System for Mobile Communications* (GSM) is the European standard for digital cellular systems operating in the 900-MHz band. This technology was developed out of the need for increased service capacity due to the analog systems' limited growth. This technology offers international roaming, high speech quality, and increased security, and advanced systems features can be developed for it. The development of this technology was completed by a consortium of 80 pan-European countries working together to provide integrated cellular systems across different borders and cultures.

5. The *Nordic Mobile Telephone* (NMT) standard was developed by the Nordic countries of Sweden, Denmark, Finland, and Norway in 1981. This type of system was designed to operate in the 450- and 900-MHz frequency bands. These are noted as NMT 450 and NMT 900. NMT systems have also been deployed throughout Europe, Asia, and Australia.

6. The *Total Access Communications System* (TACS) is a cellular standard that was derived from the AMPS technology. TACSs operate in both the 800- and 900-MHz bands. The first system of this kind was implemented in England. Later these systems were installed in Europe, China, Hong Kong, Singapore, and the Middle East. A variation of this standard, JTACS, was implemented in Japan.

7. The *Integrated Dispatch Enhanced Network* (iDEN) is the name for an alternative form of cellular communication that operates in the specialized mobile radio (SMR) band just adjacent to the cellular frequency band. The iDEN is a blend of wireless interconnect and dispatch services, which makes it very unique as compared to existing cellular and PCS systems. The iDEN utilizes a digital radio format called *QAM*, and, with the exception of the radio link, it is a derivative of the GSM.

There are several PCS systems that an operator can possibly utilize, and some operators are actually using several systems within their market in order to capture market share. Some of the more common PCS systems are listed below for quick reference:

1. The *DCS-1800* is a digital standard based upon the GSM technology except that this type of system operates at a higher frequency range, 1800 MHz. The DCS-1800 technology is intended for use in PCS systems. Systems of this type have been installed in Germany and England.

2. The *PCS-1900* is the same as the DCS-1800 and is a GSM system. The only difference between the PCS-1900, also called *DCS-1900,* and the DCS-1800 and GSM is the frequency band of operation. The PCS-1900 operates in the PCS frequency band for the United States, 1900 MHz.

3. The *Personal Digital Cellular* (PDC) system is a digital cellular standard developed by Japan. PDC systems were designed to operate in the 800-MHz band and in the 1.5-GHz band.

4. The *IS-661* is the technology platform that is being promoted by Omnipoint. It is a spread-spectrum technology that relies on *time division duplexing* (TDD).

5. The *IS-136* is the PCS standard that relies on the NADC system except that it operates in the 1900-MHz band.

6. The *CDMA* is another popular PCS platform, and it utilizes the same standard as does the CDMA for cellular systems except that it too operates in the 1900-MHz band.

Generic System Configuration

A generic system configuration is shown in Fig. 1.2. The configuration shown in the figure involves all the high-level system blocks of a cellular or PCS network. There are many components within each of the blocks shown.

In the figure, the scenario for originating or terminating a call is approximately the same regardless of the technology platform or whether it is a PCS or cellular system. In all cases the mobile unit communicates to the cell site through use of radio transmissions. It is the type of radio

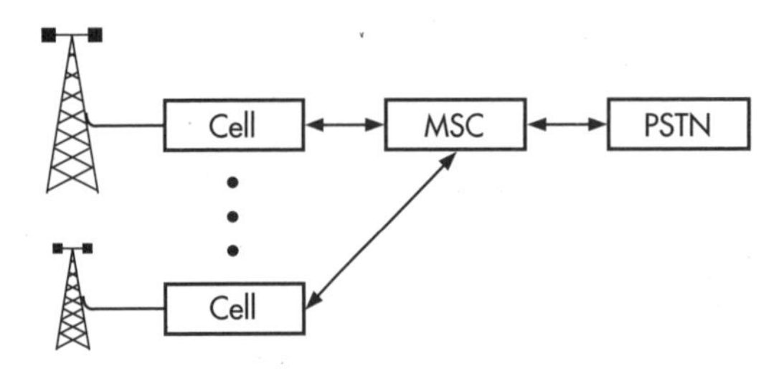

Figure 1.2 Generic wireless system. MSC = mobile switching center. PSTN = Public Switched Telephone Network.

transmission used that differentiates cellular and PCS systems.

The cell site acts as a conduit for the information transfer by which the radio energy is converted into another medium. The cell site sends and receives information from the mobile unit and the *mobile switching center* (MSC). The MSC is connected to the cell site by either leased T-1 lines or through a microwave system. The wireless system is made up of many cell sites, which are all interconnected back to the MSC.

The MSC processes the call and connects the cell site radio link to the *Public Switched Telephone Network* (PSTN). The MSC performs a variety of functions involved with call processing, and it is effectively the brains of the network. The MSC maintains subscriber data such as the individual subscriber records, the current status of the subscribers, and call routing and billing information.

Generic Cell Site Configuration

Figure 1.3 is an example of a generic cell site configuration which can be applied to any one of the various technology platforms for cellular and PCS systems. The cell site configuration shown in the figure is a picture of a monopole cell site. The monopole cell site has an equipment hut connected with it that houses the radio transmission equipment. The monopole, which is next to the equipment hut, supports the antennas used for the cell site at the very top of the monopole. The cable tray between the equipment hut and the monopole supports the coaxial cables that connect the antennas to the radio transmission equipment.

The radio transmission equipment used for a cellular base station, located in the equipment room, is shown in Fig. 1.4. The equipment room layout shown is a typical arrangement in a cell site. The cell site

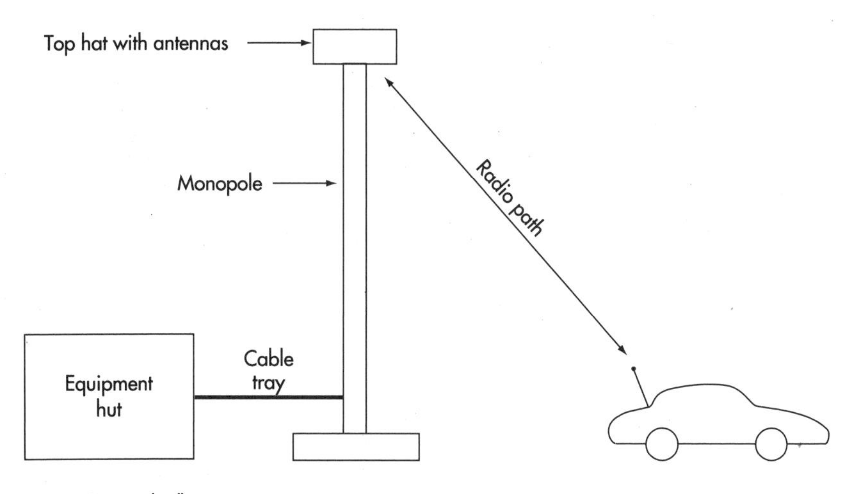

Figure 1.3 Monopole cell site.

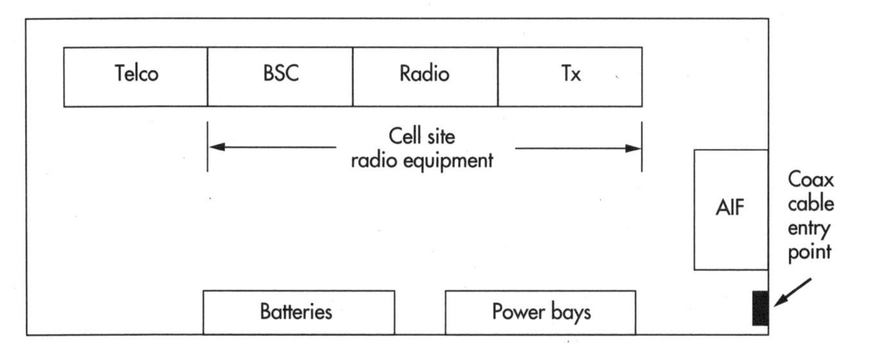

Figure 1.4 Radio transmission equipment for a cellular base station. BSC = base site controller. Tx = amplifier. AIF = antenna interface frame. Telco = T-1, microwave interconnect.

radio equipment consists of a *base site controller* (BSC), a *radio bay,* and the *amplifier* (Tx) *bay.* The cell site radio equipment is connected to the *antenna interface frame* (AIF), which provides the receiver and transmitter filtering. The AIF is then connected to the antennas on the monopole through use of the coaxial cables located next to the AIF bay.

The cell site is also connected to the MSC through the *telco bay.* The telco bay either provides the T-1 leased line or microwave radio link connection. The power for the cell site is secured through use of *power bays,* or *rectifiers,* that convert ac electricity to dc. Batteries are used in the cell site in the event of a power disruption to ensure that the cell site continues to operate until power is restored or the batteries are exhausted or replaced.

Call Setup Scenarios

There are several call scenarios that are used for all wireless systems. There currently are a few perturbations of the call scenarios discussed here, driven largely by fraud prevention techniques employed by individual operators. Moreover, there are numerous algorithms utilized throughout the call setup and processing scenarios that are not included in the diagrams presented here. However, the call scenarios selected to be shown here provide the fundamental building blocks for all call scenarios utilized in wireless communication systems. The call scenarios here are based on the AMPS system. All the other wireless systems currently in use today are improved versions of the original AMPS system. Therefore, the process for call setup is fundamentally the same for all wireless systems, and understanding one will facilitate the understanding of all the other technology platforms. (See Figs. 1.5, 1.6, and 1.7.)

Handoffs

The handoff concept is one of the fundamental principles used in wireless telephony. Handoffs enable cellular and PCS subscribers to operate at lower power levels, and they provide high capacity through facilitating frequency reuse as well as improving the amount of talk time allowed for a subscriber unit between battery charges. There is a multitude of algorithms that are invoked for the generation and processing of a handoff request and eventual handoff order. The individual algorithms are dependent upon the individual vendor for the network infrastructure and the software loads utilized.

Handing off from cell to cell is fundamentally the process of transferring the mobile unit that has a call in progress on a particular voice channel or time slot to another voice channel or time slot, all without interrupting the call. Handoffs can occur between adjacent cells or sectors of the same cell site. The actual need for a handoff is determined by

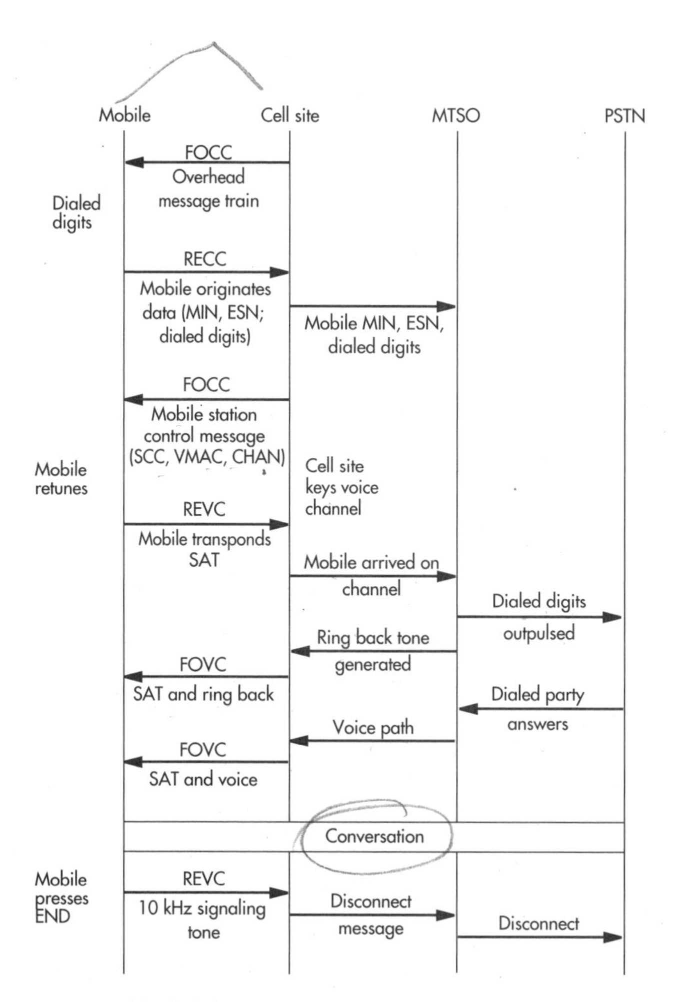

Figure 1.5 Mobile to land call.

the actual quality of the RF signal received from the mobile unit into the cell site or that received by the mobile unit itself. The decision for determining the handoff is driven by the technology utilized by the wireless operator. As the mobile unit transverses the wireless network, it is handed off from one cell site to another cell site, ensuring a high-quality call is maintained for the duration of the conversation.

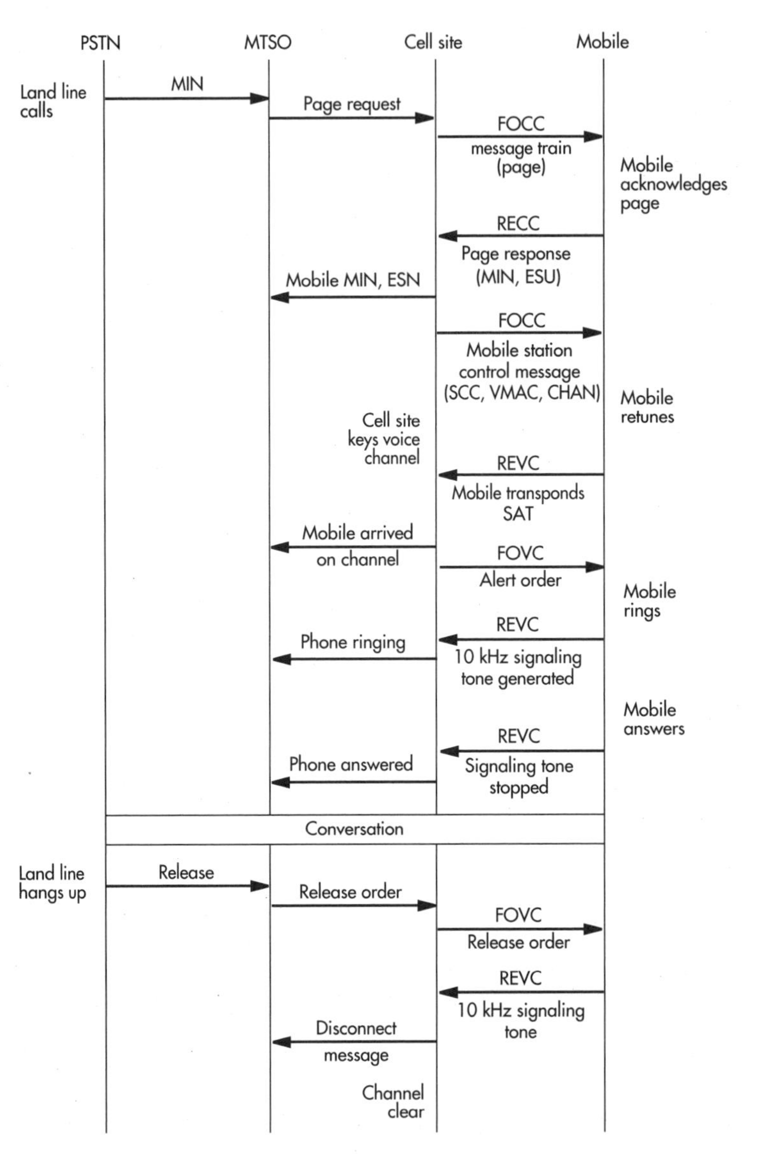

Figure 1.6 Land to mobile call.

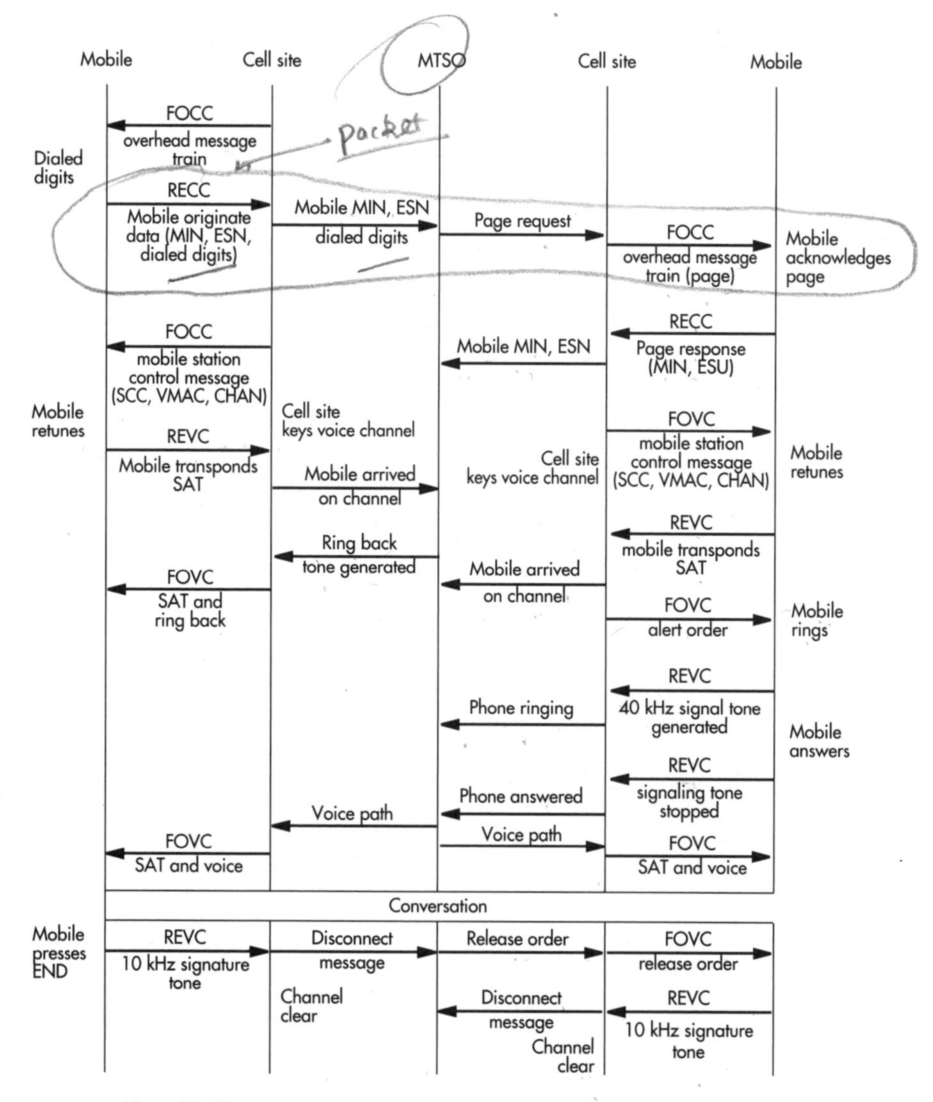

Figure 1.7 Mobile to mobile call.

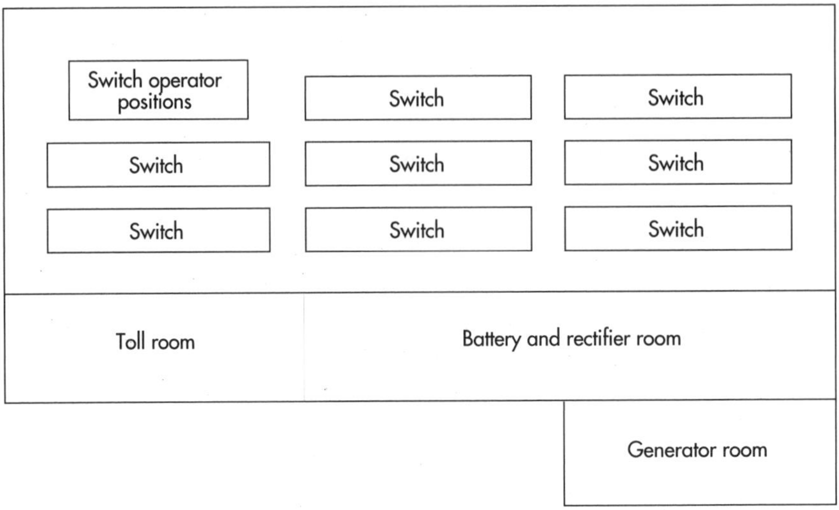

Figure 1.8 Generic MTSO configuration.

Generic MTSO Configuration

Figure 1.8 is a generic configuration for a *mobile telephone switching office* (MTSO), also called a *mobile switching center* (MSC). The MSC is the portion of the network that interfaces the radio world to the public telephone network. In mature systems there are often multiple MSC locations, and each MSC can have several cellular switches located within each building.

2

Technologies

This chapter addresses the multitude of technologies that are utilized in the wireless industry. The topics covered are meant to provide a general overview of the common questions that are often brought up regarding each of these exciting technological capabilities. Also, tables are included in which some of the technologies are compared. The comparison describes only the technical differences; it does not compare how well one performs against another. It must be stressed that every technology available has its advantages and disadvantages. It is through knowing the advantages and disadvantages and comparing them against the system services you need that you can make a rational decision as to which technology or technology platforms would work well for your system.

Q. *What is cellular communication?*

A. *Cellular communication* is a form of wireless communication that enables several key concepts to be employed:

- Frequency reuse
- Mobility of the subscriber
- Handoffs

The cellular concept is employed in many different forms. The term *cellular communication* is typically applied to either the AMPS or TACS technology. The AMPS operating frequency is in the 800-MHz band: 821 to 849 MHz for the base station receiver and 869 to 894 MHz for the base station transmitter. For TACS the frequency range is between 890 MHz and 915 MHz for the base receiver and 935 MHz and 960 MHz for the base station transmitter.

Many other technologies also fall under the guise of cellular communication, including those involving the PCS bands—both the domestic and the international bands. In addition, because the same concept is applied to several technology platforms that are currently used in the SMR band (IS-136 and iDEN), the term *cellular communication* sometimes is used to describe these as well. However, *cellular communication* usually refers specifically to the AMPS and TACS bands.

Q. *What is a handoff?*

A. The term *handoff* typically refers to the process whereby the wireless subscriber's communication link is transferred from one base station to another base station. The type of handoff used depends on several factors. However, all handoff types employ the same concept in that the subscriber's communication link is transferred from one cell site to another without the communication link being permanently disrupted. An example of how a handoff takes place is shown in Fig. 2.1.

The decision process utilized for determining the type of handoff and the execution of the handoff proceed differently for each technology utilized. The example in Fig. 2.1 is for an AMPS. Note that although subsequent technologies have attempted to improve upon it, the AMPS process remains the most commonly used handoff procedure.

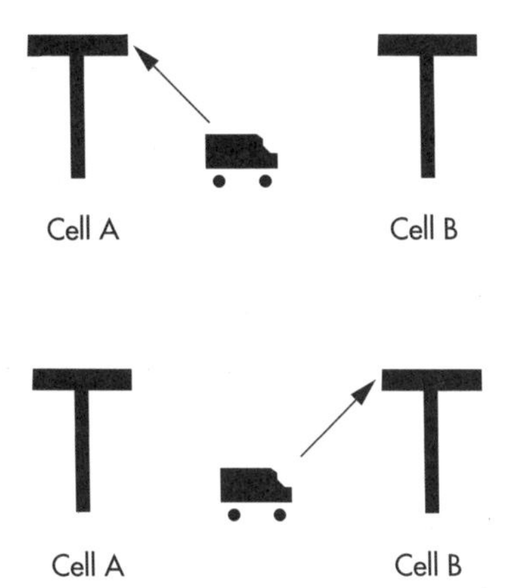

Figure 2.1 Handoff from cell A to cell B.

Q. *What is an SAT?*

A. A *supervisory audio tone* (SAT) is a series of tones that is utilized in an AMPS or TACS for identification of the mobile unit. An SAT is similar in concept to a private line (PL) tone found in a two-way system, and a PL is used for the same purpose except that in two-way systems there are no handoff or other enhanced features as found in AMPSs and TACSs.

An SAT consists of three frequencies, each spaced 30 Hz apart from each other. Each SAT frequency has a corresponding code that is referenced differently depending on the switching manufacturer being used. The table below is a quick reference for SATs:

Frequency, Hz	Code	Code, alt
5970	0	1
6000	1	2
6030	2	3

In the table, the alternative codes are given because in many cases the datafill for the cell site is populated using a 1-through-3 numbering scheme. The specification for air communication uses the 0-through-2 numbering scheme.

Q. *What does the term analog mean?*

A. The term *analog* has many meanings in wireless technology.

Analog communication refers to any communication that does not utilize a digital modulation format to convey its information. For example, a form of analog communication is the *amplitude-modulated* (AM) or the *frequency-modulated* (FM) station that you listen to in your vehicle or home.

In general, the term *analog* usually refers to an FM signal. For a cellular communication system, when a person refers to an *analog channel,* he or she means the 30-kHz AMPS channel that was used prior to the advent of digital radio platforms and that continues to be used today.

For an SMR system, the term *analog* refers specifically to a two-way push-to-talk environment.

It must be stressed that although *analog* may be used in reference to cellular and SMR systems, these systems rely on digital modulation for conveying control and subscriber information. Therefore, when a person refers to a system as "analog," he or she means the voice communication portion only.

Q. *What is an analog radio?*

A. An *analog radio* is any radio, whether a transmitter or receiver, that modulates and demodulates voice communication. The modulation used is either FM or AM; however, the typical method used in wireless communication for an analog radio is FM.

In an analog cellular system, the voice communication is digitized within the cell site itself for transport along the T carrier to the MTSO. The voice representation and information transfer utilized in AMPS cellular is analog, and it is this part in the communication link that engineers are currently trying to convert to a digital platform.

An example of an analog radio is shown in Fig. 2.2 for quick reference.

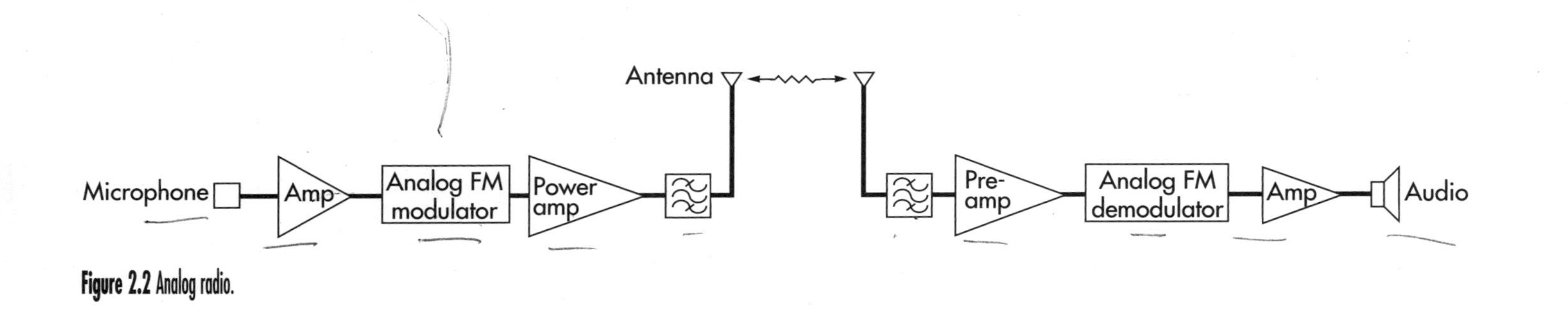

Figure 2.2 Analog radio.

Q. *What does the term digital mean?*

A. Like *analog*, the term *digital* has many meanings in the context of wireless technology.

Digital communication refers to any communication in a modulation format that relies on some type of data format for transmission. More specifically, *digital communication* refers to the process whereby the sending location digitizes the voice communication and then modulates it. At the receiver, the exact opposite is done.

Q. *Why use digital?*

A. This is a question that many operators, new and current, have had to wrestle with. Digital radio technology is deployed in cellular and PCS or SMR systems primarily to increase the quality and capacity of the wireless system over its analog counterpart.

The use of digital technology is meant to take advantage of many features and techniques that are currently not obtainable for analog cellular communication. Several competing digital technologies are presently being used, and each has advantages and disadvantages in particular applications. The major benefits associated with utilizing digital radios for a cellular or PCS environment are the following:

- Increased capacity
- Reduced capital infrastructure costs
- Reduced capital-per-subscriber cost
- Reduced cellular fraud
- Improved features
- Improvement in customer-perceived performance
- Encryption

Q. *What is a radio system?*

A. The term *radio system* has many meanings depending upon the context in which it is used. However, all radio systems are one of three types: simplex, duplex, or full duplex. The particular appli-

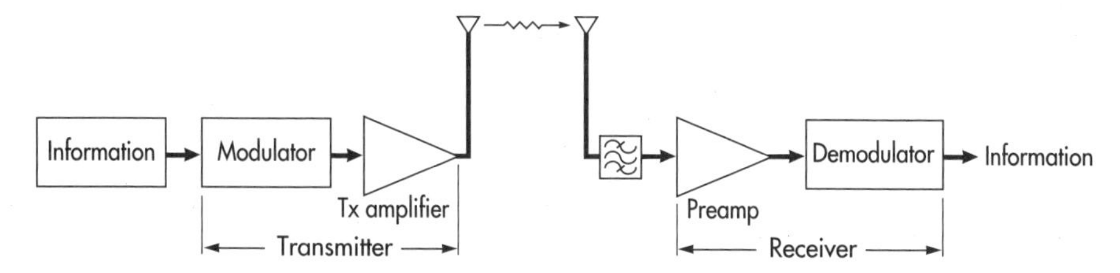

Figure 2.3 Basic radio system.

cation that a designer is trying to solve dictates the type of system that will be used and the available options. A basic radio system is shown in Fig. 2.3.

Q. *What is a simplex radio system?*

A. In a *simplex radio system,* both radio links operate on the same frequency for transmitting and receiving the information. This system is typically used for point-to-point communication equipment such as walkie-talkies. The system is very basic and relies on only one party in the communication link transmitting at any time. There can be multiple users on this type of system.

Q. *What is a duplex radio system?*

A. In a *duplex radio system,* both radio links utilize two separate frequencies for transmitting and receiving. Thus, in a duplex environment, unit 1 would send information to unit 2 on frequency F1 while unit 2 would send information to unit 1 on frequency F2.

The chief difference between half- and full-duplex systems is how the information between the two units is transferred. For example, in a *full-duplex system,* both unit 1 and unit 2 can be sending and receiving information at the same time without impeding each other. In contrast, in a *half-duplex system,* the operation is similar to that of a simplex system with the primary difference being that two discrete frequencies are used instead of just one.

Q. *What are the fundamental building blocks of a communication system?*

A. The fundamental building blocks for a communication system are effectively the same for all technology platforms. The major components consist of an antenna, filters, receivers, and transmit-

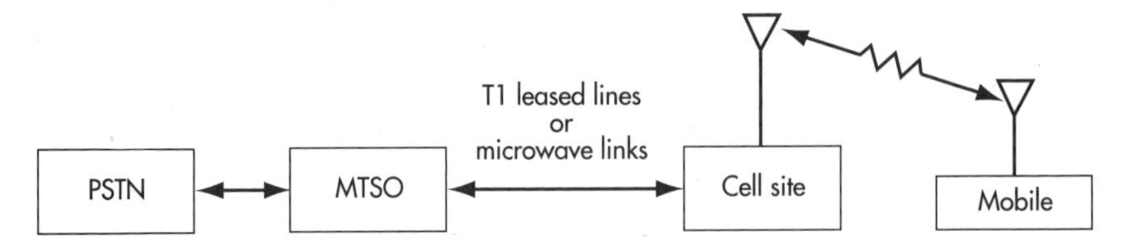

Figure 2.4 Simplified cellular system.

ters, and components that accomplish modulation, demodulation, and propagation. (See Fig. 2.4.)

Q. *What is the difference between digital and analog radios?*

A. The basic designs of a digital radio and an analog radio are shown in the block diagrams in Fig. 2.5. The diagrams can represent either the base station or subscriber unit for a variety of radio service applications, namely, cellular or PCS.

As can be seen in the diagram, for a digital radio, the initial information content, usually voice, is input into the microphone of the transmission section. The speech then is processed in a *vocoder,* which converts the audio information into a data stream utilizing a coding scheme to minimize the number of data bits required to represent the audio. The digitized data then goes to a *channel coder,* which takes the vocoder data and encodes the information even more so it will be possible for the receiver to reconstruct the desired message. The channel-coded information is then modulated onto an RF carrier utilizing one of several modulation formats covered previously in this chapter. The modulated RF carrier is then amplified, after which it passes through a filter and is transmitted out an antenna.

The receiver, at some distance away from the transmitter or placed at the same location as another transmitter, receives the modulated RF carrier via the antenna, which then passes the information through a filter and into a preamplifier. The modulated RF carrier is then down converted in the digital demodulator section of the receiver to an appropriate intermediate frequency. The demodulated information is then sent to a channel decoder, which performs the inverse of the channel coder in the

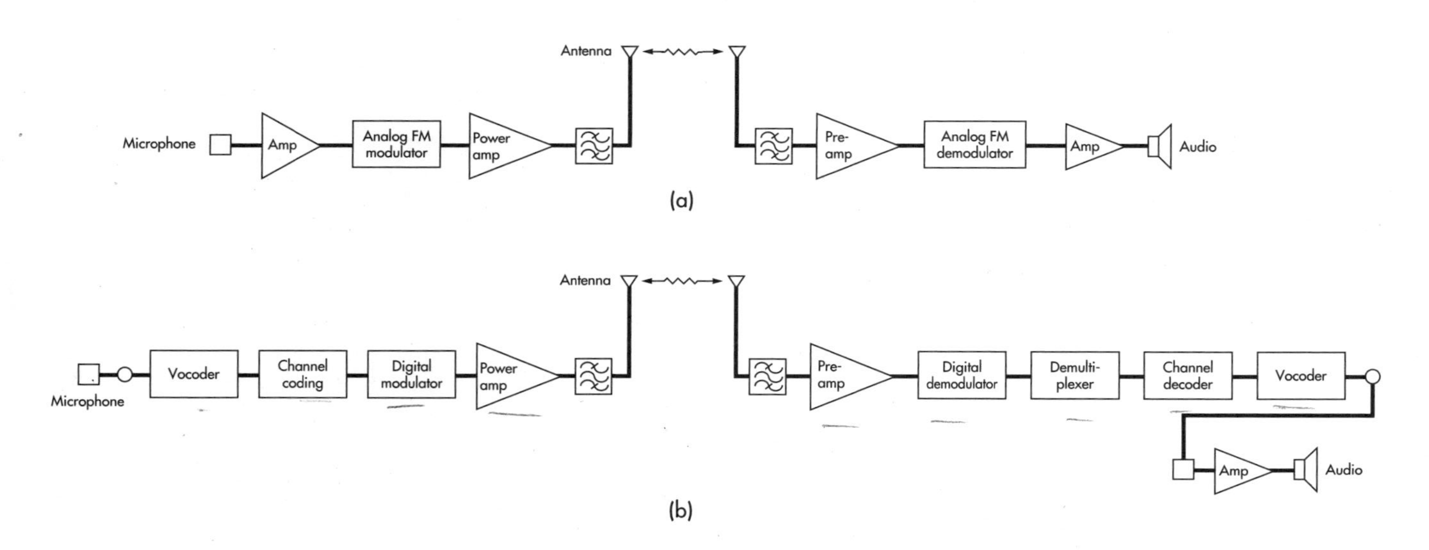

Figure 2.5 Analog (*a*) and digital (*b*) radio.

transmitter. The digital information is then sent to a vocoder for voice information reconstruction. The vocoder converts the information in digital format into an analog format, from which it is passed to an audio amplifier connected to a speaker through which the user at the other end of the communication path may listen to the message sent.

Q. *What is the E-TACS?*

A. The *Extended TACS* (E-TACS) is the Total Access Communication System that has been enlarged to include the additional frequencies later allotted by the FCC for use in cellular systems.

Q. *What is the NMT?*

A. The *Nordic Mobile Telephone* (NMT) *standard* is the cellular standard that was developed by the Nordic countries of Sweden, Denmark, Finland, and Normandy in 1981. This type of system was designed to operate in the 450- and 900-MHz frequency bands. These are noted as "NMT 450" and "NMT 900." NMT systems have also been deployed throughout Europe, Asia, and Australia.

Q. *What is a PCS?*

A. *Personal communication services* (PCSs) is a general name given to systems that have recently been developed out of the need for more capacity and design flexibility than that provided by the initial cellular systems.

Personal communication services is the next generation of wireless communications. The similarities between PCS and cellular lie in the mobility of the user of the service. The differences between PCS and cellular are in the applications for the systems and the spectrums available for their use.

The PCS spectrum in the United States is being made available through an action process set up by the FCC. (See Fig. 2.6.)

The geographic boundaries for PCS licensing are different from those imposed on cellular operators in the United States.

1850–1865	1865–1870	1870–1885	1885–1890	1890–1895	1895–1910
A (MTA)	D (BTA)	B (MTA)	E (BTA)	F (BTA)	C (MTA)

1930–1945	1945–1950	1950–1965	1965–1970	1970–1975	1975–1990
A (MTA)	D (BTA)	B (MTA)	E (BTA)	F (BTA)	C (MTA)

Figure 2.6 PCS allocation. (All frequencies in MHz.)

Specifically, PCS licenses are defined as MTA and BTAs. The MTA has several BTAs within its geographic region. There are 93 MTAs and 487 BTAs defined in the United States. Therefore, there are 186 MTA licenses available for the construction of PCS networks, and each has 30 MHz of spectrum to utilize. In addition, there were 1948 BTA licenses awarded in the United States. Of the BTA licenses, the C band will have 30 MHz of spectrum while the D, E, and F blocks will only have 10 MHz of spectrum.

Currently, there is no single standard upon which PCS operators may pick a technology platform for their networks. The choice of PCS standards is daunting, and each has its advantages and disadvantages associated with it. The current philosophy in the United States is to let the market decide which standard or standards are the best. This is significantly different from the manner in which cellular was developed, whereby every operator had a single preestablished interface for the analog system from which to operate.

A few major standards, however, have been picked by the licensees for the A-band and B-band PCS operators. The standards so far selected for PCSs are DCS-1900, IS-95, IS-661, and IS-136. The DCS-1900 utilizes a GSM format and is an upbanded DCS-1800 system. The IS-95 is the CDMA standard that will be utilized by cellular operators, except that it will be upbanded to the PCS spectrum. The IS-661 is a time-division-duplex system offered by Omnipoint Communications. The IS-136 standard is an upbanded cellular TDMA system that is currently being used by cellular operators.

Q. *What is the difference between narrow-band PCS and wideband PCS technologies?*

A. Personal communication services in the United States involve both narrow-band and wideband technologies. The narrow-band technologies are generally used in paging applications involving a combination of one-way and two-way paging. The wideband technologies that are currently being utilized and further developed involve IS-136, IS-95, CDS-1900, PACS (personal access communication system), and IS-661 standards, to mention a few. The chief differentiation between narrow-band and wideband technologies is simply the amount of spectrum that was auctioned off by the FCC to be used in PCS communications.

Q. *What is the PDC?*

A. The *Personal Digital Cellular* (PDC) *standard* was developed by the Japanese. PDC systems were designed to operate in the 800-MHz band and in the 1.5-GHz band.

Q. *What is the AMPS?*

A. The *Advanced Mobile Phone System* (AMPS) *standard* was developed for cellular use in North America, but it is now used also in South America, Asia, and Russia. This type of system operates in the 800-MHz frequency band. A typical AMPS is shown in Fig. 2.7 for reference.

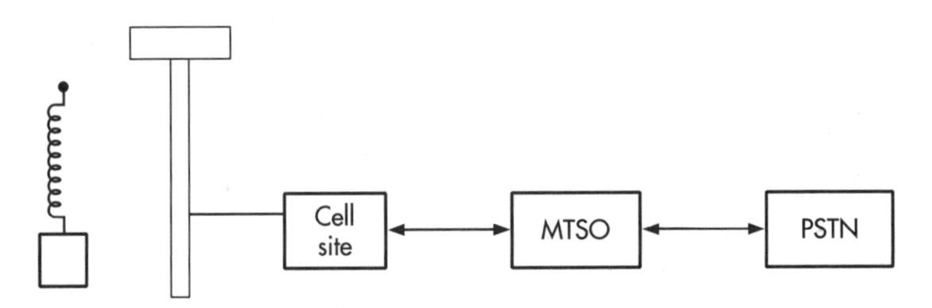

Figure 2.7 Typical cellular system.

Q. *What is the iDEN?*

A. The *iDEN system* was developed to enable the integration of several mobile phone technologies. The services that are integrated into iDEN involve dispatch, full-duplex telephone interconnect, data transport, and short messaging services.

The *dispatch system* incorporates a feature called *group call,* by which many people can engage in a conference. A user list can be preprogrammed, and a conference call can be set up just as it is done in two-way or SMR systems except that the connection can take place utilizing any of the frequencies from the pool of channels available where the subscriber is physically located.

The *telephone interconnect* and *data transport* are designed to offer conventional mobile communications. The *short messaging service* enables the iDEN phones to receive up to 140 characters for an alphanumeric message. A typical iDEN system is shown in Fig. 2.8.

Q. *What is the GSM?*

A. The *Global System for Mobile Communications* (GSM) is the European standard for digital cellular systems operating in the 900-MHz band. This technology was developed out of the need for increased service capacity due to the analog systems' limited growth. This technology offers international roaming, high-quality speech, increased security, and the capability for enhancement with as-yet-to-be-designed advanced systems features. The development of this technology was completed by a consortium of 80 pan-European countries working together to provide integrated cellular systems for use across different borders and cultures.

The GSM has achieved worldwide success. It has many features and attributes that make it an excellent digital radio standard to utilize. The GSM has the unique advantage of being the most widely accepted radio communication standard at this time.

The GSM consists of the following major building blocks: the *switching system* (SS), the *base station system* (BSS), and the *operations and support system* (OSS). (See Fig. 2.9.)

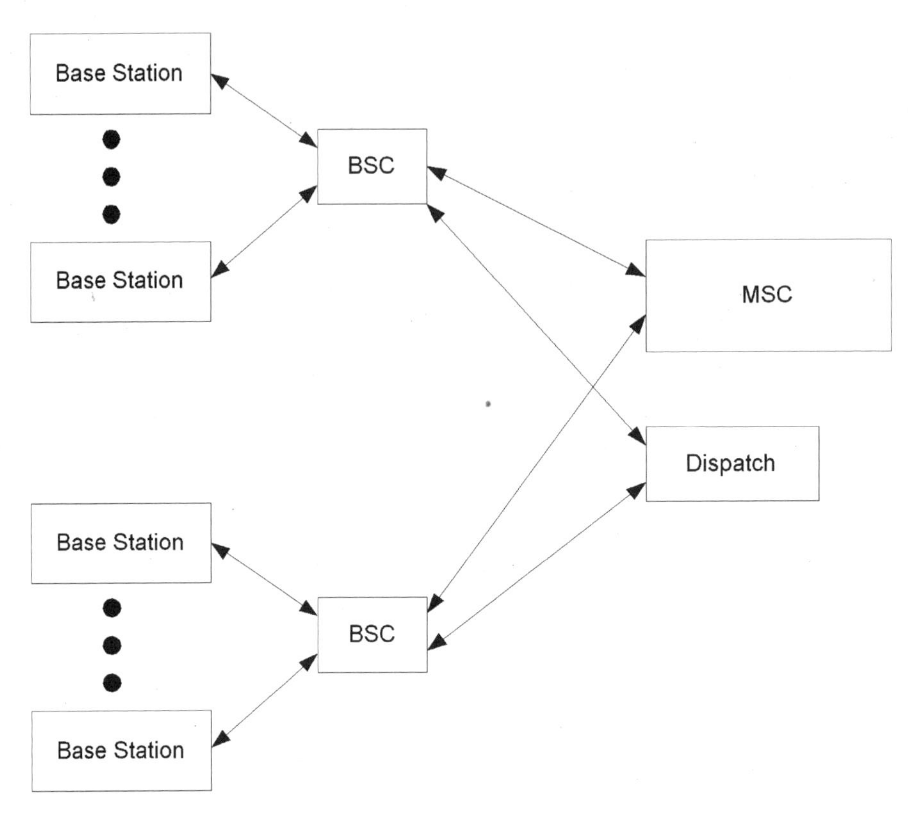

Figure 2.8 iDEN system.

The GSM shown in the figure has several subcomponents associated with each of the three major elements. The base station system comprises both the base station controller (BSC) and the base transceiver stations (BTS). In an ordinary configuration, several BTSs are connected to a BSC, and then several BSCs are connected to the MSC.

The BTS is the radio base station that contains the physical base station radios that are used to communicate with the mobile subscriber units. The BTSs are arranged in a star configuration. The BTSs are connected to the BSC via a physical link referred to as the *Abis link*.

The BSC is a high-capacity switch that is responsible for all the radio-related functions such as the management of the radio

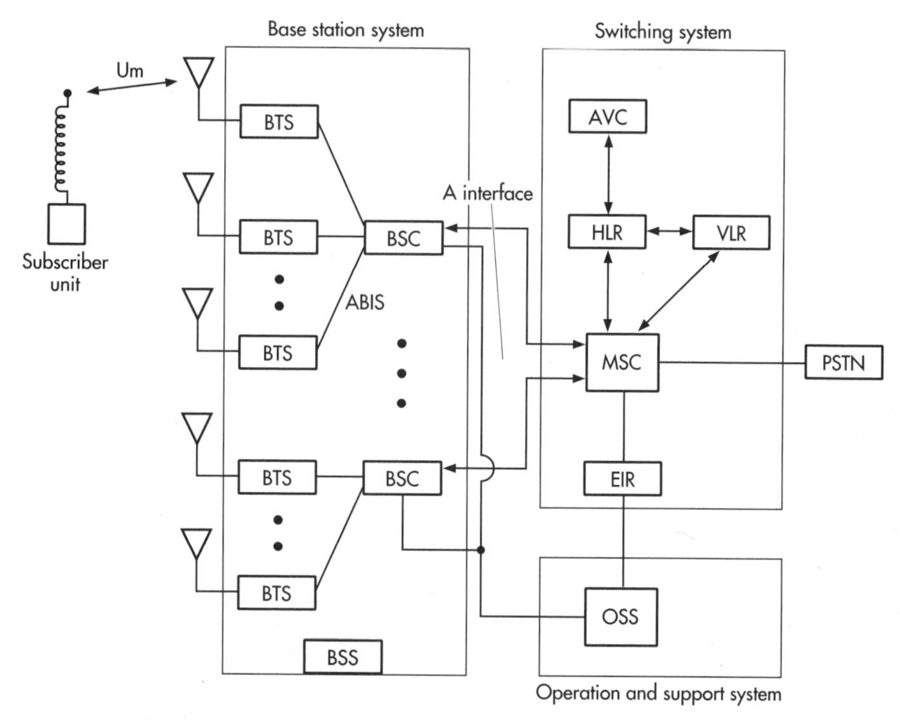

Figure 2.9 Global System for Mobile Communications.

resources, BTS parameters, and handover functions. The BSC can either be centrally located with the MSC, or it can be distributed throughout the entire network depending on the interconnect arrangement desired for the network. The BSC communicates to the MSC via the *A-link interface,* which is an open interface. The A link enables parts made by different BSC manufacturers to interconnect with parts made by different MSC manufacturers without having to resort to a special interface such as an IS-41.

The switching system consists of five main components: a mobile switching center, a *home location register* (HLR), a *visitor location register* (VLR), an *authentication center* (AUC), and an *equipment identity register* (EIR). The MSC's role in the switching system is to perform all the telephony switching functions for the network. The MSC coordinates all the traffic between the PSTN and the mobile system. Some of the functions the MSC is responsible for include SS7, network interface, and switching.

The HLR stores information about each resident subscriber. Each system has at least one HLR, and it contains permanent subscriber information like the type of subscriber and the features that they subscribe to. The HLR is an integral part of the MSC, but it is a separate node to the MSC itself.

Similar to the HLR, the VLR is a database that contains subscriber information that the MSC needs in order to properly treat each of the calls that are placed by a subscriber. The VLR is utilized by the MSC when the subscriber is not resident in the HLR.

The AUC is another node in the switching system, and it directly supports the HLR and provides the authentication and encryption parameters required for each call placed on a GSM system.

The EIR is a database that contains a variety of lists of faulty and stolen equipment. The subscriber equipment is checked against these lists as part of the call treatment for the system.

The third major component in the GSM involves the operations and support system. The OSS is the main interface with which operations personnel may monitor the network and take corrective action as required. The OSS is also the platform that is used to make any additions or deletions to the network's configuration.

Q. *What is a two-way system?*

A. A *two-way system* allows information to be both sent and received at either end of the communication link. These systems come in a multitude of configurations, each of which is based on the particular requirements of the end user and the bandwidth available for use. Two-way systems are commonly utilized for police, fire, emergency medical services (EMS), and government applications. There are also many commercial applications that utilize conventional two-way systems.

Two-way communication systems are found at all ends of the radio frequency spectrum and are not specific to any portion of the spectrum. These systems can be simplex, half duplex, or full duplex depending on the channels available.

Figure 2.10 shows a typical two-way simplex system. This system utilizes a dispatch operation whereby the base unit and the mobile unit both communicate on the same channel.

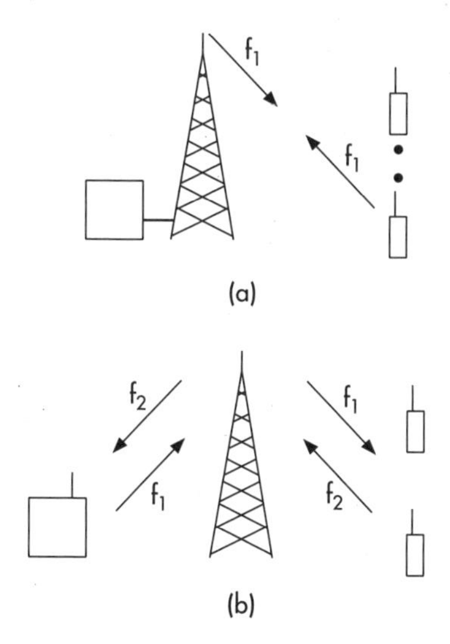

Figure 2.10 Conventional simplex (*a*) and full or half-duplex (*b*) radio systems.

The figure illustrates a duplex system that can be either half or full duplex depending on the actual configuration of the equipment. The configuration shown could also be used between two mobile units instead of between a base dispatch and a mobile unit. Operators of two-way systems could be assigned a single frequency or a frequency pair, or even multiple frequencies. The physical number of channels available determines the number of users that can be loaded onto a system. Therefore, to ensure some user selectivity on the available channels, a subaudible tone is sent along with the carrier to distinguish the desired user group.

Q. *What is a DCS-1900 system?*

A. DSC-1800 and DCS-1900 are both GSM systems. The difference between DCS-1800 and DCS-1900 is the frequency of operation. A DCS-1800 system operates in the 1800-MHz band and is normally used in European systems. The DCS-1900 is a GSM that operates in the U.S. wideband PCS frequency band at 1900 MHz.

The components that make up either a DCS-1800 or a DCS-1900 system are the same as those that make up a GSM, as described earlier, with the exception that the frequency of operation has been upbanded.

Q. *What is a DCS-1800 system?*

A. A DCS-1800 is a digital standard based on GSM technology with the exception that this type of system operates at a higher frequency range, 1800 MHz. The DCS-1800 technology is intended for use in PCN systems, such as those used commonly in Germany and England. The DSC-1800 and DCS-1900 are both GSM systems. The difference between the DCS-1800 and DCS-1900 systems is the frequency of operation. The DCS-1800 operates in the 1800-MHz band and is normally used in European systems. The DCS-1900 is a GSM system that operates in the U.S. wideband PCS frequency band at 1900 MHz.

The components that make up either a DCS-1800 or a DCS-1900 system are the same as those that make up a GSM system, described earlier, except that the frequency of operation has been upbanded. (See Fig. 2.11.)

Q. *What is the TACS?*

A. The *Total Access Communications System* (TACS) is a cellular standard that was derived from the AMPS technology. TACS operate in both the 800- and the 900-MHz bands. The first system of this kind was implemented in England. Later these systems were installed in Europe, China, Hong Kong, Singapore, and the Middle East. A variation of this standard was implemented in Japan.

Q. *What is the JTACS?*

A. The *Japan Total Access Communications System* (JTACS) is a cellular standard that was derived from the AMPS technology.

Q. *What is a NAMPS?*

A. A *Narrow-Band Advanced Mobile Phone System* (NAMPS) is used in the United States, Latin America, and other parts of the world. Specifically, the NAMP is an analog radio system that is very similar to AMPS with the exception that it utilizes 10-kHz-wide voice channels instead of the standard 30-kHz channels. The obvious advantage of using this technology is that it can deliver, under ideal conditions, three times the capacity of a regular AMPS.

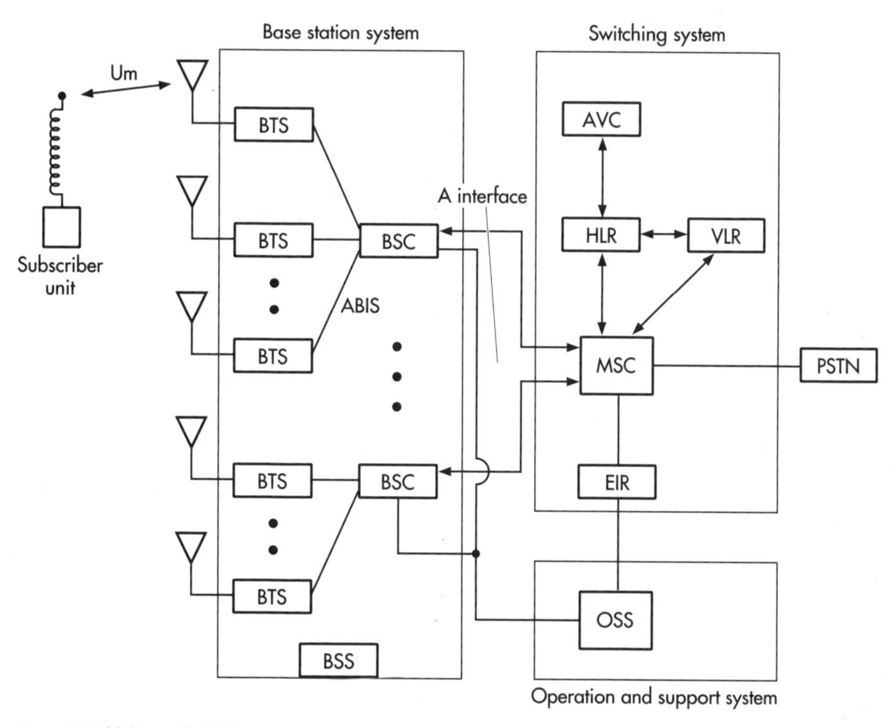

Figure 2.11 Global System for Mobile Communications.

The NAMPS can use this smaller bandwidth by changing the format, methodology, and SAT and control communications from the cell site to the subscriber unit. In particular, the NAMPS uses a subcarrier method and a digital color code in place of an SAT. These two methods make it possible to use less spectrum while communicating the same amount, or even more, information at the same time. This advantage in capacity of course requires a separate transmitter, either power amplifier (PA) or transceiver, for each NAMPS channel deployed. However, the control channel that is used for the cell site is the standard control channel, 30 kHz, that is used by the AMPS and other cellular technology platforms. Figure 2.12 is a diagram of a NAMPS. It should be stressed that, due to the narrower bandwidth channels, the carrier-to-interference ratio (C/I) requirements are different from those of a regular AMPS, which has a direct impact on the capacity of the system.

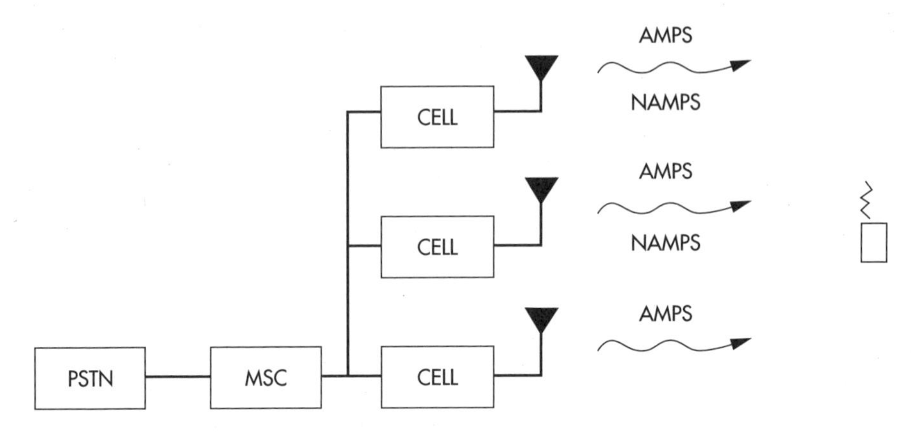

Figure 2.12 NAMPS system.

Q. *What is a D-AMPS?*

A. The *Digital Advanced Mobile Phone System* (D-AMPS) is the digital standard for cellular systems developed for use in the United States. Rather than create a completely new standard, engineers enhanced the existing AMPS standard for digital applications. This was done to accommodate rapid proliferation of analog systems.

D-AMPS is more typically referred to as IS-54A, B, or C and, of course, IS-136. There are numerous advantages of D-AMPS over AMPS, the most important being the larger overall system capacity for the same spectral bandwidth and the additional services offered to the subscribers.

Q. *What is TDMA?*

A. *Time Division Multiple Access* (TDMA) is a form of digital communication that is used for many communication platforms such as IS-136 and GSM. TDMA is also used for landline communication for DS-1, DS-3, OC-1, OC-3, or OC-12 (DS stands for digital signal and OC, optical carrier).

TDMA is another form of spread-spectrum technology that allows multiple users to occupy the same frequency spectrum through the use of time division.

In wireless communications, each radio channel occupies a certain amount of bandwidth. (See Fig. 2.13.) In the case of GSM, for example, 200 kHz is occupied in both the transmit and receive portions of the spectrum. For GSM, each radio channel is converted into eight individual time slots, or segments. The information, usually voice communication, is then placed in one of these time slots. (There is a little more to this process, but the preceding explanation is sufficient for discussion purposes.) The communication is achieved because both the sender and receiver know which particular time slot is associated with each communication.

In TDMA systems, the channel is segmented at the base radio. Thus the system requires only one digital radio at the cell site for a particular radio channel having, for example, eight time slots. However, there needs to be an individual subscriber unit for each of the time slots used, as shown in Fig. 2.13. TDMA can be used for any communication method that relies on the multiplexing and demultiplexing of information within a single communication link.

Q. *What is an IS-136?*

A. *IS-136,* also known as *IS-54,* is the digital cellular standard developed in the United States based on TDMA technology. Systems of this type operate in the same band as do the AMPSs and are used in the PCS spectrum also. IS-136 therefore can be applied to both the cellular and PCS bands and in some unique situations to downbanded IS-136, which operates in the SMR band.

TDMA technology allows multiple users to occupy the same channel through the use of time division. The TDMA format utilized in the United States follows the IS-54 and IS-136 standards and is referred to as the *North American Dual Cellular* (NADC). The IS-136 evolved from the IS-54 standard, and it enables a feature-rich technology platform to be utilized by cellular operators.

TDMA technology utilizing the IS-136 standard is currently deployed by several cellular operators in the United States. IS-136 utilizes the same channel bandwidth as does analog cellular, 30 kHz per physical radio channel. However, IS-136 enables three

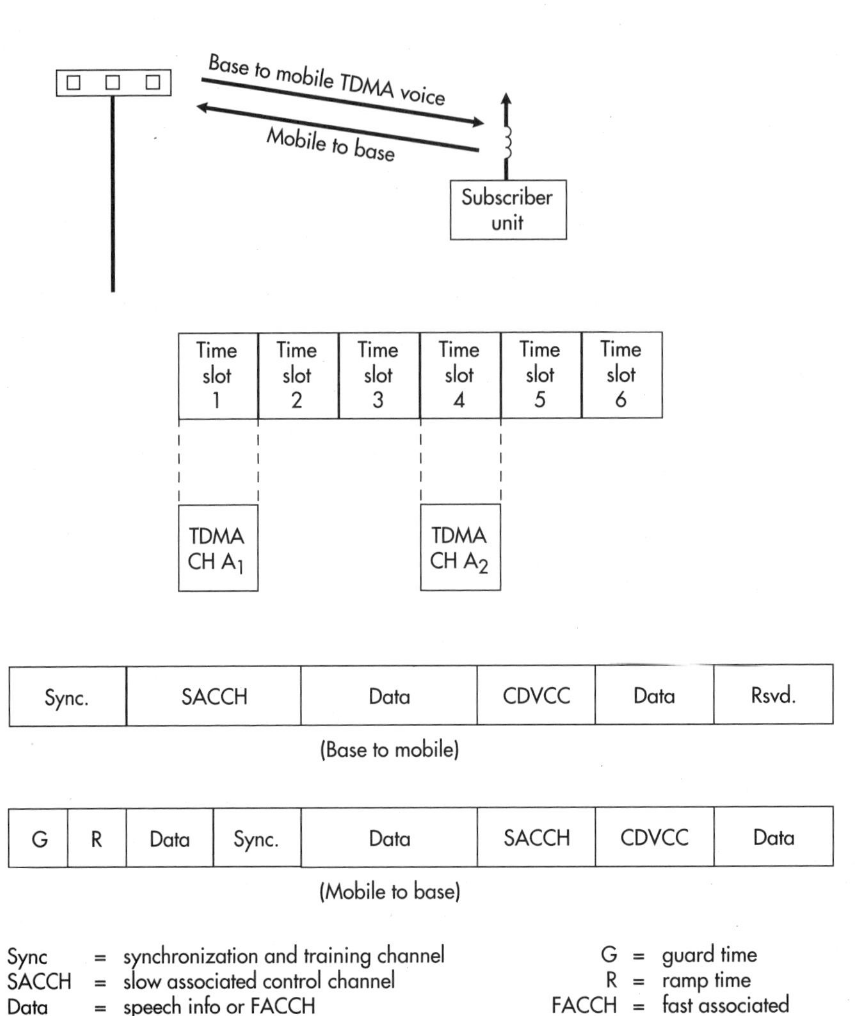

Figure 2.13 TDMA channels.

and possibly six users to operate on the same physical radio channel at the same time. The IS-136 channel presents six time slots in the forward and reverse directions. IS-136 at present utilizes two time slots per subscriber, and it has the potential to go to *half-rate vocoders,* which require the use of only one time slot per subscriber.

IS-136 has many advantages for cellular systems:

Increased system capacity, up to three times more than analog
Improved protection from adjacent channel interference
Authentication capability
Voice privacy
Less infrastructure capital needed to deploy the system
Frequency plan integration over CDMA systems
Short message paging capability

An IS-136 can be integrated into an existing cellular system more easily than into a CDMA system. The use of IS-136 in a network requires a guardband to protect the analog system from the IS-136 signal. However, the required guardband consists of only a single channel on either side of the spectrum block allocated for IS-136 use. Depending on the actual location of the IS-136 channels in the operator's spectrum, it is possible that only one or no guardband channel is necessary.

The IS-136 system has the unique advantage of affording the implementation of digital technology into a network without elaborate engineering requirements, and therefore it can be put into wide use quickly.

The implementation of IS-136 technology is further augmented by requiring only one channel, per frequency group, as part of the initial system offering. The advantage of requiring only one channel per sector in the initial deployment is that there is less capacity for reduction of the existing analog network. Another advantage of deploying one IS-136 channel per sector is the elimination of the need to preload the subscriber base with dual-mode, IS-136 handsets.

Since IS-136 operates in both the cellular and PCS bands, one of the advantages that can be exploited with IS-136 is the ability to roam and aggregate system capacity for an operator who has both the cellular and PCS spectrums.

Q. *What is an IS-54?*

A. *IS-54* is the name of the standard that is now referred to as "IS-136." IS-54C, which followed two previous versions A and B, is now called "IS-136." The channel structure and format are the same as those discussed in the preceding section on IS-136 technology. It should be noted that many operators who have had their networks for a while commonly refer to their systems as "TDMA," "D-AMPS," "NADC," or "IS-54."

Q. *What is a PCS-1900?*

A. *PCS-1900* is a common name for the GSM technology platform that uses the 1900-MHz band, which is the PCS band in the United States. The PCS-1900 system is a digital standard that operates at a higher frequency range, 1900 MHz. The PCS-1900 system is also referred to as a "DCS-1900 system." The difference between the PCS-1800 and PCS-1900 systems is the frequency of operation. The PCS-1800 operates in the 1800-MHz band and is normally used for European systems.

Q. *What is an FDMA system?*

A. A *frequency division multiple access* (FDMA) *system* uses multiple frequencies to accomplish communication—that is, to deliver information content from a sender to a receiver. FDMA systems use AMPS, IS-136, GSM, and other cellular network technologies.

The underlying concept of an FDMA system is similar to that of a TDMA with the exception that in TDMA systems, the channels are broken into individual time slots for each communication instance. In FDMA systems, multiple channels, rather than multiple time slots, are used for each communication instance. The advantage in this type of technology is that it is a flexible system, but the flexibility comes at the expense of bandwidth.

A diagram of an FDMA (analog cellular) system is shown in Fig. 2.14.

Q. *What is AM?*

A. *Amplitude modulation* (AM) is a form of modulation that is used predominantly in commercial radio systems. However, forms of AM, also known as *quadrature amplitude modulation* (QAM), have found their way into many wireless communication technologies.

AM is best shown as in Fig. 2.15.

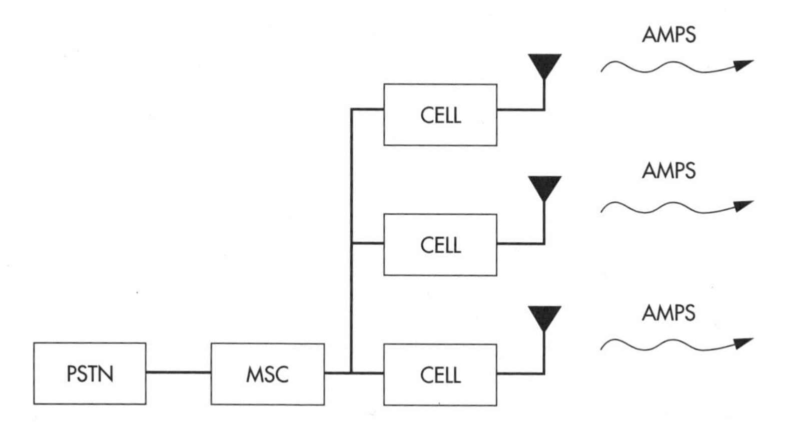

Figure 2.14 FDMA system.

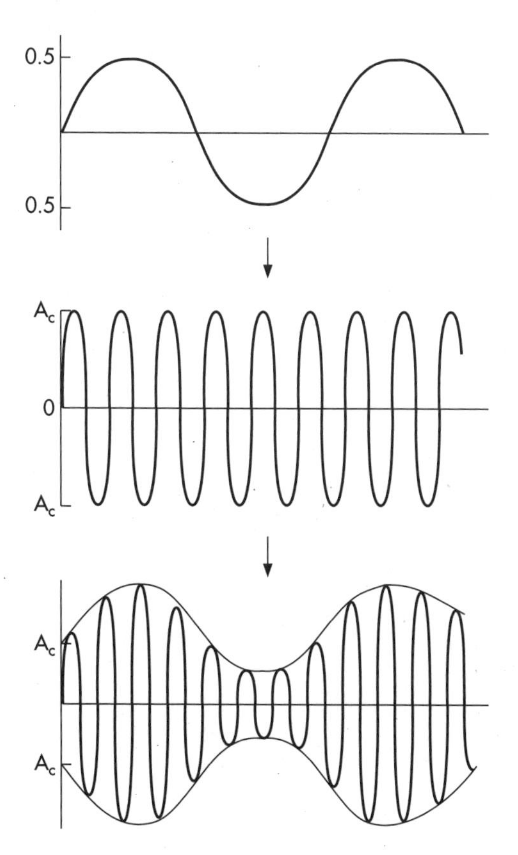

Figure 2.15 Modulation of voice or tones by varying the carrier amplitude.

The key concept to always remember when considering an AM signal for transferring information is that it works by changing the amplitude of a signal to convey information from one unit to another.

Other ways AM is used for communications involve *amplitude shift keying* (ASK), which involves changing the AM signal's amplitude, as the name applies, while keeping the frequency and phase constant. Of course, when using QAM, the phase component is also manipulated.

Q. *What is CDPD?*

A. *Cellular Data Packet Data* (CDPD) is a packetized data service utilizing its own air interface standard. The CDPD system utilized by the cellular operator is a separate functional data communication service that physically shares the cell site and cellular spectrum.

CDPD technology has many uses but is most applicable for short-burst data transfers rather than large file transfers. Examples of CDPD applications include e-mail, telemetry, credit card validation, and global positioning.

CDPD technology does not work through a direct connection between the host and server locations. Instead, it relies on the open systems interconnection (OSI) model for packet switching data communications by which the packet data are routed throughout the network. The CDPD network has various layers that make up the system: Layer 1 is the physical layer, layer 2 is the data link itself, and layer 3 is the network portion of the architecture. CDPD utilizes an open architecture and has incorporated authentication and encryption technology into its air-link standard. The CDPD system consists of several major components, as shown in the block diagram in Fig. 2.16.

The *mobile end system* (MES) is a portable wireless computing device that moves around the CDPD network communicating to the mobile database station. The MES is typically a laptop computer or other personal data device that has a cellular modem.

The *mobile database station* (MDBS) resides in the cell site itself and can utilize some of the same infrastructure that the cellular

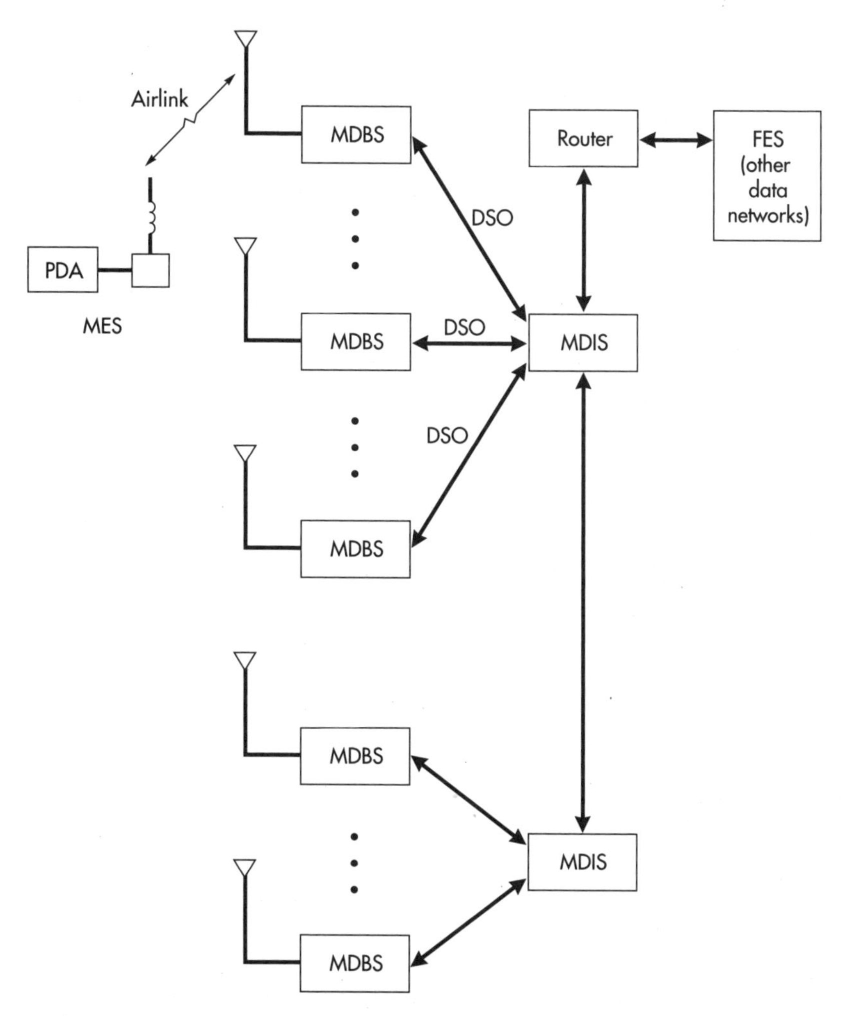

Figure 2.16 CDPD system.

system uses for transmitting and receiving packet data. The MDBS acts as the interface between the MES and the mobile database intermediate system. One MDBS can control several physical radio channels depending on the site's configuration and loading requirements. The MDBS communicates to the mobile database intermediate system via a 56-kb/s data link. Often the data link between the MDBS and the mobile database intermediate system utilizes the same facilities as the cellular system uses, but it occupies a dedicated time slot.

The *mobile data intermediate system* (MDIS) performs all the routing functions for the CDPD system. The MDIS performs the routing tasks utilizing the knowledge of where the MES is physically located within the network itself. Several MDISs can be networked together to expand a CDPD network.

The MDIS also is connected to a router or gateway that connects the MDIS to a *fixed end system* (FES). The FES is a communication system that handles layer 4 transport functions and other higher layers.

The CDPD system utilizes a *gaussian minimum shift keying* (GMSK) *method* of modulation and is able to transfer packetized data at a rate of 19.2 kb/s over the 30-kHz-wide cellular channel. The frequency assignments for CDPD can be made using one of two methods. With the first method, specific cellular radio channels are dedicated to be utilized by the CDPD network for delivering the data service. With the second method, called *channel hopping*, the CDPD's mobile database station utilizes unused channels for delivering its packets of data. Both frequency assignment methods have advantages and disadvantages.

The advantage of using a dedicated channel assignment for the CDPD system is that the CDPD system will not interfere with the cellular system with which it is sharing the spectrum. By enabling the CDPD system to operate on its own set of dedicated channels, there is no real interaction between the packet data network and the cellular voice network. However, the dedicated channel method reduces the overall capacity of the network, which, depending on the system loading conditions, might not be a viable alternative.

The advantage of using channel hopping, when this is part of the CDPD specification, is that the MDBS for that cell or sector will utilize idle channels for the transmissions and reception of data packets. In the event that the channel being used for packet data is assigned by the cellular system for a voice communication call, the MDBS detects the channel's assignment and instructs the MES to retune to another channel before it interferes with the cellular channel. The MDBS utilizes a scanning receiver, or *sniffer,* which scans all the channels it is programmed to scan to determine which channels are idle and/or in use.

The disadvantage of the channel hopping method is that it may cause interference in the cellular system. Coexisting on the same channels with the cellular system, the CDPD can create mobile-to-base-station interference. The mobile-to-base-station interference occurs because the handoff boundaries for CDPD and cellular are different for the same physical channel. The difference in handoff boundaries occurs largely because the CDPD system utilizes a BER (bit error rate) for handoff determination and the cellular system utilizes an RSSI (received signal strength indication) at the cell site in analog systems or the MAHO (mobile-assisted handoff) in digital systems.

Q. *What is a paging system?*

A. A *paging system* is a one-way message delivery system. These systems are popular and will remain so because they are cost effective. The message historically was one way in that the communication was sent to an individual who possessed a paging receiver. The paging receiver would convey the information content to the individual through a tone, voice, or numeric or alphanumeric signal.

A classic one-way paging system configuration is shown in Fig. 2.17. Paging systems are not concentrated within any band of the radio frequency spectrum, but they are usually in the 150-MHz, 450-MHz, and 900-MHz bands in the United States.

Q. *What is SMR technology?*

A. SMR and ESMR systems cover a variety of technologies and applications whose primary purpose is to offer more capacity per radio frequency than that of typical two-way systems. An SMR system is also commonly referred to as a *trunked repeater system,* and this type of system is very similar to a conventional two-way system. However, like its two-way cousin, an SMR station cannot hand off to another SMR station, which means when the extent of the coverage area has been exhausted, the conversation is then dropped. An enhanced specialized mobile radio (ESMR) system is similar to an SMR system except in an ESMR system, digital technology is used and handoffs can take place to ensure continuous communication.

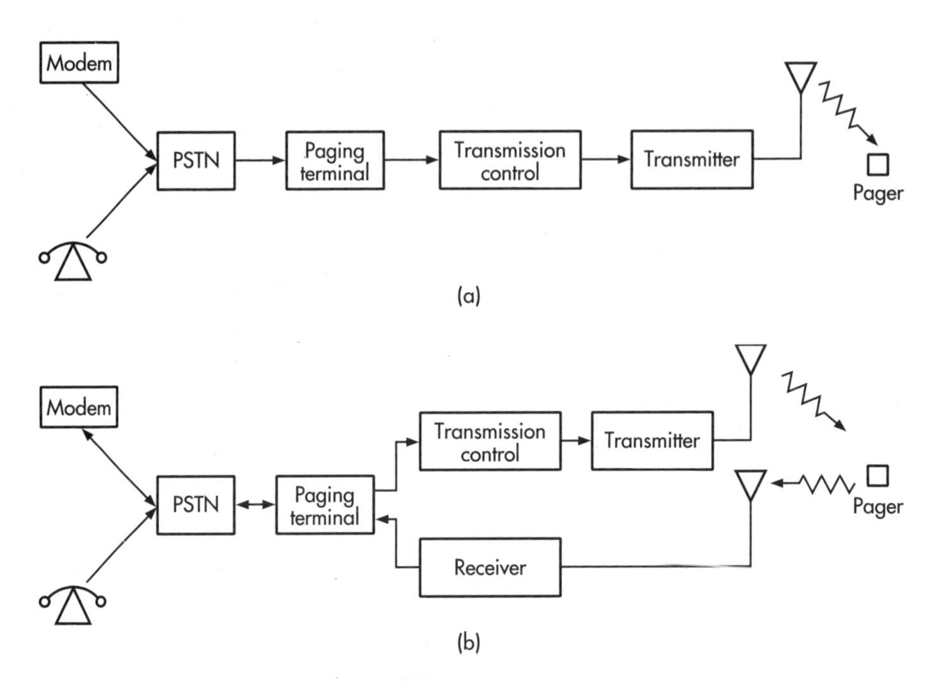

Figure 2.17 (*a*) Simplified one-way paging system. (*b*) Simplified two-way paging system.

Both SMR and ESMR systems occupy the 800- and lower 900-MHz portions of the radio frequency spectrum. The channel allocation for SMR and ESMR communication is shown in Fig. 2.18.

Q. *What is ESMR technology?*

A. ESMR technology is by definition an enhanced form of SMR. It was established to increase the capacity of existing SMR systems. An ESMR system utilizes different digital technologies to maximize the spectrum use and overall capacity for the network. There are several ESMR platforms in existence; however, two types that are commonly used involve TDMA and FDMA.

The FDMA system is utilized by GeoTek, and it is designed to increase the system capacity through *frequency hopping*. Slow frequency hopping is used by conventional cellular communication systems; however, the frequency hopping system that is employed by GeoTek involves a faster method, and the channel changing is based on a pseudo number coding sequence.

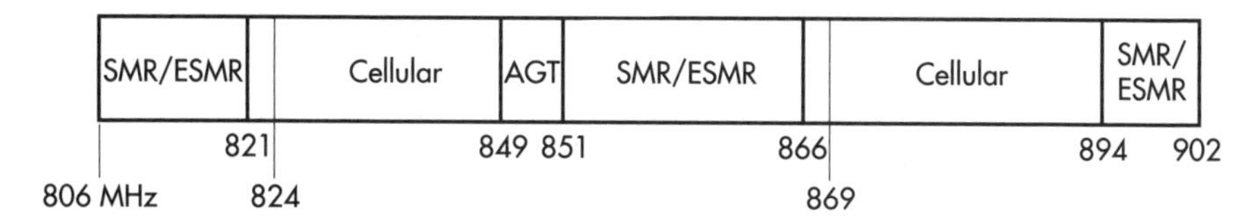

Figure 2.18 Channel allocation for SMR and ESMR communication.

The TDMA system, iDEN, is utilized currently by Nextel, and it has a similar channel structure to that of IS-136, but it occupies only 25 kHz of spectrum and utilizes QAM.

Q. *What is CDMA technology?*

A. A CDMA system uses a spread-spectrum technology platform, which enables multiple users to occupy the same radio channel, frequency spectrum, at the same time. CDMA technology has been and is being utilized for microwave point-to-point communication and satellite communication in both military and civilian applications. With a CDMA system, each of the subscribers, or users, utilizes his or her own unique code to differentiate himself or herself from the other users.

A CDMA system offers many special features including its ability to thwart interference and to improve immunity to multipath effects due to its bandwidth.

The benefits associated with CDMA technology are the following:

- More system capacity than analog and TDMA systems
- Improved protection from interference
- No frequency planning required between CDMA channels
- Improved handoffs with MAHO and soft handoffs
- Fraud protection due to encryption and authentication capabilities
- Compatibility with new wireless features

CDMA technology is based on the principle of *direct sequence* (DS) and is a wideband spread-spectrum technology. The CDMA channel utilized is reused in every cell of the system and is differentiated by the *pseudo random number* (PN) *code* that it utilizes.

However, despite the apparent advantages CDMA technology offers for cellular systems, there are several implementation concerns regarding its use in an existing system. The introduction of a CDMA system into an existing AMPS requires the establishment of a guardband and guardzone. The guardband and guardzone are required to ensure that the interference received from the AMPS does not negatively impact the ability for the CDMA system to perform well.

Q. *What is a CDMA guardzone?*

A. A CDMA *guardzone* is the physical area, outside the CDMA coverage area, that can no longer utilize the AMPS channels then occupied by the CDMA system. Figure 2.19 shows a guardzone in relation to a CDMA system coverage area. The establishment and size of the guardzone depends upon the traffic load expected by the CDMA system. The guardzone is usually defined in terms of a signal strength level from which analog cell sites operating with the CDMA channel sets cannot contribute to the overall interference level of the system. An interesting situation arises when an operator on one system wishes to utilize a CDMA system and must require an adjacent system operator to reduce his or her channel utilization to accommodate the introduction of this new technology platform.

Q. *What is the cellular spectrum?*

A. Figure 2.20 is a breakdown of the cellular spectrum and its associated frequency and channel allocation scheme. The channels defined are in increments of 30 kHz.

Q. *What are the cellular CDMA channel designations?*

A. The CDMA channel designations for cellular are shown below:

	A band	**B band**
Primary CDMA channel	283	384
Secondary CDMA channel	691	777

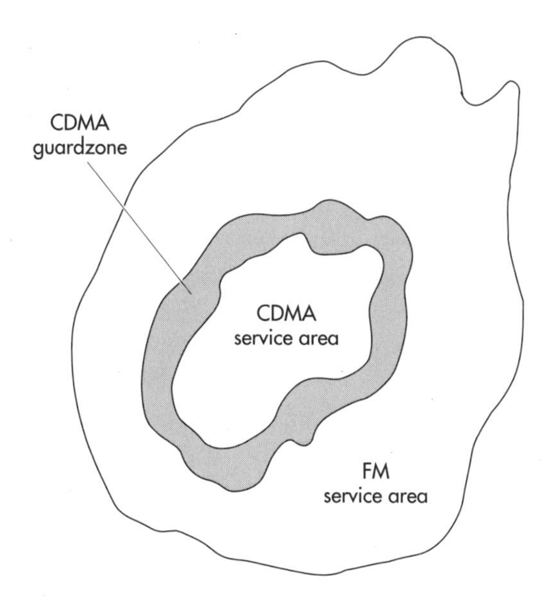

Figure 2.19 Guardzone.

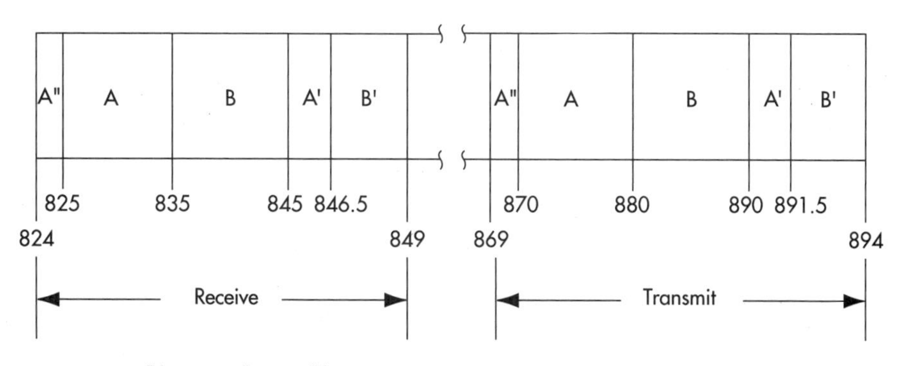

Figure 2.20 U.S. cellular spectrum allocation. (All frequencies in MHz.)

Q. *What is the SMR spectrum?*

A. Figure 2.21 shows the SMR band and its relationship with the cellular band. It should be noted that, in terms of radio frequency, its physical location is significantly away from the PCS band utilized in the United States.

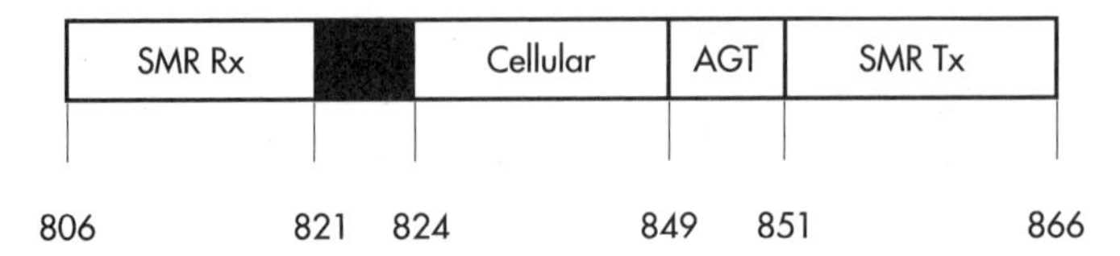

Figure 2.21 SMR spectrum (MHz).

The spectrum chart also shows the channel allocation scheme that is utilized. One interesting aspect of operating in the SMR band is the lack of contiguous channel blocks for system design and expansion. The use of contiguous channel blocks within an allocated spectrum is a critical concern when employing a cellular technology in the SMR band.

Another important consideration is that the SMR channels are 25 kHz. To utilize a technology requiring 30 kHz of bandwidth, it is necessary to aggregate two channels together, which makes the use of this technology in the SMR band without contiguous channels spectral inefficient. For example,

SMR channel = 25 kHz
AMPS channel = 30 kHz

Therefore,

2 SMR channels = 50 kHz
AMPS channel = 30 kHz
Unused spectrum = 20 kHz

If the channels were combined, then the spectral inefficiency would be minimized. However, with the diverse nature of SMR channels, the use of a non-25-kHz technology, as has been done in several systems, leads to numerous technological problems that reduce spectral efficiency.

Q. *What is the PCS spectrum?*

A. The PCS spectrum is shown in Fig. 2.22. It should be noted that it is not like cellular and SMR channels in that the frequency separation between the transmit and receive spectrums is 80 MHz. Also there are a few restrictions that are placed on the operators,

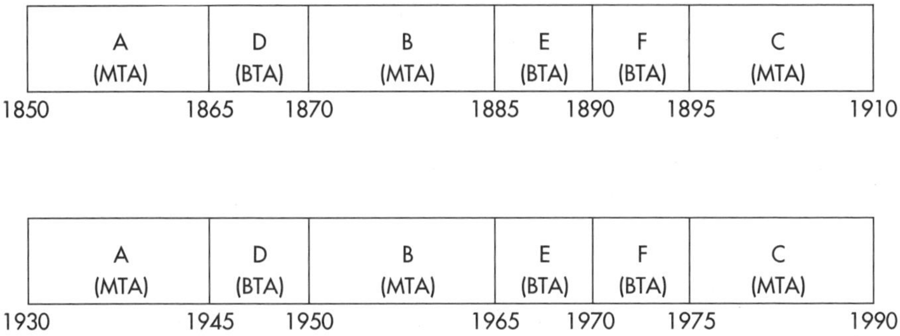

A (MTA)	D (BTA)	B (MTA)	E (BTA)	F (BTA)	C (MTA)

1850 1865 1870 1885 1890 1895 1910

A (MTA)	D (BTA)	B (MTA)	E (BTA)	F (BTA)	C (MTA)

1930 1945 1950 1965 1970 1975 1990

Figure 2.22 PCS allocation. (All frequencies in MHz.)

or rather, license holders, for each of the different PCS bands that they have. The particular issues associated with the restrictions generally apply to the service deployment and size of the network in terms of coverage area. There are some unique restrictions placed on the entrepreneur blocks, C and F, but the restrictions have changed somewhat since the inception of the auction process held in the United States.

Q. *What is an RFP?*

A. A *request for proposal* (RFP) is often used to solicit bids from vendors. However, in many cases the issuer of the RFP generates a best and final RFP, which is a polished version of the original RFP. The exception is that some operators populate the RFP with proprietary material from other vendors and share that with their competition, which is unethical. An RFP usually specifies a deadline and a format for the vendor to use in submitting a bid.

The use of a grading matrix has been an exceptional tool for making accurate comparisons among RFP responses. It cannot be overstressed that if the RFP does not specify a response format and does not establish a grading matrix, the decision process will be exceptionally difficult.

Q. *What is PCS-2000 technology?*

A. *PCS-2000* is another name for the technology platform IS-661.

Q. *What does bandwidth on demand mean?*

A. *Bandwidth on demand* describes many different technologies that are used for maximizing the spectrum available for use. Bandwidth on demand is used commonly in conjunction with local multipoint distribution services (LMDS) and provides a means for subscribers to increase the amount of bandwidth they use just at the time they need it. When they no longer need the additional bandwidth, the excess bandwidth capacity is made available for other subscribers to utilize.

Q. *What is spread-spectrum modulation?*

A. *Spread-spectrum modulation* allows communication to utilize more bandwidth than the information content actually requires. Typically spread-spectrum modulation is used primarily with wireless CDMA systems, but it is also used with TDMA and FDMA systems.

Q. *What is an unlicensed spectrum?*

A. There are numerous locations within the electromagnetic spectrum where a specific license to operate within that band is not required—hence, the name "unlicensed spectrum." The operator or service utilizing the unlicensed spectrum must, however, meet specific FCC requirements for operation, and he or she must know that there is a high likelihood that interference will occur.

Q. *What is a narrow-band CDMA system?*

A. A *narrow-band Code Division Multiple Access system* is another name for an IS-95 system. A narrow-band CDMA channel uses 1.25 MHz of spectrum to convey its information.

Q. *What is a wideband CDMA system?*

A. A *wideband Code Division Multiple Access system* is another name for third-generation wireless systems, such as IMT-2000. A wideband CDMA system utilizes 5 MHz of radio spectrum, in contrast to a narrow-band CDMA system, which uses 1.25 MHz per channel. Manufacturers report wideband CDMA features to be the following:

- Eight times more capacity than narrow-band CDMA
- Variable data transmission rates of 384 kb/s mobile and 2 Mbytes/s stationary
- Supportive of multiple wireless services
- Universal service

Q. *What is IMT-2000 technology?*

A. *IMT-2000 technology* is the standard that will be used for third-generation wireless systems to enable a truly universal platform and service. The IMT-2000 systems will utilize wideband CDMA technology, but there are two competing wideband CDMA standards, and at this time neither has been selected as the IMT-2000 standard.

Some of the basic features supported by IMT-2000 service are the following:

- High-speed data transmission
- Video-conferencing
- High-quality audio transmission

Q. *What does the term third-generation wireless mean?*

A. *Third-generation wireless* refers to the international effort to establish a universal mobile communication standard with which true international mobility can be achieved. The third-generation wireless systems will utilize IMT-2000 technology. Currently several competing standards, all based on wideband CDMA technology, will make up the IMT-2000 standard.

Q. *What does the term push to talk mean?*

A. The term *push to talk* refers to the process in dispatch and two-way communication systems whereby the sender depresses the mic key in order to talk to the other party.

Q. *What is the TCP/IP?*

A. The *Transmission Control Protocol/Internet Protocol* (TCP/IP) encompasses a variety of protocols that enable multiple computer platforms using different formats—that is, Mac and DOS—to communicate with each other. The TCP/IP is often referred to as "the technology used for the Internet."

Q. *What is the IS-41?*

A. The *IS-41* is the interswitch standard utilized for providing automatic call delivery and handoffs. The IS-41 relies on SS7 signaling, and it allows for switches from different vendors to communicate with each other. For example, if one system had Lucent equipment and the neighboring system had Ericsson, the IS-41 protocol would allow for call delivery and handoffs to occur between them.

Q. *What is the GPS?*

A. The *Global Positioning Satellite* system is a series of 24 satellites that orbit the earth for the purposes of providing position and timing information for objects. The GPS system consists of 24 satellites arrayed over six regions of the world so as to ensure that at least 4 satellites can be viewed from any single point on the earth. Visibility, of course, depends on the lack of obstructions in the viewing area on the ground.

The GPSs transmit on two frequencies using a spread-spectrum modulation technique, CDMA, using bipolar phase shift keying (BPSK). The GPS satellite frequencies are the following:

L1 = 1575.42 MHz
L2 = 1227.60 MHz

The GPS *receiver sensitivity*—that is, the minimal required signal level for proper detection—is directly dependent upon the frequency received and the code that is used:

L1 C/A code = 2160 dBW
L1 P code = 2163 dBW
L2 P code = 2166 dBW

The GPS system is used not only for MSC timing but also for iDEN and CDMA systems for the purposes of call processing and signaling between cell sites for handoffs.

Q. *What is differential GPS technology?*

A. *Differential Global Positioning Satellite* (GPS) *technology* is used to increase the positional accuracy of a regular GPS receiver. A reading made with a differential GPS system will be more accurate and within 10 m of the actual position. In contrast, ordinary GPS is accurate only to within 100 m of the true position.

Q. *What is an LMDS?*

A. A *local multipoint distribution service* (LMDS) is a broadband wireless service that uses point-to-multipoint technology to provide an alternative to telco. LMDS systems operate in frequency bands greater than 20 GHz and are available in multiple frequency bands.

Q. *What is an ADSL?*

A. An *asymmetric digital subscriber line* (ADSL) is part of the xDSL (generic DSL) technology platforms currently being deployed in the United States and around the world. ADSL technology allows for the use of the existing telephone lines connected to private residences. Data can be transmitted at rates of up to 8 Mbytes/s from the Internet to the residence and simultaneously in the uplink direction from the residence to the Internet, from 384 kb/s to 1.5 Mbytes/s. The key advantage in using an ADSL is that the *plain old telephone service* (POTS) *line* that exists at the private residence will still be usable at the same time the user is connected to the Internet, thus eliminating the need for a second telephone line. In addition, ADSL technology can be successfully applied in many other high-speed data transmission situations. An ADSL also has the potential for reducing the interconnection cost for wireless systems because data can be transported in TCP/IP packets over the Internet or a *wide area network* (WAN).

Q. *What is an HDSL?*

A. A *high-bit-rate digital subscriber line* (HDSL) is an alternative to T-1/E-1 circuits in wireless communication systems. HDSLs are less expensive than T-1/E-1 circuits, and they do not need line conditioning as do T-1/E-1 circuits, nor do they require repeating under normal situations.

Q. *What is an ATM?*

A. *Asynchronous transfer mode* (ATM) technology is used to facilitate high-speed, low-delay switching for voice, data, and video traffic. Increasing numbers of telcos and wireless companies are using some ATM systems. An ATM switch segments and multiplexes data, voice, or video traffic into small, fixed-length cells, which results in an apparently seamless connection between the sender and receiver with little delay. In addition, ATM technology can provide direct connections between remote computers.

Q. *What is an air-to-ground transmitter system?*

A. An *air-to-ground transmitter* (AGT) *communication system* operates in a frequency band next to the cellular B band. AGT systems use the 849- to 851-MHz frequency band for transmitting and the 894- to 896-MHz frequency band for receiving with respect to the base station. The subscriber units are located onboard airplanes. The AGT channel is 200 kHz wide, and it uses a GSM-like digital protocol for its communication platform.

Q. *What is an antenna farm?*

A. The term *antenna farm* describes a location in which there are many wireless operators and antennas. These locations pose a problem when colocating for potential interference between systems because they are usually located on a prime RF site.

Q. *What is beam tilting?*

A. *Beam tilting* is the process of making the peak of the vertical planes' radiation pattern of an antenna lie below horizontal for the purposes of optimizing the receiver signal strength on the ground. This technique is used by operators to reduce cochannel and adjacent channel interference. Beam tilting is also referred to as *null filling*. Beam-tilting capability can either be built into the antenna through electrically phasing the elements or provided by mechanical means.

Q. *What is a bore site?*

A. A *bore site* is simply the main axis of a directional antenna from which the antenna directs most of the radiated energy or at which the antenna can absorb the most energy.

Q. *What is a cell phone?*

A. A *cell phone* is the common term used to describe any one of a multitude of subscriber units.

Q. *What is CPE?*

A. The *customer premise equipment* (CPE) for wireless systems is the subscriber's phone, that is, the subscriber's cell phone.

Q. *What is cell splitting?*

A. RF designers use the *cell-splitting* process to handle increased traffic without enlarging the spectrum. With cell splitting, the larger cells can be divided into smaller cells so as to reuse the frequencies in a smaller geographic area.

Q. *What is a cell?*

A. A *cell* is a geographic region served by a particular cell site. This region is designed by a hexagon despite the fact that the geographic region that serves the cell site never is hexagonal in shape.

Q. *What is a channel group?*

A. A *channel group* is made up of individual voice and data channels that are used by a specific cell or sector of a cell to provide capacity to subscribers. A channel group is also referred to as a *freq group, frequency group,* and *channel set.*

Q. *What is circuit merit?*

A. The *circuit merit* (CM) is the voice quality of a wireless call. The circuit merit is subjectively rated on a scale of 1 through 5, 5 being the highest rating. The circuit merit test is similar to the MOS (mean opinion score) test.

Q. *What is a DCC?*

A. A *digital color code* is a means by which each control channel is identified uniquely. For cellular systems, there are four DCCs: 0, 1, 2, and 3.

Q. *What is the directed-retry process?*

A. When it is necessary that an originating call is redirected from the best-serving cell site to another cell site on the directed-retry list, when either the cell site does not have enough capacity or the subscriber unit accesses the system below a required signal level, the *directed-retry service* is initiated. This service is a software parameter set by engineering. It is used for a temporary capacity relief for a severely blocking cell site. There is some sacrifice of call quality, as well as some degradation in C/I due to higher levels of interference, that results from using directed retry.

Q. *What is drive testing?*

A. *Drive testing* is the process by which a subscriber unit or several units are used in a vehicle and connected to a special data collection computer so that system performance measurements can be made.

Q. *What is dual mode?*

A. The term *dual mode* refers to customer premise equipment that can work in two different protocols such as AMPS and CDMA or AMPS and TDMA.

Q. *What is dual band?*

A. The term *dual band* describes customer premise or cell site equipment that can operate in two different RF bands. Examples of dual-band operation is CPE that can operate at both 800 cellular and 1900 PCS frequencies within an IS-136 standard.

Q. *What is churn?*

A. The term *churn* refers to a subscriber's dropping service with one provider and signing up with another provider. A 2 percent churn rate for the month equates to a 24 percent rate for the year.

Q. *What are jumpers?*

A. *Jumpers* are coaxial cables that connect the feed line for the antenna system to either the antennas or the receiver multicoupler for the radio equipment. When used to connect the feed line to the antenna, jumpers are a primary maintenance problem for an antenna system. Jumpers are usually about 5 ft in length but can be longer if required.

Q. *What is an ESN?*

A. An *electronic serial number* (ESN) is a 32-bit number that uniquely identifies a particular mobile unit. This number is set at the factory and is used in conjunction with the *model identification number* (MIN) to establish the identity of the subscriber. *Cloning* involves the duplication of the ESN with the subscriber's MIN.

Q. *What is the IS-3-D?*

A. The *IS-3-D* is the TIA/EIA interim standard for the AMPS cellular mobile station and land station compatibility specifications. (TIA stands for Telecommunication Industry Association and EIA, Electronic Industries Association.)

Q. *What is a locate receiver?*

A. A *locate receiver* is a radio receiver in a Lucent series II cell site that is used to scan and perform specific call processing functions. The scan receiver is used extensively for AMPSs during the handoff process when the scan receivers at potential target cell sites scan the mobile unit to measure its RSSI to determine if the mobile unit is a valid handoff candidate for that cell.

Q. *What is the MTBF?*

A. The *mean time between failures* (MTBF) is a measure of how long a piece of equipment will last, that is, how long it is between expected failures.

Q. *What is an FOA?*

A. A *first office application* (FOA) is a term used to describe the beta load for switch and cell site software loads. Sometimes an FOA is associated with new hardware, but usually it is related to software load changes whose purpose is to fix existing problems, as well as offering new features.

Q. *What is an SID?*

A. A *system identification* (SID) is a digital identification signal used in cellular and PCS systems. An SID is transmitted to the mobile unit so that the mobile unit can determine if it is on its home system or is roaming. Each system has its own unique SID. Odd numbers in cellular identify A-band frequencies and even numbers identify B-band frequencies.

Q. *What is a transmission mask?*

A. The term *transmission mask* refers to the FCC requirement that limits the amount of RF energy which can be transmitted into an adjacent band.

Q. *What is a transportable?*

A. A *transportable* is a battery-operated, class 2 subscriber unit. It is also called a *bag phone.*

Q. *What is a UPS?*

A. An *uninterruptible power supply* (UPS) is part of a disaster recovery program for ensuring that power is maintained during a power outage of any length.

Q. *What is polarization diversity?*

A. *Polarization diversity* is a type of diversity achieved with antennas, and it is based on the concept of utilizing two different branches that are polarized in orthogonal planes. Polarization diversity lends itself to only a two-branch system in practice. There has been and continues to be work conducted with implementing this type of diversity scheme in a mobile unit environment. The main advantage that polarization diversity offers a communication system is that the number of antennas that are required for an installation is reduced, thereby providing a possible solution for difficult-to-zone areas.

Q. *What is vertical diversity?*

A. *Vertical diversity* is a diversity scheme that, when applied to a mobile unit communication field, requires the antennas be separated by a defined vertical barrier. Vertical diversity involves staking the receive antennas on top of each other to overcome fades in the network. Vertical separation can be used as a method for diversity reception for mobile communications. However, the drawback is that typically the separation of the antennas is greater for vertical planes than for horizontal planes. In addition, the multipath spread angle is usually small for a mobile unit environment, making this technique unattractive for improving the multipath environment for distant mobile units.

Q. *What is angle diversity?*

A. *Angle diversity* utilizes different antennas to achieve diversity with each antenna oriented in a different direction. This setup is meant to capture the advantage of reducing the number of antennas required for a cell site when multiple sectors are utilized. In addition, the signal from a directive antenna has less severe fading characteristics. One application of angle diversity is in a six-sector cell site where it is advantageous to reduce the number of antennas. (See Fig. 2.23.) Each sector has one antenna, and it is oriented in a different direction. The angle of arrival for the signal from the subscriber unit is where the advantages of angle diversity are gained. However, the maximum number of diversity branches that can be used with angle diversity is three.

Q. *What is the standing-wave ratio?*

A. The *standing-wave ratio* (SWR) is the measure of how much power is reflected back to the transmitter from the load, usually the antenna. Poor construction or faults with the antenna feed line or the antenna itself will result in a high SWR. The desired

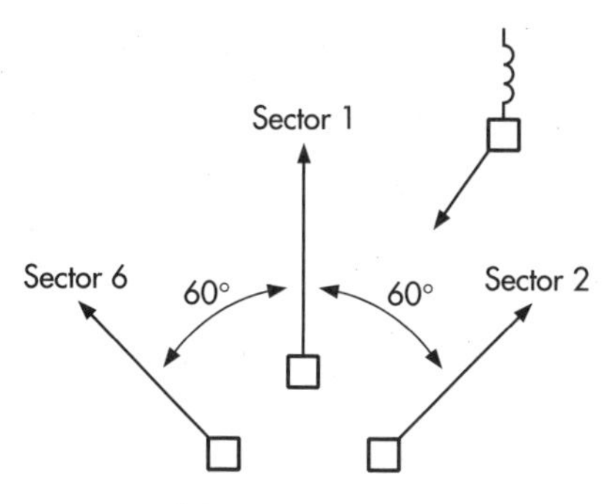

Figure 2.23 Angle diversity.

and ideal SWR is 1:1.0. A good design goal for an antenna system is an SWR of 1:1.2 or better.

Q. *What is frequency diversity?*

A. *Frequency diversity* utilizes two distinct frequencies to convey the same information content. The concept is that while one frequency is undergoing a deep fade, the other frequency is not experiencing the same fade depth. Therefore, using two frequencies will ensure that the communication link remains at the desired level of service. However, the use of frequency diversity has the unique disadvantage of requiring twice the resources both physically and also spectrally to deliver the same information content. The advantage frequency diversity has over space diversity techniques is that this method can utilize one antenna branch and therefore reduce the physical real estate on a tower. An example of frequency diversity is shown in Fig. 2.24.

Q. *What is space diversity?*

A. *Space diversity* is used extensively in wireless systems. It is achieved by physically separating the antennas either horizontally or vertically for the sole purpose of minimizing or removing the negative effects associated with fading in a radio environment. The objective is to have enough separation so that the antennas used in the system will not experience the same fade characteristics at the same instant in time. However, ordinarily diversity is used primarily at the base station and not at the subscriber unit. The reason for diversity playing a role only at the base station side of the system is that most subscriber units have only one antenna and therefore cannot possibly achieve any advantage or improvement with diversity.

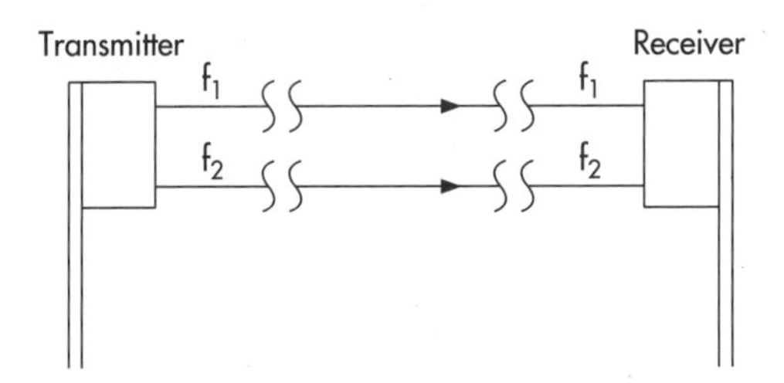

Figure 2.24 Frequency diversity.

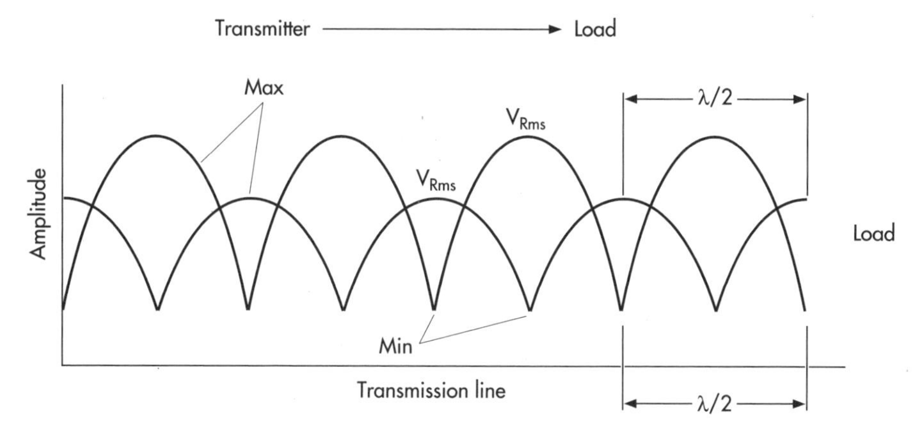

Figure 2.25 SWR, VSWR, and ISWR.

Q. *What is the relationship between the SWR, VSWR, and ISWR?*

A. The relationship between the SWR, *voltage standing-wave ratio* (VSWR), and *current standing-wave ratio* (ISWR) is shown in the following equation (see also Fig. 2.25):

$$\text{VSWR} = \text{SWR} = \text{ISWR}$$

Q. *What is the difference between the ERP and EIRP?*

A. *Effective radiated power* (ERP) and *effective isotropic radiated power* (EIRP) are the two most common references used for determining the transmit power of a communication site. The ERP and EIRP are directly related to each other, and a simple conversion can be achieved when one is known and the other is sought:

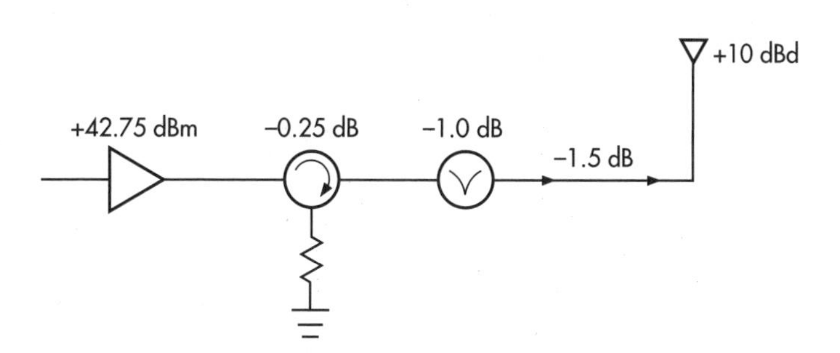

Figure 2.26 This schematic is calculated as follows: transmitter, +42.75 dBm; isolator, −0.25 dB; filter, −1.0 dB; feed line, −1.5 dB; antenna gain, +10 dBd, for a total of +50 dBm, which equals 100 W ERP or 165 W EIRP.

$$ERP = EIRP - 2.14$$

Q. *How do you calculate the ERP?*

A. Figure 2.26 contains a brief example of how to calculate the ERP for a communication site.

Q. *What types of analog systems are used for cellular?*

A. Table 2.1 lists various analog cellular systems that are used throughout the world.

Q. *What types of digital systems are used for cellular?*

A. Table 2.2 lists various digital cellular systems that are used throughout the world.

Q. *What types of PCS systems are used for cellular?*

A. Table 2.3 lists various PCS systems that are used throughout the world and in particular the United States.

Q. *What are some of the more common mobile data systems used?*

A. Table 2.4 lists some of the more common mobile data systems that are used in the United States.

Q. *What are the AMPS subscriber power levels?*

A. The following table lists various AMPS subscriber power levels:

Station class	ERP (max)
1	6 dBW (4 W)
2	2 dBW (1.6 W)
3	−2 dBW (0.6 W)

Q. *What are the IS-136 subscriber power levels?*

A. Table 2.5 lists various IS-136 subscriber power levels.

Q. *What are the various PCS specifications in common use?*

Table 2.1 Various Analog Cellular Systems Used throughout the World

	AMPS	NAMPS	TACS	NMT-450	NMT-900	C-450
Base Tx	869–894 MHz	869–894 MHz	935–960 MHz	463–468 MHz	935–960 MHz	461–466 MHz
Base Rx	824–849 MHz	824–849 MHz	890–915 MHz	453–458 MHz	890–915 MHz	451–456 MHz
Multiple-access method	FDMA	FDMA	FDMA	FDMA	FDMA	FDMA
Modulation	FM	FM	FM	FM	FM	FM
Radio channel spacing	30 kHz	10 kHz	25 kHz	25 kHz	12.5 kHz	20 kHz (b) 10 kHz (m)
Number of channels	832	2496	1000	200	1999	222 (b) 444 (m)
CODEC	NA	NA	NA	NA	NA	NA
Spectrum allocation	50 MHz	50 MHz	50 MHz	10 MHz	50 MHz	10 MHz

TABLE 2.2 VARIOUS DIGITAL CELLULAR SYSTEMS USED THROUGHOUT THE WORLD

	IS-136	IS-136*	IS-95	GSM	iDEN
Base Tx	869–894 MHz	851–866 MHz	869–894 MHz	925–960 MHz	851–866 MHz
Base Rx	824–849 MHz	806–821 MHz	869–894 MHz	880–915 MHz	806–821 MHz
Multiple-access method	TDMA/FDMA	TDMA	CDMA/FDMA	TDMA/FDMA	TDMA
Modulation	Pi/4 DPSK	Pi/4 DPSK	QPSK	0.3 GMSK	16 QAM
Radio channel spacing	30 kHz	30 kHz	1.25 MHz	200 kHz	25 kHz
Users per channel	3	3	64	8	3/6
Number of channels	832	600	9 (a) 10 (b)	124	600
CODEC	ACELP/VCELP	ACELP	CELP	RELP-LTP	
Spectrum allocation	50 MHz	30 MHz	50 MHz	50 MHz	30 MHz

*Special application for Hawaii and Brazil only.

Table 2.3 Various PCS Systems Used in the United States and throughout the World

	IS-136	IS-95	DCS-1800	DCS-1900	IS-661
Base Tx	1930–1990 MHz	1930–1990 MHz	1805–1880 MHz	1930–1990 MHz	1930–1990 MHz
Base Rx	1850–1910 MHz	1850–1910 MHz	1710–1785 MHz	1850–1910 MHz	1850–1910 MHz
Multiple-access method	TDMA/FDMA	CDMA/FDMA	TDMA/FDMA	TDMA/FDMA	TDD
Modulation	Pi/4 DPSK	QPSK	0.3 GMSK	0.3 GMSK	QPSK
Radio channel spacing	30 kHz	1.25 MHz	200 kHz	200 kHz	5 MHz
Users per channel	3	64	8	8	64
Number of channels	166/332/498	4–12	325	25/50/75	2–6
CODEC	ACELP/VCELP	CELP	RELP-LTP	RELP-LTP	CELP
Spectrum allocation	10/20/30 MHz	10/20/30 MHz	150 MHz	10/20/30 MHz	10/20/30 MHz

Table 2.4 Mobile Data Systems Used Commonly in the United States

	CDPD	RAM-Mibotex	Ardis-RD/LAP
Base Tx	869–894 MHz	935–941 MHz	851–869 MHz
Base Rx	824–849 MHz	896–902 MHz	806–824 MHz
Multiple-access method	FDMA	TDMA/FDMA	TDMA/FDMA
Radio channel spacing	30 kHz	12.5 kHz	25 kHz
Number of channels	832	480	720
Spectrum allocation	50 MHz	12 MHz	32 MHz

A. There are numerous PCS specifications. However, some of the more prevalent PCS standards are listed below for reference:

- PCS-1900—upbanded GSM cellular
- TIA IS-136—upbanded TDMA digital cellular
- TIA IS-95—upbanded CDMA digital cellular
- TIA IS-88—upbanded NAMPS narrow-band analog cellular
- TIA IS-91—upbanded analog cellular
- J-STD-014—PACS
- TIA IS-661—Omnipoint composite CDMA/TDMA
- TIA IS-665—wideband CDMA

Q. *What are some commonly used decibel reference values?*

A. Table 2.6 can be used to make a general comparison for various decibel references. The decibel reference table is not complete, but it does cover most, if not all, of the decibel references that are encountered in the wireless industry.

Q. *What are some commonly used standard distance measurement conversions?*

A. Table 2.7 lists many of the standard conversions needed for various measurements encountered in a wireless system. Often the

Table 2.5 Various IS-136 Subscriber Power Levels

Station class	ERP (max)
1	6 dBW (4 W)
2	2 dBW (1.6 W)
3	22 dBW (0.6 W)
4	22 dBW (0.6 W)

	Nominal transmit power (dBW) per mobile unit station power class			
Mobile unit power level	**1**	**2**	**3**	**4**
0	6	2	−2	−2
1	2	2	−2	−2
2	−2	−2	−2	−2
3	−6	−6	−6	−6
4	−10	−10	−10	−10
5	−14	−14	−14	−14
6	−18	−18	−18	−18
7	−22	−22	−22	−22
8	—	—	—	−26
9	—	—	—	−30
10	—	—	—	−34

more common conversions deal with converting from metric to U.S. Customary System (USCS) units and vice versa.

Q. *What are some commonly used standard temperature conversions in wireless systems?*

A. The table at the top of p. 67 lists the standard temperature conversions needed for a wireless system.

TABLE 2.6 DECIBEL REFERENCE TABLE

Decibels	Reference	Comment
dB	None	
dBm	1 mW	Standard wireless value
dBs	1 mW	Japanese wireless system reference
dBc	None	Referenced to the carrier power
dBW	1 W	1 watt reference
dBk	1 kW	1 kilowatt reference
dBμ	1 microvolt	Standard wireless value
dBV	1 V	1 volt reference

TABLE 2.7 STANDARD CONVERSIONS FOR WIRELESS SYSTEM MEASUREMENTS

From	To	Multiply by
Meters	Feet	3.28
Feet	Meters	0.3048
Miles	Kilometers	1.609
Kilometers	Miles	0.6214
Kilometers	Feet	3281
Feet	Kilometers	0.0003408
Liters	Gallons	0.2642
Gallons	Liters	3.785
Rods	Feet	0.06061
Yards	Feet	3
Yards	Meters	1.094
Inches	Centimeters	2.54
Centimeters	Inches	0.3937
Feet	Centimeters	30.48
Centimeters	Feet	0.03281

From	To	Multiply by
Fahrenheit	Kelvin	(F + 459.67)/1.8
Celsius	Fahrenheit	(C · 9/5) + 32
Fahrenheit	Celsius	(F − 32) · 5/9
Celsius	Kelvin	C + 273.1

Q. *What are the major parts of a fixed network design associated with a wireless system?*

A. The following is a brief list of the elements of a fixed network design associated with a wireless system. There are obviously more elements to a network design, including parts to facilitate coordination with adjacent markets as well as with the various vendors. However, the following list should prove helpful in establishing the resources and time frame needed for a fixed network design to be successful.

Network design criteria
MTSO design
Network operation center (NOC) design
Acceptance test procedures
Data network plan
Define network software
Switch database
HLR database
Signaling control point (SCP) database
Signaling transfer point (STP) database
Voice mail system
Short messaging
Authentication
BSC database
BTS database
System dialing plan
System routing plan
Interconnect plan
Telephone code management
STP/SCP design
Call delivery testing
Billing systems

Alternative interconnect
System integration

Q. *What should the format be for a scope-of-work outline?*

A. The format that should be used for a *scope-of-work outline* (SOW) is given below. Obviously, the topics below can and should be modified to meet your particular requirements. However, the format presented should provide good guidance in establishing a well-defined scope-of-work outline.

Scope of Work

Overview

Briefly describe the project itself. This is where you describe what the project is in a paragraph or two—no doctoral dissertations unless you feel compelled to do so.

Subprojects

Provide a bulleted list of subprojects with a brief description of each subproject. For example, a subproject under "MTSO design" would be "Network Equipment Specifications," and it might have another subproject under it like "Network Equipment Design and Layout."

Project Duration

Indicate the time frame in terms of days, or hours, in which each of the subprojects will be completed. Keep in mind that this is an estimate, or best guess. You will most likely need to complete this timetable in conjunction with several other items listed next.

Starting Time

Indicate when this project needs to start. For most projects, the earliest starting time is today. However, when considering the starting time, divide projects into two types. The first is a project that does not require that another project be completed first. In this case, try to pick the time that you believe it should begin. The second type of project, or subproject, requires that another project be completed or begin at the same time. Indicate which one it is. You might have a project that can begin at any time but a subproject that requires a specific project be completed first. Also, note that subprojects do not have to be serial in nature. There are bound to be situations in which parallel processing can take place.

Input Required

Indicate what information you will need to accomplish a particular project. All projects will require several forms of data to complete. Indicate the types of information and the potential source of the information. The information required could be a design from another project in the fixed network portion or an RF design. Keep in mind that to get a project moving, all the information is not needed at once. Therefore, indicate at what time, for which subproject, you will require the information to be available.

Personnel Resources Required

Indicate what the human resource requirements will be for each subproject. This data will be used to help secure more people, the right people, to help complete the project. When putting together the human resource requirements, try to indicate the skill level needed. The suggested format for arriving at the skill sets needed is as follows:

1. Senior engineer
2. Engineer
3. Associate engineer
4. Technician

When listing the personnel needed, indicate the estimated number of days each would be involved with the project or subproject. Expand on the list above as you need to, and indicate a particular skill set required.

Training

Indicate any special training you or other project team members may need to complete this project. If you do not know the name of the particular training class you or others need, just indicate the type of training. For example, the type of training could be "Nortel switch architecture."

Equipment Required

Indicate what the equipment requirements are or are expected to be. Break down equipment requirements into several major groups: fixed network, test equipment, and engineering. The fixed-network equipment needed for the project could be a digital access cross-connect system (DACS) or "voice mail system." The test equipment needed for the project could be a protocol analyzer or T Bird. The engineering equipment needed could be an engineering DOS and/or UNIX workstation, software, plotters, and printers. Provide as much detail as you can.

The actual procurement policy and procedure that will be followed is not defined at this time. An *engineering recommendation* (ER) will be generated later for all major capital purchases above a certain threshold.

It is recommended that the format for the equipment required is as follows:

1. Equipment: Date needed
2. Testing Equipment: Date needed
3. Engineering: Date needed

Output

Indicate what the output will be from the project and the subprojects itself. Keep in mind that some of your subproject outputs will be input requirements for other projects. For example, the output could be a design specification, equipment ordered, or equipment integrated successfully into an existing network.

3

Network System

The following section addresses many of the commonly asked questions pertaining to network systems. The term network system is used differently from company to company within the wireless industry. In this book the term refers to the switching environment, or rather, for our discussion purposes, the part of the system that does not involve actual radios. However, note that microwave backbones for transport can also be included with network systems. It is interesting to note that many of the issues covered in this section also apply directly to the wire-line arena.

Q. *What is an MSC?*

A. A *mobile switching center* (MSC) houses the brains of the wireless network. (See Fig. 3.1.) The MSC coordinates all the cell sites it is responsible for and performs all the call delivery functions and auxiliary platform functions. There can be multiple MSCs within a network, as is the common case with large networks. Usually an NOC is used to coordinate all the activities of the various MSCs. The MSC is also called the *mobile telephone switching office* (MTSO), although that term is currently out of vogue.

Q. *What is 900 service?*

A. *900 service* is a telephone service that has 900 as its area code. The service is free to the company or person who receives the call and is charged to the person making the call. 900 service is billed through the telephone company (telco), and the fee charged per call or minute is determined by the receiving party. Some immediate examples of 900 services involve weather information, but the service is more commonly associated with pornography. Most, if not all, wireless operators *block service,* that is, prevent their subscribers from using 900 services. The reason for this is that the wireless company is liable for paying the service costs since it delivers the calls.

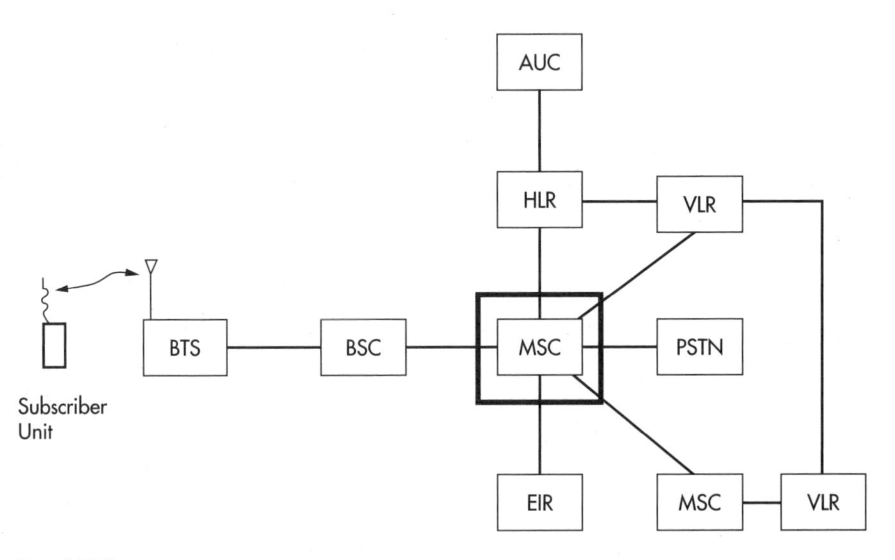

Figure 3.1 MSC.

Q. *What is an AIN (or IN)?*

A. An *advanced intelligent network* (AIN), also called an *intelligent network* (IN), enables carriers to create and manage a vast array of telecommunication services within a common architecture. An AIN gives the communication network the ability to determine how to route a call and how to treat a call based on the calling pattern of the user. With an AIN, the customers can have their calls routed the way they want. AIN technology is making its way into the wireless arena because it offers additional services and roaming capabilities for delivering calls.

An AIN's specific format and methodology depends on the manner in which it is implemented in each company. However, in general, an AIN is used to route calls based on a preestablished set of conditions on a real-time basis using defined rules instead of classical switch routing techniques. Another way it is used is to allow the calling party or receiving party to alter the call's delivery or reception.

An AIN utilizes a database for making the routing, delivery, or reception alterations that are different from classic telephony. The database that holds the customer-specific information that is used to route the incoming or outgoing calls is the SCP and is part of the SS7/CC7 network to which the operator has access.

Q. *What is an ANI?*

A. The term *automatic number identification* (ANI) has become synonymous with *caller ID*. This feature allows the party at the receiving end, or termination point, to know who the caller is before he or she answers the call. Caller ID has become a standard feature associated with digital systems for the wireless community.

Q. *What is an HLR?*

A. A *home location register* (HLR) maintains all the permanent information on the subscribers that it is responsible for regardless of their physical location. In wireless systems, the HLR coordinates all subscriber identification activities when the mobile unit roams between systems or between MSCs within the same system and provides all the necessary identification information to not only ensure that the subscriber is a valid subscriber but also to provide information on how his or her calls are to be treated and the associated privileges that he or she is allowed.

The HLR does not have to be physically part of the switch that is processing the subscriber's call. The HLR is usually a separate platform that is located at a central location within the network, or it can be part of the existing MSC. However, having the HLR reside within an existing switch is not a common application due to the volume of subscribers that exist on many networks.

The HLR is a key element in the functioning of an IS-41 system. The HLR updates and sends subscriber information to the *visitor location register* (VLR) of the serving MSC for the subscriber. The location of the HLR within a typical wireless system is shown in Fig. 3.2.

Q. *What is a VLR?*

A. There is a *visitor location register* (VLR) in every switch used in a wireless application. Therefore, if there are three switches within a network, then there should be three VLRs. The VLR is used when the subscriber is roaming or does not have its database resident

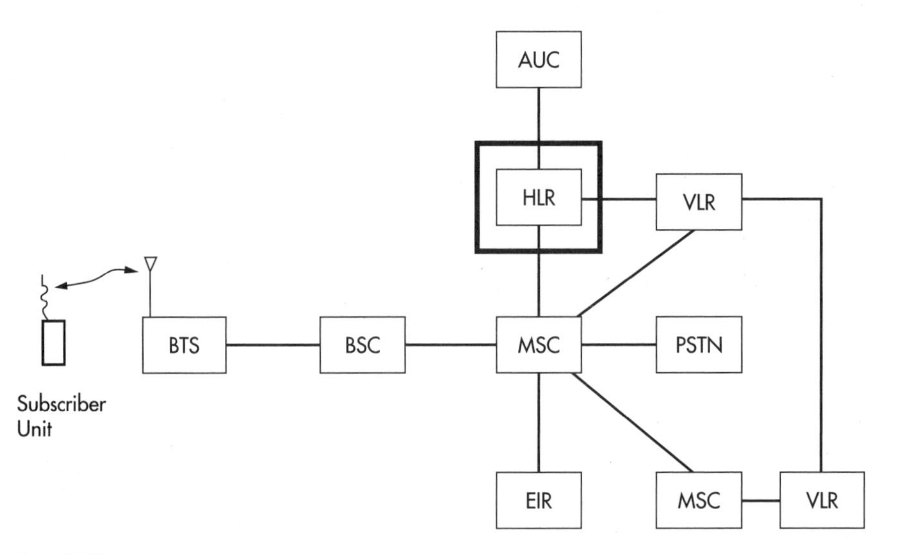

Figure 3.2 HLR.

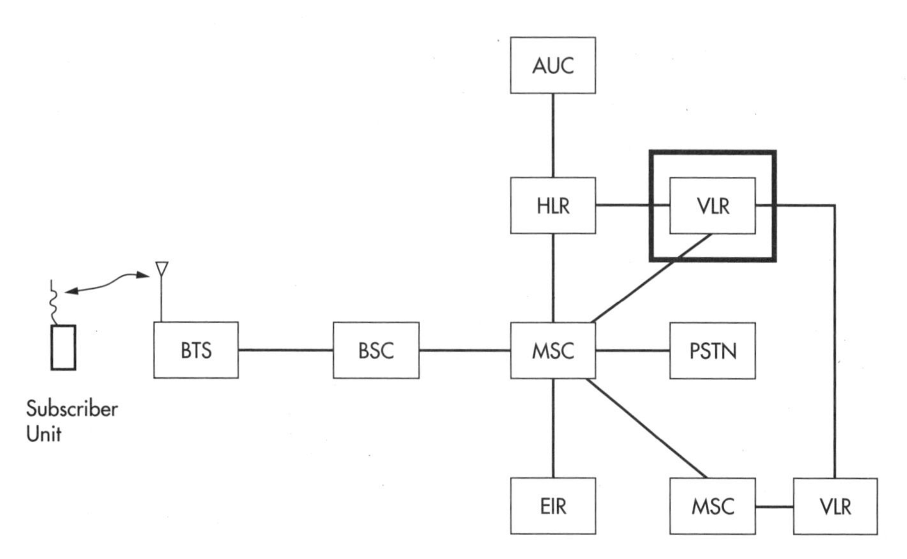

Figure 3.3 VLR.

on that particular switch. In a wireless system each switch has its own VLR, and each VLR interacts with the HLR to maintain the current information regarding the subscriber. Therefore, as the subscriber turns on or off his or her phone or moves from one switch to another, the VLR associated with the switch that the subscriber is currently associated with updates the HLR regarding the status of the subscriber and his or her location. Additionally, the VLR administers the TMSI (temporary mobile station identifier) and MSRM and any handover information. The location of the VLR is shown in the diagram of a typical switch in Fig. 3.3.

Q. *What is an EIR?*

A. An *equipment identity register* (EIR) is used to deny access to the wireless system by unapproved, faulty, and stolen wireless phones. The EIR is a valuable tool for fraud prevention. The location of an EIR in a typical wireless application is shown in Fig. 3.4.

Q. *What is an AUC?*

A. An *authentication center* (AUC) prevents access to the wireless network from an unauthorized source. The AUC is normally part of the HLR where it performs unique functions and authenticates the subscriber unit before it allows access to the system. (See Fig. 3.5.)

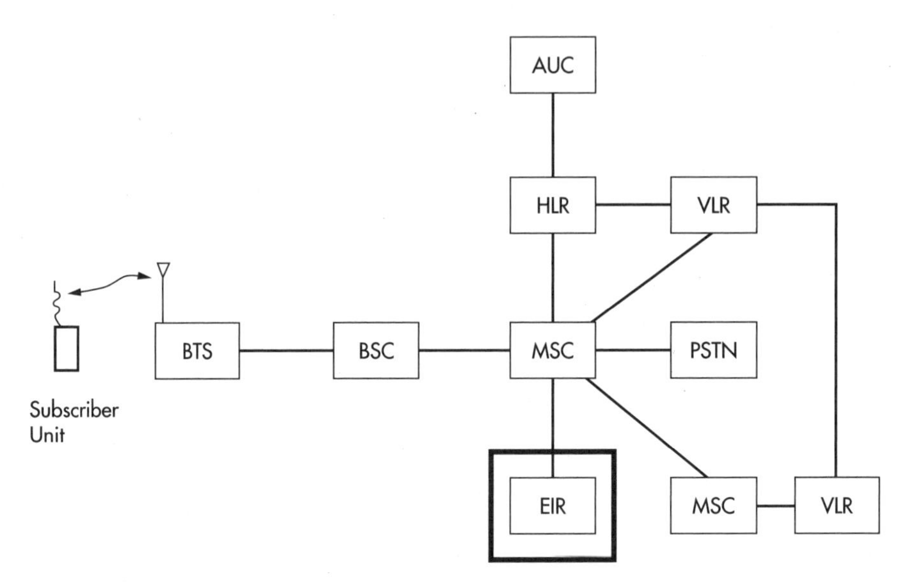

Figure 3.4 EIR.

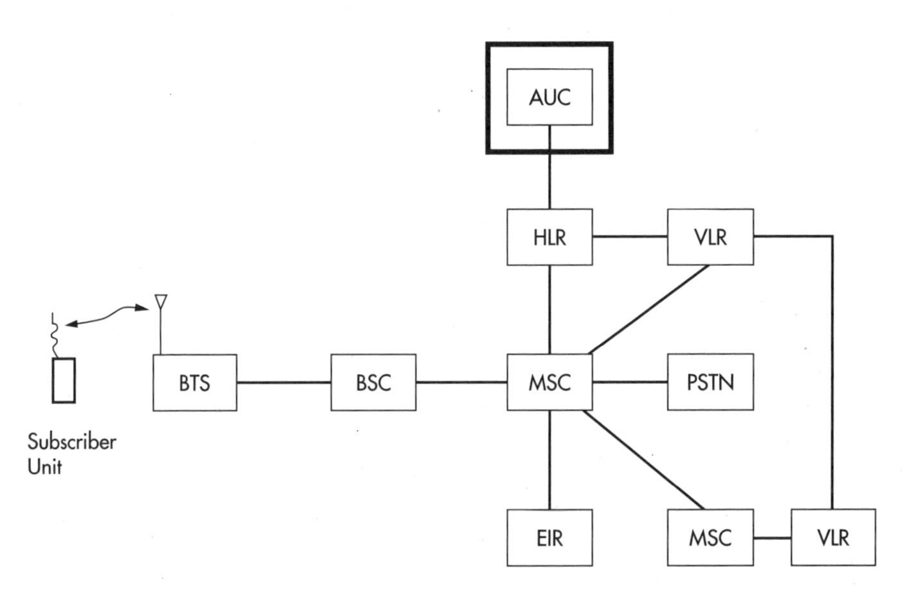

Figure 3.5 AUC.

Q. *What is an SMS?*

A. A *short-message service* (SMS) *center*, provided by a wireless company, offers capabilities similar to those of a paging system. If an SMS is properly installed and operational, it allows for the integration of the telephony functions with paging, thereby eliminating the need to carry two devices. Some of the features or functions of an SMS system are the following:

- Short-message, text and numeric, transmission
- Short-message broadcasting
- Confirmation of short-message delivery
- Delivery acknowledged back to sender, two-way paging

The SMS used by a wireless operator can vary and there are a multitude of different vendors. The location of the SMS in a wireless application is shown in Fig. 3.6.

Q. *What is an A interface?*

A. An *A interface* is a component called for in the GSM standard, and its purpose is to assure intervendor compatibility. Specifically, the A interface is the signaling link between the

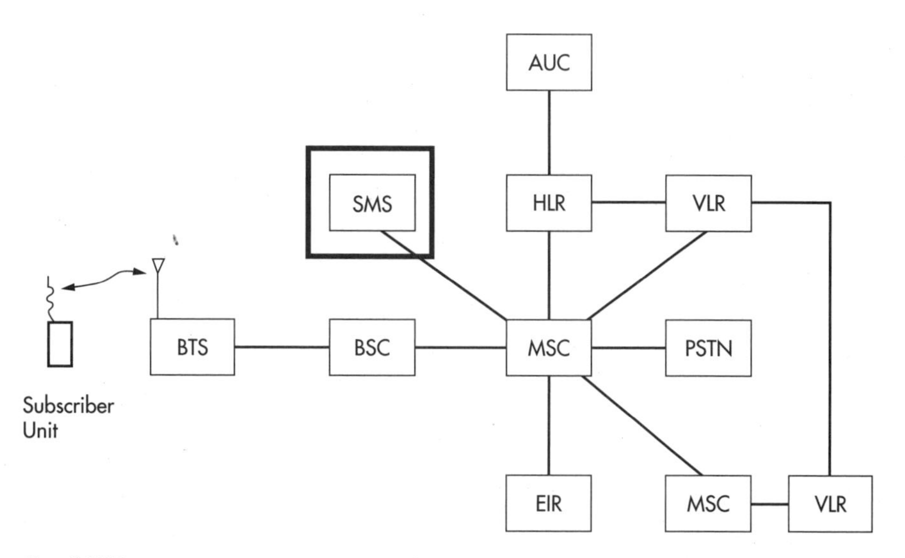

Figure 3.6 SMS.

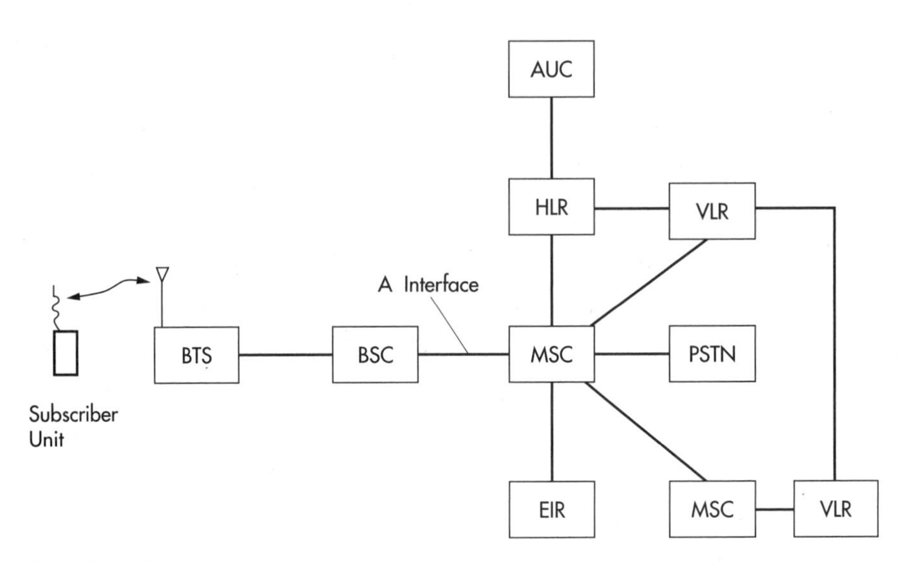

Figure 3.7 A interface.

vocoder portion of a GSM system and the main switch. For example, using Nortel equipment, the A interface is positioned between the TCU and the MSC; when using Ericsson equipment, the A interface is between the BSC and the MSC. The Ericsson BSC also performs the Nortel TCU function, but it is combined into one platform instead of being separated. The standard MSC location of the A interface is shown in Fig. 3.7.

Q. *What is a BSC?*

A. The term *base station controller* (BSC) has different meanings to different wireless vendors. Usually, the BSC is the common name for the processor at the cell site. However, for a network, the BSC is part of the GSM network or iDEN where there are several cell sites that connect to the BSC, which then connects to the MSC. The location of the BSC is shown in the network in Fig. 3.8.

Q. *What is the U_m interface?*

A. The *U_m interface* is the air interface for a GSM through which is passed the communication to and from a subscriber unit. The U_m interface utilizes a LAPDm (link access procedure—D mobile) protocol. The location of the U_m interface is shown in Fig. 3.9.

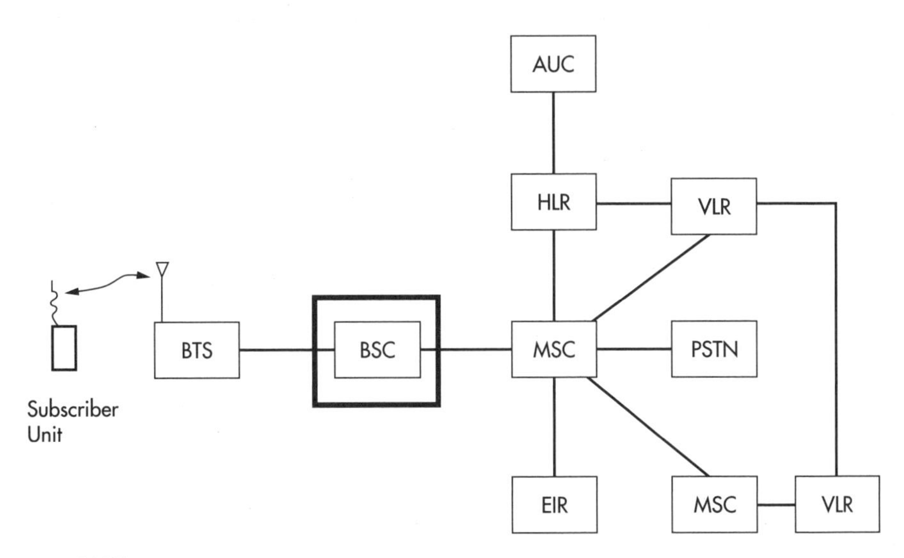

Figure 3.8 BSC.

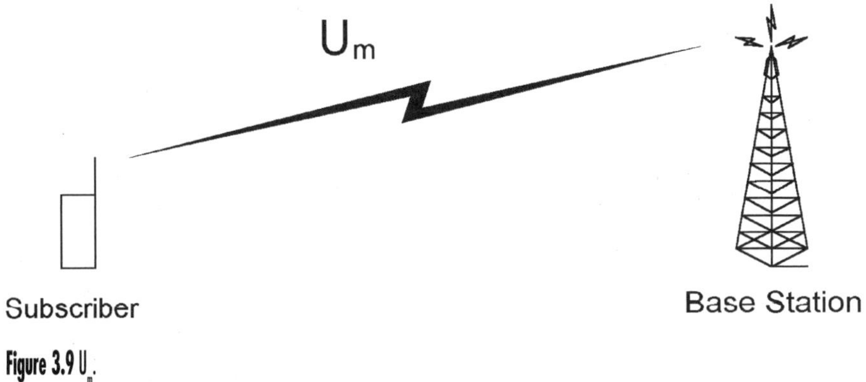

Figure 3.9 U_m.

Q. *What is the Abis interface?*

A. The *Abis interface* lies between the BSC and the individual BTSs of a GSM system. Communication passes from the BSC to the BTSs and vice versa through the Abis interface in which it is subject to a uniform protocol, thus ensuring vendor compatibility. The location of the Abis interface is shown in Fig. 3.10. The Abis utilizes a T-1/E-1 using *pulse code modulation* (PCM) to transport the information.

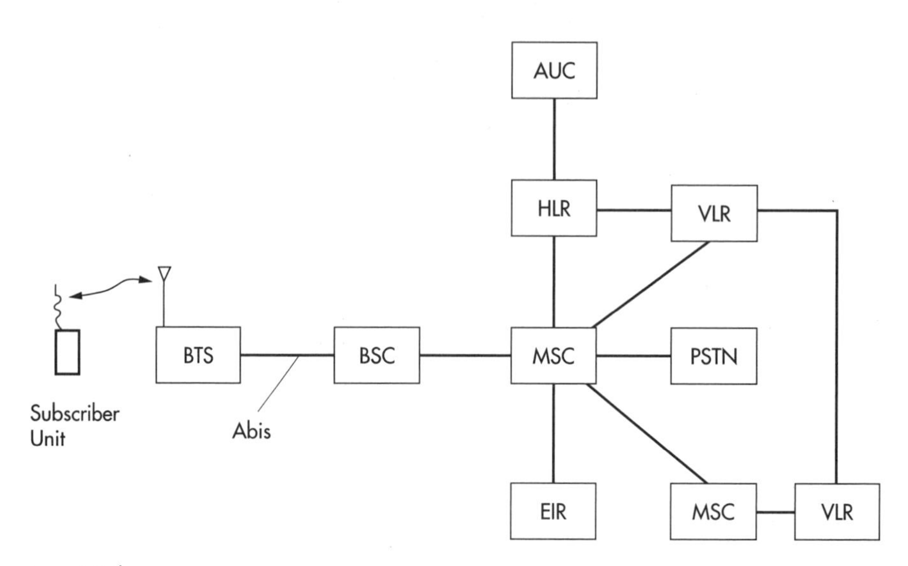

Figure 3.10 Abis.

Q. *What is the Ater interface?*

A. The *Ater interface* is a component that is used in GSMs to facilitate communication between a BSC and the TCU, or rather, vocoder units. The signaling between the devices utilizes the LAPD protocol. However, between the BSC and the MSC, CCS7 signaling is utilized. (See Fig. 3.11.)

Q. *What is X.25?*

A. The *X.25 protocol* is used in packet switching applications to provide a direct connection to a packet switched network. The X.25 protocol utilizes the OSI modem, and it has three layers: physical, data link, and network. A wireless system utilizes X.25 signaling in various parts of its network to ensure error-free communication with a guaranteed delivery of the information.

Q. *What is ADPCM?*

A. *Adaptive differential pulse code modulation* (ADPCM), also known as *bit crunching* or *compressing,* is used in wireless applications to reduce the facility costs when a second DS-1 is added to the network for connecting a cell site to the MSC. The

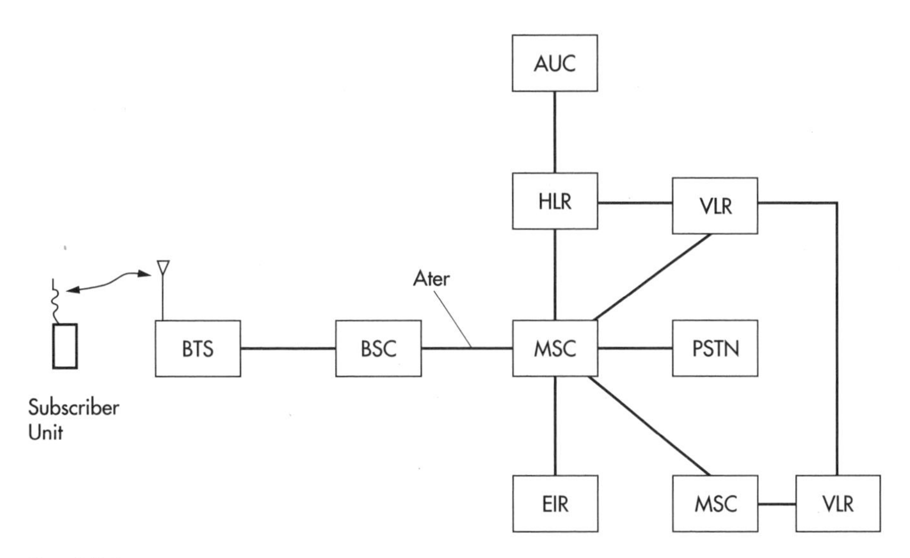

Figure 3.11 Ater.

ADPCM is accomplished by compressing the PCM code that is contained in a normal DS-1 from 8 kb/s to 4 kb/s, thereby allowing 48 conversations to be transported over a single DS-1 instead of relying on two DS-1s to do the same job. The ADPCM process requires a device at both ends in order to encode and decode the material. However, it should be noted that some wireless systems require that the control information that is sent from the MSC to the cell site cannot be compressed and therefore requires the use of variable-rate ADPCM.

Q. *What is variable-rate ADPCM?*

A. *Variable-rate ADPCM* performs the same task as ADPCM, but it provides for clear channels, that is, channels that are not compressed. The use of clear channels is necessary for certain wireless applications for control signaling between the cell site and the MSC.

Q. *What is an area code?*

A. An *area code* is a specific three-digit code that identifies a particular toll center using the North American dialing plan. The area code is also commonly referred to as the *numbering plan area*

(NPA), and it is part of the dialing sequence that is used to deliver a call. The dialing pattern takes on the following format: NPA-NXX-XXX, where the NPA is the area code portion. To be more specific, in the number 914-987-1787, the 914 represents the NPA of 914 and refers calls to the toll center for 914 for delivery.

The NPA references a potential 1000 exchanges that could be within its domain. Typically a wireless operator is not granted his or her own NPA for use. However, there are NPAs dedicated to specific services such as 800 and 888 for toll-free dialing, 900 for calling-party-paying calls, and 500 for universal services.

Q. *What is a dialing plan?*

A. A *dialing plan* is an engineering document, which is used by both vendors and operators, that defines how calls are to be processed within a wireless or landline telephony network. For example, a dialing plan would specify which NPA-NXXs reside in the network. Also, the dialing plan would specify whether permissive dialing is allowed and how it is achieved. For any wireless network, the dialing pattern is frequently modified. It is essential that all changes be recorded in the dialing plan.

Q. *What does the term asynchronous mean?*

A. *Asynchronous* describes communication that is not synchronized. Wireless systems use a combination of synchronous and asynchronous communication. When communication devices transmit in their own time frame rather than transmit the moment the sender initiates the transmission, it is referred to as *asynchronous communication.* Some simple examples of asynchronous communication equipment are modems, faxes, and packet data TCP/IP switches.

Q. *What is B8ZS line coding?*

A. *Binary 8 zero substitution* (B8ZS) is a type of line coding. This format is used to prevent too many consecutive zeros from being sent in a row, which would negatively impact the timing that is derived from the signal. B8ZS line coding is used for DS-1 communication, which is the primary interface between cell sites and the MSC.

Q. *What is back-haul technology?*

A. The term *back haul* describes the routing method that is used by an operator to deliver a call service between the cell site and the MSC, at the least expensive method. This term is commonly used in the wireless industry to describe the method used to deliver. The term is a misnomer in the sense that the signal may take a longer route than necessary as it has to be routed past its destination and then rerouted back to where it should be delivered.

Q. *What is a backbone?*

A. In any wireless communication system, a *backbone* refers to the fixed-network portion that connects all the major nodes. Many times, the term *backbone* in wireless systems refers more specifically to the wire-line portion of the communication network.

Q. *What is the BER?*

A. The *bit error rate* (BER) is the ratio of errors to nonerrors that are received on a bit-by-bit basis. The BER for a system or subsystem depends on the communication system requirements and thus varies from system to system.

Q. *What is a BERT?*

A. A *bit-error-rate test* (BERT) is one method that is used to measure the quality, or rather the integrity, of a communication system with respect to its digital component. A BERT for wireless operations can be conducted on the landline portion of a communication link—that is, the T-1 or T-3 lines—or on the radio environment. When a bit-error-rate test is conducted, the objective is to determine that the line or path being tested is performing at or better than the defined *bit error rate* (BER) that has been established for acceptable communication. The actual BER used by the operator is different for each device tested due to different modulation formats and error correction codes. The BERT determines the ratio of errors to nonerrors that are received on a bit-by-bit basis. Therefore, a BERT that shows a BER of 0.01 percent means that the path used has 1 error for every 10,000 bits sent. Depending on the communication system requirements, this value can be acceptable or not.

Q. *What is a CDR?*

A. A *call detailed record* (CDR) stores information about how a call transversed through the system, including how it was treated and its duration. Billing information is extracted from the CDR so that the caller can be billed. Wireless operators use CDRs not only for billing purposes but also for resolving billing disputes that may arise. The CDR can also be used to help isolate and troubleshoot a system problem because it contains the records of how a call was treated. A typical CDR will contain the originating point for the call, the treatment of the call, the duration of the call, the facilities it utilized, and the manner in which the call was terminated.

Q. *What is caller ID?*

A. *Caller ID* is a service available in an advanced intelligent network (AIN). It allows a caller's phone number to be sent and displayed to the person receiving the call; its sole purpose is to identify callers. Caller ID is a common feature now available on wireless systems, and it is used to screen incoming calls.

Q. *What is clear channel coding?*

A. *Clear channel coding* (CCC) is a service provided for data transmission via a T-1 line. Specifically, CCC involves using the T-1 for *out-of-band signaling,* which means that the entire 64 kb/s is available for each DS-0. The T-1 uses out-of-band signaling so that there is no bit robbing, and signals like dial tones, hook flashes, and DTMF (dual-tone multifrequency) are sent over the 24th DS-0 of the T-1. Thus 23 DS-0s are capable of 64-kb/s data rates instead of their usual 56 kb/s. To increase the data rate in this way, the 24th DS-0 is dedicated to just signaling.

Q. *What is the central office code?*

A. The *central office code* is also commonly referred to as the NXX in the numbering pattern used for the North American dialing patterns. The NXX is also called an *exchange number* because it refers to a particular exchange within an NPA code. The exchange number, or NXX, is used to represent a geographic area for a central office. However, with number portability, this relationship no longer exists. Thus the assignment of NXX addresses is almost a random process and tends to not follow any logical pattern, which is unfortunate.

In the North American dialing plan, the number NPA-N*XX*-*XXXX*, or rather using a direct example, 914-987-1787, the N*XX* refers to the exchange that houses the 987 codes. The exchange code represents a total of 10,000 numbers, from the *XXXX* code that is assigned to each NXX.

Q. *What is a centrex system?*

A. A *centrex system* is a service that is offered by telephone companies for the purpose of providing private branch exchange (PBX) functions without the customer's having to install a PBX system. The centrex service allows the customer to have all the features and benefits of a PBX system without incurring the capital costs or dealing with the implementation and maintenance issues associated with the installation of a PBX system. Centrex service is not typically offered by wireless service providers at this time, but it could be bundled with a *competitive local exchange carrier* (CLEC).

Q. *What is a cesium clock?*

A. A *cesium clock* is used to synchronize a communication network to a given reference. It is called a "cesium clock" because it uses the vibrations of the atomic element cesium to derive its timing. One example of cesium clock use is in a synchronous optical network (SONET).

Q. *What is a circuit ID?*

A. A *circuit ID* is the identification code given to a particular circuit. It is used by wireless communication services in the ordering of spans, T-1s, from a local telephone company or a CLEC. The circuit ID uniquely identifies a circuit for troubleshooting purposes when initially establishing service or restoring service. The circuit ID is also used for billing purposes.

Q. *What is a CID code?*

A. A *circuit ID* (CID) *code* refers to the particular name that is assigned to an SS7 or a CCS7 circuit. The CID is important for an SS7 link because it is a unique identification of the link, which is needed for the link to connect and operate correctly.

Q. *What is a class-of-service designation?*

A. A *class-of-service* (COS) *designation* describes the type of service or line that is leased or purchased from a telephone company, *local exchange network* (LEC), or even a CLEC. There are many types of COS designations, and the following chart elaborates on each one.

COS	Comments
1FR	This means *one-flat-rate residential service,* and it is typically what subscribers have for their home telephone service.
1MR	This means *one-measured-rate residential service.* This service has a defined low monthly rate for a certain number of local calls. When the preset volume of local calls is exceeded, then a fee is attached to additional calls.
1FB	This is a *one-flat-rate business service,* and it is typically what businesses have for their telephone service.

Q. *What is the CLLI (silly) code?*

A. The *common language location identifier* (CCLI) *code*, also referred to as the *silly code,* is used to identify the physical location of a telephone communication network. A CCLI code is an 11-digit alphanumeric code that every wire-line and wireless operator needs to process telephony calls. Every wireless operating system has a CLLI code, which is needed for a variety of reasons starting with billing.

Q. *What is a conditioned circuit?*

A. A *conditioned circuit* is any twisted-pair electric cable that is modified to carry digital information instead of analog voice. The use of conditioned circuits is needed for the T-1 circuit that connects the cell sites to the MSC. The difference between a conditioned circuit and nonconditioned circuit is that the conditioned circuit has noise filters placed on it to improve the *signal-to-noise*

(S/N) *ratio* of the signal. However, the unconditioned circuit usually has a load circuit associated with it, which helps reduce noise but, because it is a choke, also restricts the bandwidth of the circuit, impairing the transmission of digital signals.

Q. *What is a CSU?*

A. A *channel service unit* (CSU) is used to provide a demarcation point, line coding, and alarm capabilities for the operator. The CSU is used in conjunction with a T-1 and is usually located at the cell site or customer premise location. The CSU provides the demarcation point from the telco or CLEC to which the operator delivers the service. From that point, the telco is responsible for providing a specified quality of signal. The line formatting that the CSU also offers allows for the conversion between the telco and the customer premise equipment. Often the CSU is referred to as a *CSU/DSU,* which is another name for *channel service unit/data service unit.* The CSU/DSU's location in a wireless application is shown in Fig. 3.12.

Q. *What is a DACS?*

A. A *digital-access cross-connect system* (DACS) is also referred to in some circles as the *DXX* and can be seen in other literature as a *DCS,* all of course meaning the same function. What a DACS does is best illustrated in Fig. 3.13.

Q. *What is a demarc?*

A. A *demarcation point* (demarc) is the point at which the telephone company or CLEC is no longer responsible for the circuit's performance. Specifically, the demarc separates the customer's equip-

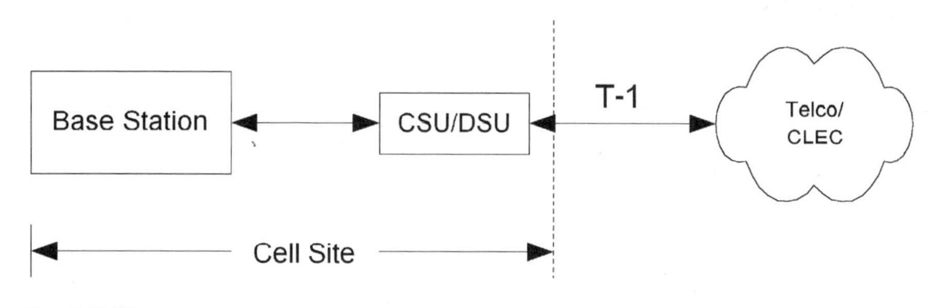

Figure 3.12 CSU.

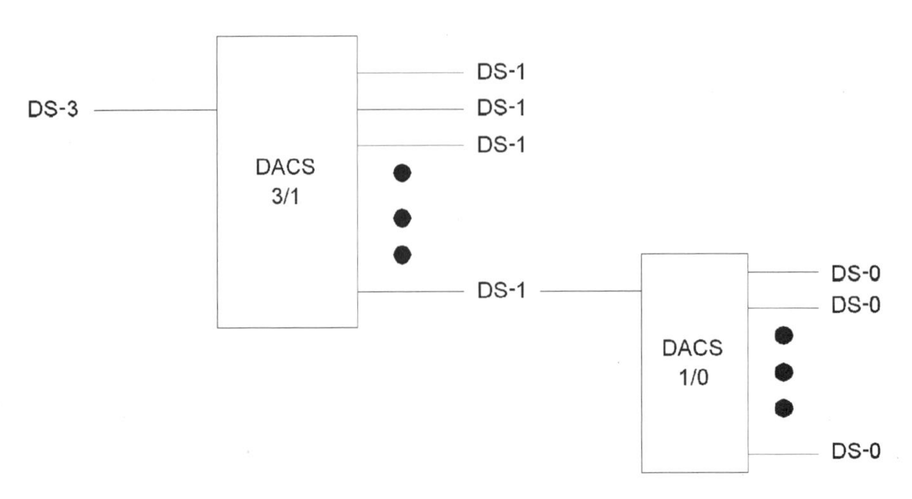

Figure 3.13 DACS.

ment from that of the telephone company. The demarc for a cell site is usually the CSU, and for the MSC it can be the DACS.

Q. *What is a drop?*

A. A *drop* is a name given to the service wire that is used by the telco company for physically connecting its system to the demarc point. The drop can be either fiber-optic cable or copper wire, depending on the situation and service available.

Q. *What is AWS?*

A. *Aerial wire service* (AWS) refers to the method the telco uses to deliver its physical cable or wire to a facility for providing service. Specifically, it refers to telephone service delivered to the location from a telephone pole. Service provided to the MSC should not be provided by an AWS; instead, it should be provided by buried-line methods.

Q. *What is BWS?*

A. *Buried-wire service* (BWS) refers to a method by which telephone service is delivered to a particular location. Specifically, it refers to telephone service delivered via underground wires. BWS is the common method used for delivering service to an MSC. In fact, there are usually two different BWS routes used so as to hasten disaster recovery.

Q. *What is a dry span T-1?*

A. A *dry span T-1* is a span that does not have −135 Vdc battery voltage.

Q. *What is a wet span T-1?*

A. A *wet span T-1* is a span that has −135 Vdc battery voltage.

Q. *What are the DS-1 formats?*

A. See Table 3.1.

Q. *What is a DS-1c?*

A. A *DS-1c* is a format that aggregates two DS-1s together. (See Table 3.1.)

Q. *What is E&M signaling?*

A. *Ear and mouth* (E&M) *signaling* uses four wires with one pair dedicated to the mouth signal (sending) and the other pair to the ear signal (receiving). In some cases, a third pair of E&M wires is used for additional control and signaling methods. E&M signaling is typically not used anymore, but it may be encountered when interfacing with older PBXs associated with wireless PBX applications that use the customer's equipment to handle some of the wireless traffic.

TABLE 3.1 DS-1 FORMATS

Line conditioning	Format	Signaling	Application
AMI	SF/D4	Inband	24 voice or modem channels
AMI	ESF	Inband	24 voice or modem channels
AMI	ESF	Out of band	23 voice or modem or digital or data channels
B8Zs	SF/D4	Inband	24 voice or modem channels
B8ZS	ESF	Inband	24 voice or modem channels
B8ZS	ESF	Out of band	23 voice or modem or digital or data channels

Q. *What is an end office?*

A. An *end office* is the wireless communications equivalent of a central office that connects customers to the telephone system.

Q. *What is a fractional T-1?*

A. A *fractional T-1* is a service that can be used when more than one DS-0 is needed for bandwidth but a full T-1 may be way too excessive or expensive. The fractional T-1 enables the aggregation of several DS-0s for use by the client. The use of a fractional T-1 may be applicable in those cell site applications that do not require a full T-1 (and that would benefit from reduced facility costs).

Q. *What is a ground start?*

A. A *ground start,* or rather *ground start trunk,* refers to a type of phone system that requires the phone line to use a ground instead of a short circuit for determining when the phone goes off hook. This technology is applicable to wireless applications in which older PBXs are interfaced. In these cases, when trying to provide an alternative interconnection method, a ground start can provide the correct interface circuit to achieve the conversion.

Q. *What is inband signaling?*

A. *Inband signaling* is a method of proving signaling of information that is being transmitted within the channel being used. An example of an instance in which inband signaling would be used is a typical residential phone that carries *dual-tone multifrequency* (DTMF) *signaling* along with the voice- and line-sensing signaling. An inband signaling for a T-1 line would have all the signaling information contained within the DS itself. The advantage of inband signaling is the simplicity of its use. The disadvantage of inband signaling is that bandwidth, and thus throughput, is wasted due to the overhead of carrying the inband signaling information. Inband signaling typically reduces a DS-0 from a potential of 64 kb/s to 56 kb/s, and it also reduces trunking efficiency.

Q. *What is out-of-band signaling?*

A. *Out-of-band signaling* is a method of proving signaling of information that is being transmitted outside the channel being used.

Out-of-band signaling is used by telephone companies and businesses that handle large volumes of traffic and data. With out-of-band signaling, the signaling information for a T-1 line is contained in the 24th DS-0 while the remaining 23 DS-0s operate at the maximum bandwidth possible, 64 kb/s. This method allows for a greater overall throughput as well as increased trunking efficiency.

Q. *What is a pedestal?*

A. A *pedestal* is a demarcation point, or rather a box, that is used by the telephone company to house the terminals and to serve as a splice point. Often in wireless applications, for exterior cabinet installation the pedestal could be a box that is physically attached to the wireless cabinet and needs power provided to it.

Q. *What is the POTS?*

A. The term *plain old telephone system* (or *service*) (POTS) refers to the separate phone line used in a cell site so that technicians can communicate with the MSC when conducting maintenance at the cell site. A POTS line is an added expense, but it is needed when communication via wireless is unobtainable usually because the serving cell site is off the air for a maintenance reason.

Q. *What is the voice band?*

A. The *voice band* associated with telephony covers the frequency range of 500 to 3500 Hz, which means that there are 3000 Hz of voice band available.

Q. *What is circuit switching?*

A. *Circuit switching* is a type of switching architecture that is utilized by both the telephone companies and wireless operators for the processing and routing of the telephony calls made by their subscribers. Input and output lines are connected to each other within a switch. The connection of such lines provides a single dedicated voice path over which a phone call may take place. The input and output line circuits remain connected for the duration of the call. Upon termination of the call, the connection is broken and the lines return to the idle mode, awaiting another call to process.

Q. *What is packet switching?*

A. *Packet switching* is a type of switching architecture that is used in many telecommunication applications. Many communications systems that utilize data transmissions among various nodes and computers in a network will utilize packet switching. Packet switching is well suited for the short-burst transmissions of the data network. Packet switching involves the sorting of data packets from a single line circuit and the switching of them to other circuits within the network. These sorting and switching functions are determined by an imbedded network address within the data packet itself. There are numerous applications for packet switching for wireless systems—for example, SMS or TCP/IP communications.

Q. *What is PCM?*

A. *Pulse code modulation* (PCM) is a signaling format used extensively in the telecommunications industry. This technology enables analog voice signals to be digitized, and it is therefore used widely in T-carrier systems. This technique is based upon the Nyquist sampling theorem, which requires a sampling rate of 8000 signals a second. Figure 3.14 shows a functional block diagram of the PCM process. In this digitizing process, the first step is to choose and measure points on the analog speech curve. The actual measured values are called *samples,* and the process itself is called *sampling.* These samples produce a series of pulses that represent the amplitudes of the various signals at specified times. The sampled amplitudes are then quantized so that they can be represented by a binary digital code. This is the second step in the PCM process (*quantizing function*). The final step (*encoding*) takes the digital codes from the quantizing process and then combines them into a serial data stream, which can then be transmitted over a single transmission line. This process is then reversed at the opposite end to yield the original analog signal.

Q. *What is an IXC?*

A. An *interexchange carrier* (usually IXC but also referred to as an IEC) is a telecommunication operator that provides service between specified *local access transport area* (LATA) boundaries. The IXC offers a toll-call service for the delivery of long-distance calls.

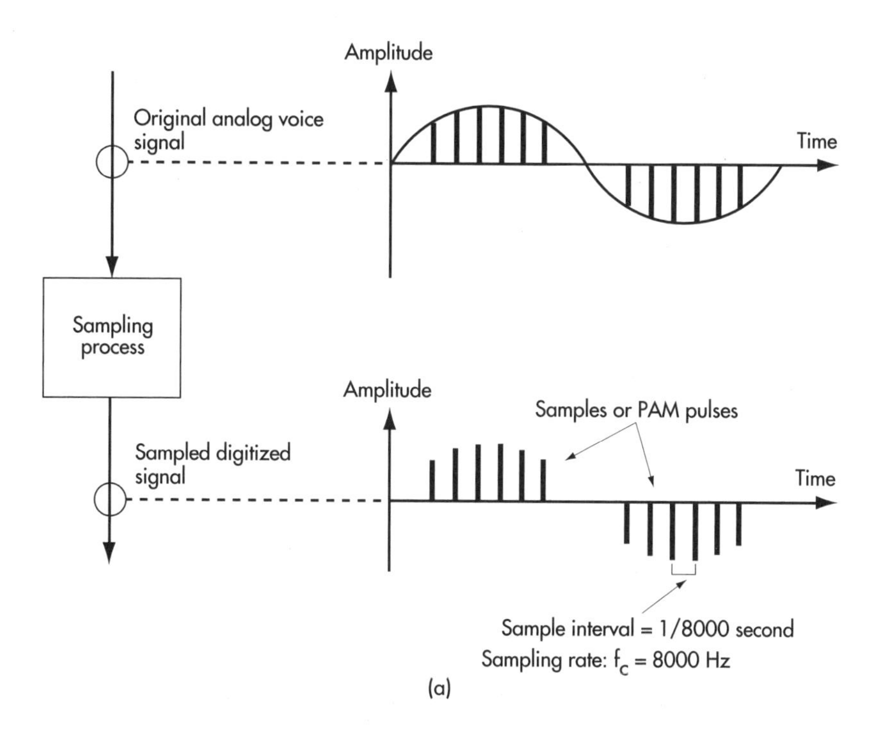

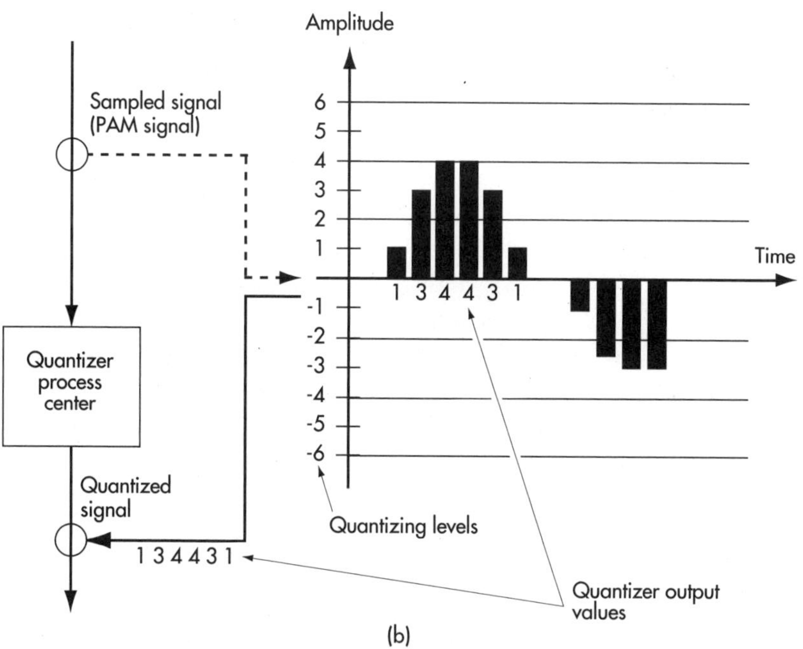

Figure 3.14 (*a*) PCM sampling process. (*b*) PCM quantification process.

Most wireless operators have many IXCs they interface with, and they can deliver long-distance service based on the company's ability to use the least-cost routing, which increases their margin, or by advising the subscriber to utilize a specific IXC.

Q. *What is time division switching?*

A. *Time division switching* refers to a method of transferring voice or data signals or a combination of the two from one location to another. In time division switching, it is the data of the individual time slots or PCM channels that are interchanged within the switch matrix between incoming line circuits and outgoing line circuits. Time division switching involves the switching of actual digitized voice data within the individual time slots.

Q. *What is space division switching?*

A. *Space division networks* establish a physical link through the network for connecting originating and terminating users to one another. Thus, in *space division switching,* an input line is connected to an output line within the switch, building a complete physical path through the matrix so that a call may take place.

Q. *What are the classes of switches?*

A. There are five classes of switches within the *North American Public Switched Telephone Network* (PSTN). These classes can be subdivided into the central office class 5 switch and the remaining tandem-type switches of classes 1 through 4. The basic difference between these two categories of switches is that a class 5 local switch will be directly interconnected to subscriber terminal equipment while a tandem switch will be interconnected to other switching equipment. The local class 5 switch has the ability to terminate a call to one of its subscriber units while a tandem switch can only route calls to other destined nodes, and it can never act as a final call delivery point.

In wireless systems, the switch that is utilized is a hybrid of a class 5 and a class 4 switch. The reason they are combined is that in most applications the wireless switch does not provide answer supervision, and while it performs many of the same functions as a class 5 switch, it is still a gateway. (See Fig. 3.15.)

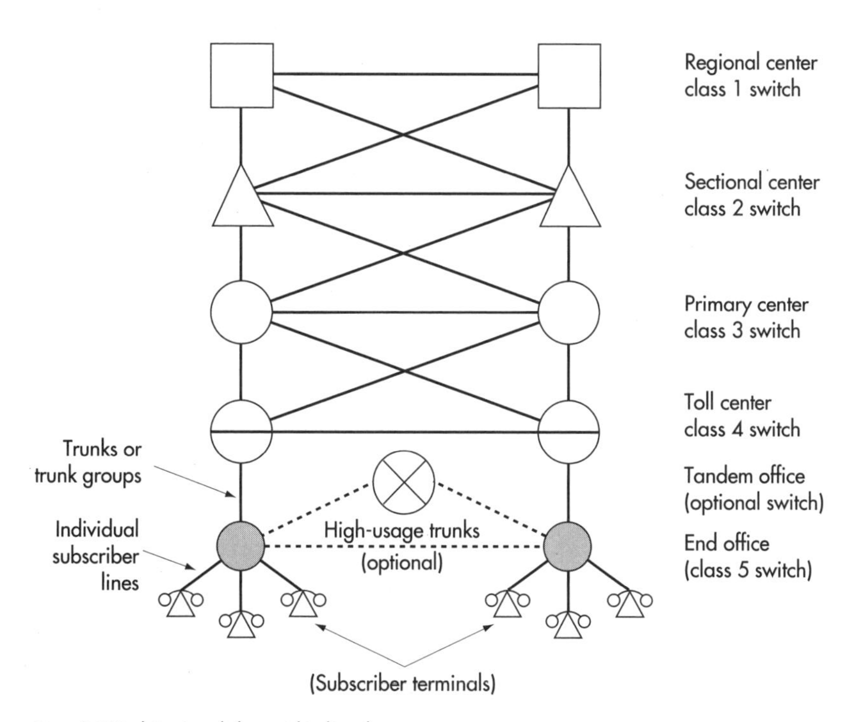

Figure 3.15 North American telephone switching hierarchy.

Q. *What is a T-1 carrier?*

A. The *T-1 carrier* is also called a "T-1," a "DS-1," or even a "span." The T-1 carrier transmits digital, two-way voice, data, or video signals over a single high-speed circuit. The actual data transmission rate of the T-1 carrier system is based on the bandwidth utilized to transmit one digitized voice signal. This signal is referred to as a *digital signal, level zero* (DS-0). The bandwidth of this DS-0 signal is equal to 64,000 b/s. The initial T-1–based systems transmitted 24 voice channels over two pairs of twisted cables at a rate of 1,536,000 b/s. Later systems introduced a necessary control bit for synchronizing T-1 multiplexing equipment. This additional bandwidth brought the T-1-carrier transmission rate up to 1,544,000 b/s (also denoted as 1.544 Mb/s). This is now known as DS-1, the first signal level of a T-1-carrier system. Table 3.2 gives the various signal levels and transmission rates for T-1-carrier–based systems.

TABLE 3.2 VARIOUS SIGNAL LEVELS AND TRANSMISSION RATES FOR T-1-CARRIER–BASED SYSTEMS

Signal level	Carrier system	Number of T-1s	Voice circuits	Megabits per second
DS-1	T-1	1	24	1.544
DS-1c	T-1c	2	48	3.152
DS-2	T-2	4	96	6.312
DS-3	T-3	28	672	44.736
DS-4	T-4	168	4032	274.760

Q. *What types of PSTN interconnections are there?*

A. There are numerous types of PSTN interconnection formats used throughout the industry. The type of interconnection that is optimal for a particular application depends upon the situation. Table 3.3 lists facilities and their potential applications.

Q. *What is an LEC?*

A. A *local exchange carrier* (LEC) is a wire-line phone company that is designated to provide telecommunication service within a specified LATA. The LEC is normally the local telephone company for the area and is the company the wireless operators normally interface with to obtain facilities from to connect the cell sites to the MSC. However, the landscape for this is changing due to deregulation, and the emergence of *competitive local exchange carriers* (CLECs) is substantially increasing the number of interconnection options.

Q. *What are the SS7 and CCS7 (or C7) data networks?*

A. *Signaling system 7* (SS7) is a data communications transfer protocol used to provide out-of-band signaling in processing calls within the PSTN in North America. The *common channel signaling 7* (CCS7 or C7) *protocol* is the international version of the SS7 protocol. The SS7 protocol has three major advantages over inband signaling. The first advantage is its improved postdial delay for faster call setup times. This network improvement is very noticeable to the end user. The second advantage is its ability to signal

Table 3.3 PSTN Facilities and the Possible Interconnection Applications

Facility user type	Signaling	IXC-LEC circuit type	Application
A	DTMF	Line circuit	Switch to switch
B	DTMF	Trunk	950 service
C	DP/DTMF	Trunk	AT&T interconnection only
D	DP/DTMF	Trunk	Equal access

in the reverse direction, improving communication among various nodes in the network. And a third advantage of SS7 over conventional inband signaling is that SS7 offers better network fraud protection because with SS7 the data (voice) messages and signaling messages are sent over separate routes. (See Fig. 3.16.)

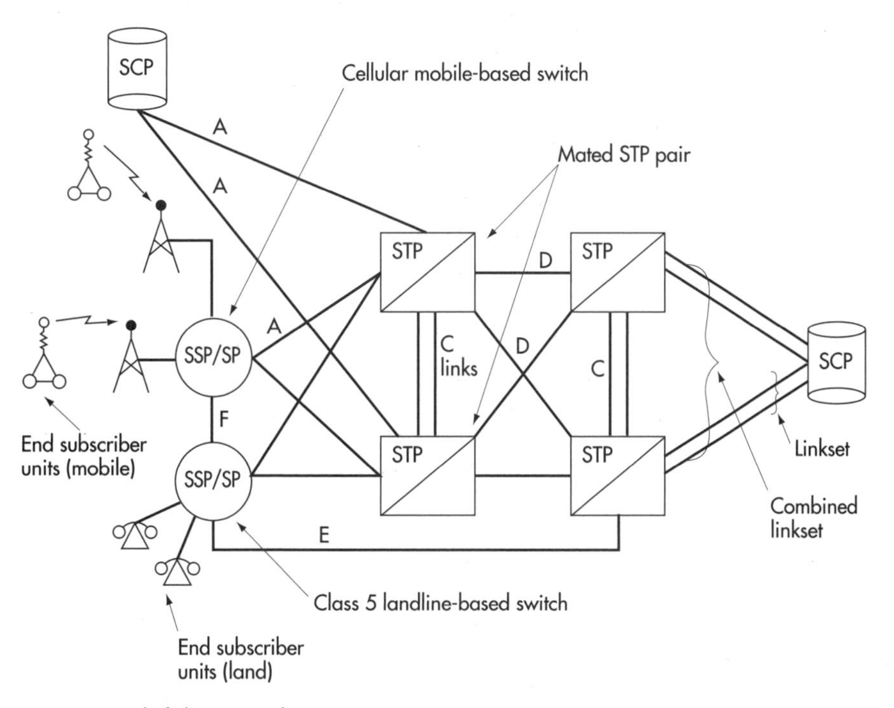

Figure 3.16 Example of a basic SS7 network.

Q. *What are signaling points?*

A. *Signaling points* (SPs) and *signaling service points* (SSPs) are part of an SS7 or CC7 network and serve as the connecting points for end subscriber units such as landline and mobile phones. These nodes perform call processing on calls that originate, terminate, or tandem at the location.

Q. *What is an STP?*

A. A *signal transfer point* (STP) is used in an SS7 or CC7 network. Specifically, the STPs transfer SS7 messages between interconnected nodes based on information contained in the SS7 address fields.

Q. *What is an SCP?*

A. A *signaling control point* (SCP) is used in an SS7 or CC7 network. SSPs, SPs, and SCPs operate similarly but only SCPs perform database functions. An SCP typically houses the subscriber records and information about how a call should be treated.

Q. *What are the SS7 links?*

A. The *SS7 links* typically include A, B, C, D, E, and F links, each of which has a specific function and application within the SS7 network. What follows is a brief description of each of these links. The link names are the same when describing a CC7 network.

A links are *access links* that carry SS7 messages between SSPs and STPs and between STPs and SCPs.

B links are *bridge links* that carry SS7 messages between STPs in different regions of the same network.

C links are *control links* that carry SS7 messages between mated STPs.

D links are *diagonal*, or *quad*, *links* that carry SS7 messages between STPs in networks.

E links are *extended links* that carry SS7 messages between SSPs and remote STPs.

F links carry SS7 messages between SSPs.

Q. *What is a linkset?*

A. A *linkset,* as the name implies, is a set of two or more links that connect adjacent nodes and share the same routing information. Linksets are associated with SS7 networks.

Q. *What types of data network protocols are supported in an SS7 network?*

A. There are a number of protocols that can be supported by and utilized in an SS7 network. While the SS7 network protocol itself provides the reliable transport of data messages among network nodes, there exists even more protocols used to build and decode these various messages based upon the particular application in the network. For instance, SS7 is the transport protocol used to carry IS-41–based messages among network nodes. However, the IS-41 protocol is yet another specification used to produce and translate IS-41–based messages. These messages are used to conduct mobile calls between switches of different vendor types. Another example of a network protocol is ISUP (integrated services user part). This protocol is used to provide ISDN (integrated services digital network) services between the mobile unit network and the PSTN.

Q. *What is a validation system?*

A. *Validation systems* are required by cellular operators as a means of checking the validity of the mobile unit subscriber. If the subscriber is valid (that is, he or she is a paying customer with good credit), then he or she is allowed to use the system. If the mobile unit subscriber is not a valid user, then he or she will be denied service on the system.

There are two basic types of validation systems: *postcall* and *precall.* The early cellular systems used postcall validation systems for determining if the mobile unit subscriber was authorized to use the system. Once a mobile unit user made a call on the system, the system could then begin to analyze the data about the subscriber and determine his or her status in the system. These systems worked well in the beginning, but soon a more advanced system was needed. The next generation of validation systems took advantage of the SS7 networking protocol and registration process of the mobile unit system to validate a user prior to his or her making a call. This system has helped reduce the number of unauthorized calls from mobile units in the system.

Q. *What is voice mail?*

A. *Voice mail* is a service that most wireless systems offer to their subscribers. Voice mail is basically an electronic answering service made available to mobile unit subscribers. This service has quickly become very popular because it is useful to mobile phone customers; however, like any system, it must be engineered, maintained, and expanded when its use begins to reach a critical upper limit. Often when switching from one voice mail platform to another, the code that the subscriber used previously is changed, leading to much frustration. It is suggested that, when upgrading or changing the voice mail platform, the user codes should remain the same.

Q. *What are value-added services?*

A. *Value-added services* are the additional features offered by wireless carriers to their mobile unit subscribers. Examples of these services are listed below:

- Information line for traffic and sports reports, trivia games, and so on
- Data services for using wireless modems in the system
- 911 service
- 611 customer service
- Caller ID
- Voice mail
- SMS
- Four-digit dialing
- Telemetry

Q. *What is the network grade of service?*

A. In the ideal network, the facilities would be available 100 percent of the time, and there would be no blocking of any call traffic. However, as nice as this sounds, the facilities costs would be too great to construct such a network. Any company attempting to do so would probably go bankrupt. Therefore, a level of acceptable blocking must be set, and the network is then built around that value to obtain a practicable design. Using the Erlang B formula or referring to the Erlang B table, an engineer can find the required number of facilities for a given traffic load and *grade of service* (GOS). This method is easy and used widely throughout the telecommunications industry. Note that the grade of service for network facilities is typically better (lower blocking probability value) than the level set for the RF portion of a cellular system. Some typical values are $P=0.001$ GOS for the network (landline) facilities and $P=0.02$ GOS for the RF system channels. Again, the basic idea is that for a given grade of service and traffic level, one could actually obtain the required number of facilities.

In practice, the landline interconnection is based on an Erlang C model while the wireless portion of the system, that is, the radios, utilizes an Erlang B model. The particular GOS that is chosen, as mentioned above, is driven by customer needs and expectations, as well as by monetary considerations.

Q. *What is network diversity?*

A. The term *network diversity* refers to available alternate routes built into a network design. The extent of network diversity depends on many factors such as the type of facilities and equipment available for use, the number, location, and distances of the mobile units in the area being served by the central offices, the number of nodes in the wireless system along with their method of interconnection, and budgetary constraints. Depending on the importance of the cell site, diverse routing may also be provided for those facilities to which there are two different physical routes into the building leading from two different central offices.

Q. *What is a cross bar?*

A. A *cross bar* is an analog switch that was used in older telephone systems as a mechanical relay with which telephone calls were connected. Cross-bar switches are not used in wireless

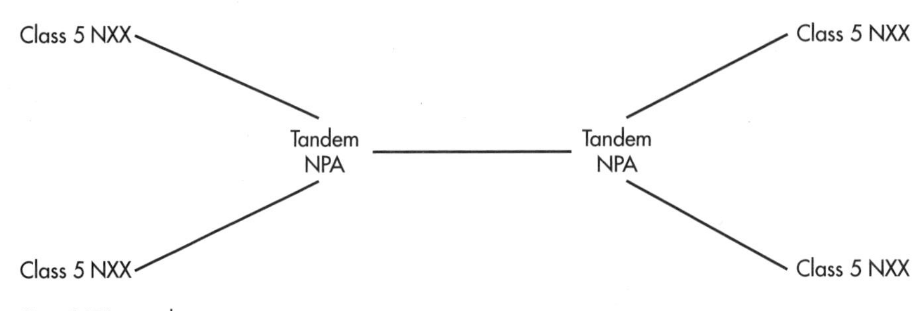

Figure 3.17 Access tandem.

applications; instead the switching and transferring of calls are accomplished by digital matrixes. However, the term *cross bar* is used sometimes in wireless situations in reference to mechanical relays. The digital matrix device is used instead of a cross bar because the digital matrix is much smaller and faster and requires little maintenance.

Q. *What is an access tandem?*

A. An *access tandem* is a class 4 switch through which inter- and outer-area-code traffic is processed. (See Fig. 3.17.)

Q. *What is a span?*

A. In the wireless community, a *span* is another name for a DS-1 circuit.

Q. *What is the difference between DS-1 and T-1 circuits?*

A. There is no difference between DS-1 and T-1 circuits. Needless to say, the use of two different names for the same type of circuit has led to some confusion. A DS-1 (or T-1) circuit occupies 1.544 MHz of bandwidth, and it is provided on a pair gain circuit. The DS-1 (or T-1) circuit has 24 individual DS-0s. Each of the DS-0s occupies 64 kHz of bandwidth. A DS-1 (or T-1) circuit is also referred to as a *span*.

Q. *What is a telco?*

A. The term *telco* is a nickname for "local telephone company." However, in wireless applications the term *telco* often is slang for the facilities that are provided by the telephone company. In many instances the term *telco* on a planning chart refers to local telephone company facilities available to the cell site, that is, the DS-1.

Q. *What is a toll (or telco) room?*

A. A *toll room,* or *telco room,* is part of the mobile switching center. Wire-line operators use the term *toll room* while non-wireline operators use the term *telco room* to refer to the same room or function. In particular, the *toll room* refers to the physical portion of the MSC where the interconnection equipment is located for connecting to the PSTN, LEC, CLEC, ILEC (incumbent local exchange carrier), IXC, and so on. It is at this point, where the network interfaces to an outside system, that an additional charge is incurred—hence, the term *toll.* Typically, the telco or toll room will house equipment such as DACS and intercept.

Q. *What is a BWM?*

A. A *broadcast warning message* (BWM) is part of the Lucent process of proving patches to their software without having to make a major revision, that is, a new software load. The use of BWMs is widely accepted, and they are especially valuable for revising an existing wireless operation. When using the BWMs, however, extra care should be taken to ensure that installing the BWMs has not introduced additional problems into the network.

Q. *What is a point of presence, or point of interconnection?*

A. A *point of presence* (POP), also known as a *point of interconnection* (POI), is the wireless operator's demarcation point for delivering interconnection calls with a local telephone company. Typically a POP, whether physical or virtual, is needed for every NPA that is obtained for that market. The POP can be a cell site or an MSC, but it must exist within the given NPA. Usually the POP is chosen to reduce the interconnection costs associated with call delivery to minimize the mileage component of the rate structure. For example, if the market were to have mobile unit numbers with the NPAs 914 and 630, then there would be two POPs associated with the system, one in the 914 area and the other in the 630 area.

Q. *What is CCS?*

A. The term *centa call seconds* (CCS) stands for *100 call seconds* of traffic. The unit of measure centa call seconds is a common switch metric that expresses the amount of carried traffic within an MSC or wireless network. If there are 36 CCS of usage per hour, then there is 1 erlang of usage.

$$1 \text{ erlang} = 36 \text{ CCS}$$

Q. *What is a 5ESS, or 5E?*

A. A *5ESS,* or *5E,* is the name given to a class 5 switch manufactured by Lucent, formerly AT&T. The term *5ESS* can refer to either a class 5 switch for landline applications for the end office or to the switching matrix of an autoplex switch. In wireless applications a reference to a 5ESS, or 5E, means an autoplex switch from Lucent.

Q. *What is a switch questionnaire?*

A. A *switch questionnaire* is a form generated by switch vendors to be filled out by their customers. The purpose of the questionnaire is to ascertain customers' requirements for current and future system use. The questionnaire usually is filled out by the fixed-network engineering group and is an essential if not a critical element in the building or expansion of any wireless network.

Q. *What are translations?*

A. Some vendors such as Lucent use the term *translation* to describe the mechanical process for entering in recent change commands to a switch, and some vendors use the term to describe the treatment of a call within a wireless system. Lucent uses the term *translation* to refer specifically to the different switch forms that are used to enter data into the switch to instruct the switch and the cell site in how to function.

Q. *What are recent change commands?*

A. *Recent change commands* (RCCs), also called *man-machine interface* (MMI) *commands,* refer to the data entry forms required to program a switch and its peripherals to perform in a specified fashion. An example of an RCC could be a neighbor list change or the activation of call forwarding for a subscriber.

Q. *What is a class 5 switch?*

A. A *class 5 switch* refers to an end office switch that provides answer supervision and other similar functions. The normal application for a class 5 switch is to provide residential service.

Q. *What is a dialing plan?*

A. A *dialing plan* is an engineering design that specifies how calls are to be treated within a wireless network. A dialing plan needs to be generated for every wireless system, and it must contain details as to how every call is to be treated and every feature is to be processed. Some examples of processes covered by a dialing plan are call forwarding, outpulsing to a PSTN, and four-digit dialing.

Q. *What are DMS and DMX?*

A. A *digital multiplex switch* (DMS) is a generic name for a switch device that connects and routes calls within a mobile switching center. Motorola calls the same component a *DMX* (digital mobile exchange).

Q. *What is an FER?*

A. *FER* is the abbreviation for the synonymous terms *frame erasure rate* or *frame error rate*. The *frame erasure* (or *error*) *rate* is a performance measurement of a digital communication system that transmits voice and data in packets. On the receiving end of the communication path, the frames must be first decoded and then reconstructed to obtain intelligible information. The 1s and 0s that make up the data in each frame are often corrupted for a variety of reasons due to impairments acquired along the communication path. Depending on the digital communication system, different levels of FER can be tolerated. However, as the FER increases, the quality of the network decreases, leading to performance and quality-of-service issues.

Q. *What is an MIN?*

A. A *mobile identification number* (MIN) is a 34-b number that represents the 10-digit directory telephone number assigned to the wireless subscriber. An MIN, sometimes along with an *electronic serial number* (ESN), is used by wireless operators for a variety of functions, one of which being to distinguish one mobile unit from other mobile units in a particular network. The MIN is transmitted as an encrypted code by the subscriber's unit when it registers for use.

Q. *What are wire-line and non-wire-line carriers?*

A. A *wire-line carrier* is a name that usually refers to a particular *regional Bell operating company's* (RBOC's) cellular operation. *Wire line* can also refer to the telephone company. However, corporate mergers and acquisitions have blurred somewhat the distinction between wire-line and non-wire-line carriers. Therefore, the term *wire line* is commonly used for the B-band operator in a market. For example, at this time in one particular market, Bell Atlantic operates the A band while SNET operates the B-band systems. Thus, the wire-line operator is SNET while the non-wire-line operator is Bell Atlantic. However, in other markets Bell Atlantic operates the B band and is referred to there as the *wire-line operator.*

Q. *What is a digroup?*

A. A *digroup* is the digitally multiplexed group of 24 DS-0s on a T-1. For wireless applications, a digroup is the channel bank, usually associated with analog service, that multiplexes and demultiplexes the 24 DS-0s on the T-1 for use by the operator.

Q. *What is a B-band carrier?*

A. A *B-band carrier* is the wire-line cellular operator to whom the FCC has allocated the B-band spectrum in a particular area.

Q. *What is an A-band carrier?*

A. An *A-band carrier* is the non-wire-line cellular operator to whom the FCC has allocated the A-band spectrum in a particular area.

Q. *What is the PSTN?*

A. The *Public Switched Telephone Network* (PSTN) refers to the traditional landline telecommunications systems and/or the central telephone company office.

Q. *What is an RBOC?*

A. A *regional Bell operating company* (RBOC) initially was one of the seven telephone companies that made up the AT&T system. The seven companies that existed in 1982 at divestiture were NYNEX, Bell Atlantic, Bell South, Southwestern Bell, Pacific Telesis, US West, and Ameritech. Some of these companies no longer exist due to mergers and acquisitions.

Q. *What is the S/N ratio?*

A. The *signal-to-noise* (S/N) *ratio* is a common parameter used to gauge the quality of an analog system associated with either a wireless or landline telephone system. As rule of thumb, a low S/N ratio indicates that the line or path has been degraded.

Q. *What is an erlang?*

A. An *erlang* is a unit of measure used to express traffic volume. It is important to note that an erlang is dimensionless. However, in the wireless industry, one erlang of usage commonly means 1 hour or 60 minutes of use for a circuit or facility. Depending on the grade of service and the probability model used—Erlang B, Erlang C, or Poisson—the number of facilities or circuits can be calculated based on the number of erlangs used.

Q. *What is a DID circuit?*

A. A *direct inward dialing* (DID) *circuit* is a feature that enables subscribers or stations to receive incoming calls directly, without the assistance of an attendant. DID circuits are used extensively in wireless applications.

Q. *What is the 911 service?*

A. The *911* service, which all wireless operators *must* provide their subscribers, is an emergency number. In emergency situations such as car accidents, the subscriber dials 911. Upon receiving the call, the wireless operator routes it to the appropriate 911 center for the area in which they are physically located. In many cases calls made within individual sectors of cell sites are routed to a specific 911 center. The 911 database is constantly maintained and updated based on the site build program to ensure accurate system response. (Note that there cannot and should not be any attempt by a wireless operator to restrict a 911 call. It should always be assumed that a 911 call is a legitimate emergency call because 911 cannot be used for any other purpose.)

Q. *What are 3/1/0, 3/1, and 1/0 capabilities?*

A. The term *3/1/0* refers to a type of DACS equipment that enables operators to accept DS-3, DS-1, or DS-0s into a unit. Having all three primary functions increases the flexibility of operators to offer unique services and groom their facilities for maximum cost savings. The 3 in the term refers to the ability to interface with a DS-3, while the 1 refers to the ability to interface with a DS-1. The 0 means that the DACS can interface with a DS-0 and thus can remove or add information at the DS-0 level, which provides drop and insert capabilities. Of course, there are limits as to how the DACS can be configured, and the unit must be provisioned properly to ensure that the appropriate number of DS-3, DS-1, and DS-0s can be handled.

The term *3/1* refers to DACS equipment that can accept either DS-3 or DS-1s into a unit. The 3 in the term refers to the ability to interface with a DS-3, while the 1 refers to the ability to interface with a DS-1. The inclusion of 3/1 capability in a DACS enables wireless operators to have a DS-3 for transmitting into and out of the MSC. The reason for including a 3/1 capability is that it will be necessary to multiplex and demultiplex DS-1s into and out of the DS-3.

The term *1/0* refers to DACS equipment that enables operators to perform drops and inserts into a DS-1. Being able to drop and insert allows operators to groom their facilities to maximize the use of the DS-1. For instance, if only six DS-0s were being used for four DS-1s, it would be possible to multiplex all four together and fill all 24 time slots of a typical DS-1, thereby eliminating the need for three DS-1s. The 1 in the term refers to the ability to interface with a DS-1, while the 0 refers to the ability to interface with a DS-0.

Q. *What is a DS-0?*

A. A *digital signal, level 0* (DS-0) is an individual time slot on a DS-1. The number of DS-0s differs depending on the format that is used, either T-1 or E-1. For instance, in a T-1 there are 24 DS-0s, while in an E-1 there are 30 usable DS-0s. A DS-0 is also called an *individual voice circuit* even though it may transport data.

Q. *What is a T coder?*

A. A *T coder,* which is another name for an ADPCM or variable-rate ADPCM device, compresses the signals for a DS-1 to enable one DS-1 to handle twice the traffic load. Specifically, a DS-1 has 24 DS-01s. With a T coder, it is possible for a DS-1 to transport 48 DS-0s over the same facility, thereby reducing the overall operating cost of the network.

Q. *What is a protocol analyzer?*

A. A *protocol analyzer* is a piece of test equipment that is used to analyze the signals between two machines. The protocol analyzer enables the technician to monitor and capture specific information that can be used to help troubleshoot a problem. The protocol analyzer is a critical tool for an operations group as well as any network engineering group. For example, on location, a protocol analyzer can be used to troubleshoot an SS7 link.

Q. *What is a T berd?*

A. A *T berd* is a piece of test equipment that is used to test DS-1 and DS-3 facilities. The term *T berd* is also commonly used to mean any test equipment used for facilities management. There are many manufacturers of T berds, and thus there are many models available.

Q. *What facilities are commonly used in wireless communication applications?*

A. The different facilities and some of their characteristics that are typically used in a wireless application are shown in Table 3.4. The table can be used to determine which circuit may fit the application at hand.

TABLE 3.4 SOME TYPICAL WIRELESS FACILITIES AND THEIR CHARACTERISTICS

Signal level	Carrier system	Number of DS-1 systems	Megabits per second
DS-0	DS-0	1/24	0.064
DS-1	T-1	1	1.544
DS-1c	T-1c	2	3.152
DS-2	T-2	4	6.312
DS-3	T-3	28	44.736
DS-4	T-4	168	274.76
OC-1	OC-1	28	51.84
OC-3	OC-3	84	155.52
OC-12	OC-12	336	622.08
OC-48	OC-48	1344	2488.32

Q. *What is an OC-12 facility?*

A. An *OC-12 facility* is the equivalent of 4 OC-3s, 12 DS-3s or rather 336 DS-1s. The service relies on fiber-optic cable, which is better than copper wire in its performance and reliability. In addition, the use of fiber-optic cable enables the system to utilize a SONET ring, thereby affording a better level of redundancy and disaster recovery.

Q. *What is an OC-48 facility?*

A. An *OC-48 facility* is the equivalent of 16 OC-3s, 4 OC-12s, 48 DS-3s or rather 1344 DS-1s. The service relies on fiber-optic cable, which is better than copper wire in its performance and reliability. In addition, the use of fiber-optic cable enables the system to utilize a SONET ring, thereby affording a better level of redundancy and disaster recovery.

Q. *What is an OC-1 facility?*

A. An *OC-1 facility* is the equivalent of 1 DS-3 or rather 28 DS-1s. The service relies on fiber-optic cable, which is better than copper wire in its performance and reliability. In addition, the use of fiber-optic cable enables the system to utilize a SONET ring, thereby affording a better level of redundancy and disaster recovery.

Q. *What is an OC-3 facility?*

A. An *OC-3 facility* is the equivalent of 3 DS-3s or rather 84 DS-1s. The service relies on fiber-optic cable, which is better than copper wire in its performance and reliability. In addition, the use of fiber-optic cable enables the system to utilize a SONET ring, thereby affording a better level of redundancy and disaster recovery.

Q. *What is the difference between an OC-1 facility and an OC-3 facility?*

A. The difference between an OC-1 facility and an OC-3 facility is simply the amount of bandwidth that each can handle. The OC-1 is the equivalent of a DS-3, while the OC-3 enables an operator to have three times that capacity, leading to a reduction in the facilities costs. The more bandwidth, the more circuits, and thus the more capacity the service can offer. It is important to note that when designing a network, diverse routing should be considered in addition to SONET ring capability. Also, the entrance facilities need to be reviewed.

Signal level	Carrier system	Number of DS-1 systems	Megabits per second
OC-1	OC-1	28	51.84
OC-3	OC-3	84	155.52

Q. *How do OC-1, OC-3, and OC-12 facilities differ?*

A. OC-1, OC-3, and OC-12 facilities differ simply in the amount of bandwidth that each can handle. The more bandwidth, the more circuits, and thus the more capacity the service can offer. It is important to note that when designing a network, diverse routing should be considered in addition to SONET ring capability. Also, the entrance facilities need to be reviewed.

Signal level	Carrier system	Number of DS-1 systems	Megabits per second
OC-1	OC-1	28	51.84
OC-3	OC-3	84	155.52
OC-12	OC-12	336	622.08
OC-48	OC-48	1344	2488.32

Q. *What is a wire center?*

A. A *wire center* is another name for a central office.

Q. *What is a T-1 circuit?*

A. A *T-1* is a DS-1 in the North American Digital Cellular system, and it has 24 DS-0s. A T-1 is also called a *T-1 carrier.* It occupies 1.544 MHz, and it can transmit 1.544 Mb/s, with each DS-0 having a maximum throughput of 64 kb/s. The T-1 circuit is very common in wireless systems.

Q. *What is a T-2 circuit?*

A. A *T-2 circuit* is a special T carrier that is used when an operator wishes to aggregate traffic onto a larger pipe. The T-2's bandwidth and traffic-handling capabilities enable it to transmit four T-1s or four DS-1s worth of traffic. Another name for a T-2 is DS-2.

Q. *What is a T-3 circuit?*

A. A *T-3 circuit* has the capability of handling 28 DS-1s. The T-3 is used when an operator wishes to aggregate traffic onto a very large pipe. The T-3s bandwidth and traffic-handling capabilities enable it to transmit 28 T-1s' or 28 DS-1s' worth of traffic. Another name for a T-3 is DS-3.

Q. *What is a DS-1 circuit?*

A. A *DS-1 circuit* can be either a T-1 circuit or an E-1 circuit. A DS-1 circuit is a single pair gain circuit that can carry either 24 DS-0s if it is a T-1 or 32 DS-0s if it is an E-1. The DS-1 is one of the circuits more commonly used in a wireless system because it can connect cell sites to an MSC. The amount of bandwidth used by a DS-1 differs for T-1s and E-1s. The T-1 has 1.544 Mb/s, while the E-1 has 2.048 Mb/s.

Q. *What is a DS-3 circuit?*

A. A *DS-3 circuit* can be either a T-3 or an E-3 circuit. A DS-3 circuit is a single pair gain circuit. The number of DS-0s that are associated with the DS-3 will, of course, depend on whether the underlying service is T-1 or E-1. For example, 28 T-1s multiplexed together make up a DS-3, while 16 E-1s are required to make up a DS-3. Operators use DS-3s when they wish to aggregate traffic onto a very large pipe. The DS-3's bandwidth and traffic-handling capabilities enable it to transmit 28 T-1s' or 16 E-1s' worth of traffic. Other names for a DS-3 are T-3 in North America and E-3 in Europe.

Q. *What is the difference between a DS-3 circuit and a T-3 circuit?*

A. There is no difference between a DS-3 circuit and a T-3 circuit used in the United States. The name T-3 is often interchanged with the name DS-3.

Q. *What is an E-1 circuit?*

A. An *E-1 circuit* is a DS-1 outside the North American Digital Cellular system. The E-1 has 32 DS-0s that are contained in it. If the E-1 is associated with voice traffic, typically only 30 DS-0s are utilized; however, if it is for a primary rate interface (PRI), then 31 DS-0s are utilized. The E-1 circuit is very common in the wireless systems outside of the United States.

Q. *What is the difference between a T-1 circuit and an E-1 circuit?*

A. The difference between a T-1 circuit and an E-1 circuit in the wireless industry is the number of DS-0s that the pair gain can contain. Specifically, a T-1 has 24 DS-0s while an E-1 has 32 DS-0s—30 for voice or 31 for a PRI. Therefore, an E-1 has typically six more DS-0s to use than a T-1, or rather it has 25 percent more capacity when used in a similar configuration. The signaling, however, is different between the two, and they are not directly compatible without an intermediate device in between. Of course, there is also the obvious issue of the other six DS-0s, which are not an issue when going from T-1 to E-1 but are an issue when going from E-1 to T-1.

The difference between T-1 and E-1 is important when reviewing the switch port count in that if the switch is set up for E-1 and there is a T-1 interface, then there is a port utilization problem requiring a preconversion with a DACS.

Q. *What is an E-3 circuit?*

A. An *E-3 circuit* is a DS-3 that uses E-1s as its DS-1 format. An E-3 is a single pair gain circuit that has 16 E-1s multiplexed onto it. The DS-3 is chosen by operators when they wish to aggregate traffic onto a very large pipe. The DS-3's bandwidth and traffic-handling capabilities enable it to handle 16 E-1s' worth of traffic. Another name for an E-3 is DS-3.

Q. *What is a DS-4 circuit?*

A. A *DS-4 circuit* is a T-4, and it is a pair gain circuit that has 168 DS-1s multiplexed onto it. The DS-4 is chosen by operators when they wish to aggregate traffic onto a very large pipe. For wireless applications, the use of DS-4s is not common because the volume of traffic from the DS-4s is so great that it would flow over a single pipe.

Q. *What is a stratum clock?*

A. A *stratum clock,* or really *timing source,* refers to four different timing sources. Each of the four levels of stratum, 1 through 4, represents different timing requirements, and each level is directly associated with the type of electronics it is applied to in the telecommunications industry. Stratum 1, the master reference for the network, is the highest and most accurate, and stratum 4 is the least accurate.

Q. *What is stratum 1?*

A. *Stratum 1* is the highest level in the stratum hierarchy, and it is the *primary reference source* (PRS) that serves as a master clock for the entire network. The stratum 1 clock utilizes a traceable timing reference. Of all the stratum levels, stratum 1 has the highest holdover period in that, when it loses its source, it will keep its accuracy for the longest period of time. Furthermore, it is accurate to within 1×10^{-11} bit per second or better. In many wireless systems, a GPS and/or a Loran C clock source is used as the stratum 1 source. However, GPS and Loran C are not stratum 1 sources since they are steerable, that is, they achieve their timing from an outside source. Even though neither a GPS nor a Loran C is classified as a stratum 1 source, they still are classified as PRSs within the network and are designated stratum 1E.

Q. *What is stratum 2?*

A. *Stratum 2* is the second-highest level in the stratum hierarchy. Stratum 2 clocks are normally found in tandems and STPs. A stratum 2 clock is typically found in an MSC, and it is sometimes referred to as a *transit node clock*.

Q. *What is stratum 3?*

A. *Stratum 3* is the third-highest level in the stratum hierarchy. Stratum 3 clocks are normally associated with DACS and digital switches utilized for wireless and landline applications. Many central offices utilize stratum 3 clock references.

Q. *What is stratum 4?*

A. *Stratum 4* is the lowest level in the stratum hierarchy. The use of stratum 4 clocks is normally associated with PBX, channel bank, echo cancellors, or similar network elements.

Q. *What is the NACN?*

A. The *North American Cellular Network* (NACN) is used for roaming, feature transparency, and consistency among different cellular and wireless markets. The NACN was initially associated with the A-band cellular operators in North America.

Q. *What is the D4 signaling format?*

A. A D4 signal channel bank is used to convert analog signals into digital PCM signals for transport along a T-1 circuit. The *D4 signaling format* was, and in many places still is, the predominant signaling format utilized by many telcos for their DS-1 circuits.

A D4 signal comprises 24 DS-0s and has a bandwidth of 1.544 MHz, or rather 1.544 Mb/s of throughput. The D4 was the predecessor to both the superframe and the extended superframe formats that are used in today's telecommunication systems. The D4 signaling format is structured as follows:

$$24 \text{ DS-0s} \times 8 \text{ bits} = 192 \text{ bits}$$

and

$$1 \text{ frame bit (F bit)} = 1 \text{ bit}$$

Thus,

$$192 \text{ bits} + 1 \text{ bit} = 193 \text{ bits per frame}$$

Since each DS-0 is sampled at 8000 per second, the bandwidth is arrived at as follows:

$$193 \text{ bits per frame} \times 8000 \text{ samples per second} = 1.544 \text{ megabits per second}$$

The disadvantage of D4 is that it is not as robust for maintenance and troubleshooting as the superframe or extended superframe formats are. To troubleshoot a problem, the D4 circuit, DS-1, must often be removed from service.

Q. *What is the superframe signaling format?*

A. The *superframe* (SF) *signaling format* is a logical extension of D4 formatting. Superframe formatting was developed to improve the diagnostics and the performance of the DS-1 circuit itself. The formatting of the DS-1 is such that it contains 12 DS-1 frames that are transmitted in succession. Many wireless operations utilize superframe formatting for their DS-1.

Q. *What is the extended superframe signaling format?*

A. The *extended superframe* (ESF) *signaling format* is a type of DS-1 formatting that enables the better collection of far-end signaling information. ESF is the latest in the evolution of DS-1 maintenance and monitoring facilities. ESF allows for in-service monitoring, sectorized troubleshooting, and far-end signal analysis—none of which are supported with D4 signaling. An ESF consists of 24 DS-1 frames, but it also restructures the 8000 bits per second.

Q. *What is a CLEC?*

A. A *competitive local exchange network* (CLEC) is a telecommunications company that tries to compete directly with the local exchange (LEC) telephone company. There are numerous CLECs presently in existence with new ones emerging every day. To maintain its license to operate, a CLEC must comply with many regulatory constraints, or rather requirements. For example, a CLEC must offer services typically associated with a local telephone company such as 911 and directory assistance.

Q. *What is an LATA?*

A. A *local-access transport area* (LATA) is a geographic area in which an LEC is allowed, or rather authorized, to provide local exchange service. When an LEC wants to provide service outside its authorized LATA, it needs to obtain a waiver.

The LATA boundaries were initially set up as part of the modified final judgment (MFJ), and in many instances the LATA was associated with an NPA. However, with subsequent divisions of the NPAs, this simple distinction has been lost in some cases.

LATA boundaries are used to determine how telecommunications companies may provide service. Depending on the type of company, the configuration of LATAs can either help or hinder the company's ability to compete successfully in certain markets.

Q. *What is a corridor circuit?*

A. A *corridor circuit* is used to transport voice and data traffic between LATAs. A corridor circuit is normally introduced when the LATA boundaries have been set up in such a manner that they defy common sense and actual traffic patterns. An example of a corridor circuit is the delivery path for calls between the borough of Manhattan in New York City and northern New Jersey.

Q. *What are inter-LATA areas?*

A. Inter-LATA spaces are outside LATA boundaries.

Q. *What are intra-LATA areas?*

A. Intra-LATA spaces are within LATA boundaries.

Q. *What is a local loop?*

A. A *local loop* is a physical wire line, or copper pair, that connects a central office to its subscriber. It is the last leg of the telecommunication network.

Q. *What is a trunk?*

A. A *trunk* is a communication channel or link between two switching platforms. The trunk can be a connection between two class 5 switches or between a class 5 switch and a class 4 tandem switch.

Q. *What is a trunk group?*

A. A *trunk group* is a group of trunks that can be used between two switching platforms. For example, in wireless systems, an MSC switch has numerous trunk groups. Thus a trunk group for a wireless system could be used in connecting one switch to another within the same network.

Q. *What is a data circuit?*

A. A *data circuit* is network equipment that is capable of providing some type of digital service.

Q. *What is a voice circuit?*

A. A *voice circuit* is network equipment that is capable of delivering some type of voice service.

Q. *What is an interconnection?*

A. An *interconnection* in the wireless industry is a telephony call made from a mobile unit to a landline.

Q. *What is an intercept?*

A. An *intercept* in the wireless industry is the legal use of wiretapping of mobile unit communications for the sole purpose of law enforcement. The request for the intercept, or wiretap, is made by a court order to the wireless company. Wiretap equipment is then used to record and register the information that will be used by the law enforcement organization in their investigation of criminal activity.

Q. *What is fraud?*

A. *Fraud* is a deception intended to induce someone to part with something of value, and it is a very large problem in wireless communications. Fraud can be perpetrated not only from an external source but also from an internal source. The deception is usually accomplished with cloned wireless phones which are used for making calls that cannot be billed to a valid subscriber. Fraud can also be perpetrated by wireless company employees who allow invalid users to make calls. Fraud is a major problem for wireless systems companies, and constant vigilance is necessary to minimize its negative financial impacts.

Q. *What is cloning?*

A. In the wireless industry, *cloning* refers to the process of duplicating a valid subscriber's phone number and ESN onto another phone either for the purpose of having two phones with the same number for legitimate reasons or for the purpose of stealing wireless service. There are many devices that have been sold and continue to be sold that can be used to obtain a subscriber's phone number and ESN so that they may be cloned.

Q. *What is four-digit dialing?*

A. *Four-digit dialing* is a feature and service offered by many wireless operators to their subscribers. Basically, four-digit dialing allows the wireless subscriber to enter only four digits to reach someone via wireless communications. The four digits are usually set up to be an extension of the company's PBX system. The use of four-digit dialing enhances the mobility and accessibility of the subscriber.

Q. *What is a smart jack?*

A. A *smart jack* is a commonly used piece of electronic demarc equipment that is placed between the DS-1 and the CSU/DSU. The smart jack is installed by the telephone company delivering the DS-1 service. A smart jack enhances the troubleshooting capabilities of the telephone company by providing diagnostic feedback. This equipment can also loop back a signal so that troubleshooting can be performed without a technician's having to be at the physical location of the DS-1.

Q. *What is a 635B jack?*

A. A *635B jack* is a commonly used, passive piece of demarc hardware. Specifically, the device is wired so that when the CSU/DSU is unplugged, a signal is looped back to the telephone company to assist them in troubleshooting and maintenance. However, to achieve the loop back, a representative of the telephone company has to be physically on the premises where the DS-1 service is located. The 635B jack was the precursor to the smart jack.

Q. *What is a progress tone?*

A. A *progress tone* is an audible signal that is used to provide a subscriber with the feeling that a call is still being processed. The objective of using a progress tone is to prevent the caller from hanging up prematurely. There are numerous progress tones; however, the more common ones are listed below for reference:

Dial tone
Ring back
Busy signal
Reorder (all trunks busy)
Special information tone (SIT), that is, the "bong"
Recorded announcement

Q. *What are supervisory signals?*

A. *Supervisory signals,* also called *line signals,* are used as part of the normal call processing of voice and data services within a wire-line or wireless communication system. The supervisory signals consist of the following:

- Idle (on hook)
- Seizure (off hook)

- Start dial (wink and delay)
- Answer
- Clear back
- Disconnect
- Blocking
- Glare

Q. *What types of signaling are used for PSTN connections?*

A. The types of signals, or rather address signals, that are normally available to wireless operators are listed below, along with their facility interface issues:

- Dial pulse
- DTMF (Touch-Tone)
- *Multifrequency* (MF), R1
- *Multifrequency compelled* (MFC), R2
- *Common channel signaling* (CCS7) and *switching system signaling* (SS7)

Q. *What is DTMF signaling?*

A. *Dual-tone multifrequency* (DTMF) *signaling,* also called Touch-Tone, is a common method subscribers use to dial another phone number. DTMF signaling utilizes the ability of each of the 12 keys on a phone or wireless handset to be represented by two discrete signals. In many digital wireless communication systems, the dialed number is not sent by DTMF but is instead encoded and then decoded for use as DTMF signals after it transverses the airwaves. Table 3.5 gives the DTMF signaling structure. The outlined portion represents a normal handset keypad, and the frequencies listed in the figure, when combined, represent the button depressed. Therefore, the 2 key with the letters A, B, or C is represented by 1336 and 697 Hz.

TABLE 3.5 DTMF SIGNALING STRUCTURE

1209 Hz	1336 Hz	1477 Hz	1633 Hz (spare)
1	ABC 2	DEF 3	697 Hz
GHI 4	JKL 5	MNO 6	770 Hz
PRS 7	TUV 8	WXY 9	852 Hz
*	OPER 0	#	941 Hz

Q. *What is fiber?*

A. In the wireless industry, the term *fiber* refers to optical fiber, which is used for DS-1 and any of the OC-1 through OC-196 facilities.

Q. *What is a nailed connection?*

A. A *nailed connection* is a circuit dedicated to deliver only one particular service. A nailed connection is established permanently and is never torn down by the system for potential reuse. A nailed connection for a wireless system is commonly used for separating control information associated with a cell site from the rest of the information associated with the voice system.

Q. *What is a fiber-loop converter?*

A. A *fiber-loop converter* is a piece of demarc equipment that is installed by the telephone company or service provider for the sole purpose of converting fiber, or light, into an electrical signal for transmission through copper wire. For wireless applications a fiber-loop converter may be installed in a cell site or an MSC.

Note that even though DACS equipment also has the ability to perform the function of a fiber-loop converter, it is not commonly used for this purpose because it complicates troubleshooting and maintenance of the wireless system.

Q. *What does the term roam mean?*

A. In the wireless telecommunications industry, to *roam* means to move among different wireless companies' market areas. Thus, when a subscriber based in one market (his or her home market) utilizes the wireless facilities in another market, he or she is roaming. For example, if a Chicago Ameritech subscriber travels to the New York area, he or she would have to use the Bell Atlantic system to place cellular calls.

Q. *What is roamer validation?*

A. In the wireless telecommunications industry, *roamer validation* is the process used to verify that a subscriber requesting service on the system is indeed a valid subscriber and is authorized to utilize the system. Roamer validation is used to prevent fraud in the wireless industry. For roamer validation to work properly, the subscribers' records need to be up to date and need to include their status for billing. This information is usually contained within the HLR, from which the roamer validation may access it for the verification process.

Q. *What is a dark fiber?*

A. A *dark fiber* is a strand of fiber dedicated to providing only one service. The wireless operator is responsible for lighting the fiber and providing equipment at both ends of the fiber strand. The provider of the dark fiber is responsible for ensuring a defined amount of attenuation and delay on the fiber itself. In wireless applications dark fiber is usually used in conjunction with microcells that convert RF signals to optic signals. (Note that for a microcell application using dark fiber, two strands are needed, one for each direction.)

Q. *What is a dim fiber?*

A. A *dim fiber* is a strand of fiber that provides more than one service. One purpose for which dim fiber would be used is to back haul signals from a hub location.

Q. *What is a DMS?*

A. A *digital multiplex switch* (DMS) is a switch manufactured by Nortel. One type of Nortel DMS is the DMS-100.

Q. *What is an AXE?*

A. An AXE is a switch manufactured by Ericsson. One type of Ericsson AXE is the AXE-10.

Q. *What is an Autoplex?*

A. An *Autoplex* is a switch manufactured by Lucent. The Lucent Autoplex 1000 is the most commonly used model, and it now comes with a 5ESS switching matrix. Previous Autoplex 1000s came with a DCS matrix.

Q. *What is an EMX?*

A. An *EMX* is a switch manufactured by Motorola. One type of Motorola EMX is the commonly used EMX-2500.

Q. *What is a channel bank?*

A. A *channel bank* is a device that is used to multiplex and demultiplex a DS-1 for the purpose of interfacing with analog equipment.

Q. *What is multiplexing?*

A. Used widely in the wireless and landline telecommunications industry, *multiplexing* is the aggregating, or concentrating, of signals from multiple sources into one signal. An example of multiplexing in wireless networks is the aggregation of DS-1 signals into DS-3 signals. Multiplexing is normally accomplished via a DACS owned by a wireless operator. Telephone companies can also perform the same function, but at a cost and loss of configuration management.

Q. *What is demultiplexing?*

A. *Demultiplexing* is the exact opposite of multiplexing. Basically, demultiplexing is the disaggregation or expansion of one signal into many signals. Demultiplexing is normally accomplished via a DACS owned by the wireless operator. Telephone companies can also perform the same function, but at a cost and loss of configuration management.

Q. *What is a hub?*

A. In a wireless network, a *hub* is a location used for traffic concentration. Often a *hub* is established in a large metropolitan area in which a particular operator does not have a wireless switch. The objective of the *hub* is to save facilities costs by concentrating the DS-1 signals in a certain area onto a DS-3 circuit for connection to the area's MSC.

Q. *What is a DOD circuit?*

A. A *direct outward dialing* (DOD) *circuit* is a feature that enables subscribers or stations to make outgoing calls directly, without the assistance of an attendant. DOD circuits are used extensively in wireless communications.

Q. *What is a drop wire?*

A. A *drop wire* is a service wire, that is, a wire used to deliver the T-1/T-3 service in a wireless system. The term *drop* can be used to describe an aerial or underground wire that runs from the demarcation point to the customer premise equipment. The drop wire's length is different depending on where the demarcation point is relative to the customer equipment, the CSU.

Q. *What is the difference between a virtual colocation and a physical colocation?*

A. The difference between a virtual colocation and a physical colocation lies in where the customer's equipment is maintained. If the customer's equipment is actually on the wire center, end office, or tandem premises, then the customer has a physical colocation. If the customer's equipment is not on the telco's premises but the telco bills the customer as though the equipment were there (a POI situation), then the customer has a virtual location.

Q. *What is an ACELP vocoder?*

A. An *algebraic code excited linear predictive* (ACELP) *vocoder* was an improved version of the VCELP vocoder commonly used in IS-136 systems. The ACELP vocoder enabled systems to relax their C/I requirements from 19 to 21 dB C/I to 17 dB C/I.

Q. *What is a VCELP vocoder?*

A. A *vector-sum excited linear predictive* (VCELP) *vocoder* is a type of linear predictive vocoder that was used in wireless systems. One system that used the VCELP was the IS-136, and it was the original vocoder that was used for the launch of TDMA systems. The C/I for a VCELP system for the IS-136 was usually between 19 and 21 dB C/I. This C/I was high, and thus the VCELP was ultimately replaced by the improved ACELP vocoder, which had a lower C/I of 17 dB.

Q. *What is a RELP vocoder?*

A. The *residual excited linear predictive* (RELP) *vocoder* is used in GSMs.

Q. *What is answer supervision?*

A. *Answer supervision* is the monitoring of a telephone line to determine when it goes off the hook for the purpose of providing billing information to the telephone company.

Q. *What is an access charge?*

A. An *access charge* is the fee that a local telephone company charges a CLEC or *long-distance* (LD) *company* to connect to their system for delivery of a call.

Q. *What is a gateway?*

A. A *gateway* is a demarcation point between different networks. An example of a gateway is a class 5 CO which is used to connect two systems, say, a North American system to an international system.

Q. *What is fiber splicing?*

A. *Fiber splicing* is the process of fusing fiber-optic fibers together for the purpose of extending the physical length of a fiber cable or repairing fiber damage that sometimes occurs to the line as a result of accidents like a back-hoe fade.

Q. *What is a back-hoe fade?*

A. A *back-hoe fade* is a situation in which underground or aerial telco facilities are severed during routine construction projects.

Q. *What is the CCITT?*

A. The *Consultative Committee on International Telegraphy and Telephony* (CCITT) is a subdivision of the *International Telecommunications Union* (ITU). The purpose of the CCITT is to make recommendations for telecommunications equipment and interoperability standards.

Q. *What is a switch?*

A. The term *switch* can mean an MSC or a central office. (And sometimes the opposite occurs when an MSC is called a "switch.") The term *switch* can also be a generic reference to the wide range of types and classes of switches, available for electronic applications.

Q. *What is a WSP?*

A. A *wireless service provider* (WSP) is an operating company that provides PCS or cellular telecommunications services.

Q. *What is a calling customer (or calling party)?*

A. A *calling customer* (or *calling party*) is simply a customer that originates a call in a wireless or telephony system.

Q. *What is a called customer (or called party)?*

A. A *called customer* (or *called party*) is the customer who receives the call that originated from the calling party. The called customer is usually identified by his or her mobile unit or directory number.

Q. *What types of telco circuits are used in a wireless system?*

A. The types of circuits used in a wireless system are type 1, type 2A, type 2B, type 2C, type 2D, and type S.

Q. *What is a type 1 circuit?*

A. A *type 1 circuit* is a circuit that is ordered from the local telephone company (LEC) for the purpose of connecting the wireless system to an end office for the purpose of establishing a connection between the wireless system and the LEC. Type 1 circuits are used for delivering calls to NXXs within the LEC's local network as well as for obtaining directory assistance, LEC operator assistance, time service, 0- and 0+ services, and other services offered by the LEC. Type 1 circuits utilize MF signaling for both incoming and outgoing signaling, and they employ line treatment.

Q. *What is a type 2A circuit?*

A. A *type 2A circuit* is a circuit that is ordered from the local telephone company (LEC) for the purpose of connecting the wireless system to the LEC's tandem, which enables the wireless system to connect not only to the LEC but also to other carriers. A type 2A circuit has no line treatment. A wireless operator using a type 2A circuit can deliver calls through an LEC tandem less expensively than he or she can deliver calls using a type 1 circuit.

Q. *What is a type 2B circuit?*

A. A *type 2B circuit* is a circuit that is ordered from the local telephone company (LEC) for the purpose of connecting a wireless system to a specific end office. For example, if a high volume of the wireless system's traffic is processed through a particular end office, it might be better to utilize a type 2B circuit than a type 2A circuit because the type 2A circuit connects to a tandem. Typically a type 2B circuit is used to augment the type 2A circuits utilized in a wireless network.

Q. *What is a PSAP?*

A. A *public safety answering point* (PSAP) is a location to which 911 calls are delivered.

Q. *What is a B911 service?*

A. A *basic 911* (B911) *service* is an emergency call service.

Q. *What is an E911 service?*

A. An *enhanced 911* (E911) *service* is an emergency call service that has more capabilities than are available from a basic 911 service.

Q. *What is a type 2C circuit?*

A. A *type 2C circuit* is a circuit that is ordered from the local telephone company (LEC) for the purpose of connecting a wireless system to emergency services by way of the LEC. Type 2C circuits are used to transport 911 calls.

Q. *What is a type 2D circuit?*

A. A *type 2D circuit* is a circuit that is ordered from the local telephone company (LEC) for the purpose of connecting the wireless system to the LEC's tandem. This circuit allows various operator services to be provided such as calling-card processing, directory assistance, and general assistance.

Q. *What is a type S circuit?*

A. A *type S circuit* is a circuit that is ordered from the local telephone company (LEC) for the purpose of establishing an SS7 connection between the LEC and the wireless operator. The type S circuit is used to connect TCAP circuits as well as SS7 ISUP circuits. A type S circuit, however, does not provide any applications and is simply a physical connection between the wireless operator and the LEC.

Q. *What is a TCAP?*

A. A *transactions capabilities application part* (TCAP) is a type of switching signal.

Q. *What is a wink?*

A. A *wink* is an off-hook signal, or rather a pulse.

Q. *What is a feature group?*

A. A *feature group* (FG) is the group of services available to cellular phone customers who select a particular switching arrangement offered at a particular access tariff. Wireless telecommunication companies offer four general types of feature groups: A, B, C, and D.

Q. *What is feature group A?*

A. *Feature group A* (FGA) is a line-side connection that is designed for interstate or intra-LATA calls. Feature group A provides three types of service:

- Resale [mobile telephone service (MTS) and *wide-area telephone service* (WATS)]
- No resale (foreign exchange services)
- Off-network access (originating or receiving inter-LATA calls within a private network)

Other services provided by an FGA include hunt groups, call denials, and service code blocking.

Q. *What is feature group B?*

A. *Feature group B* (FGB) is a trunk-side switch connection that is accomplished via an access tandem. FGB provides subscribers with a telephone number that can be used nationwide and that follows the North American numbering plan. The number is assigned in the 950-*XXXX* format. Feature group B uses two-dial-tone operation for originating calls. FGB also provides the following:

- A universal number for gaining access to services
- Verification of subscriber's credit card

Q. *What is feature group C?*

A. *Feature group C* (FGC) is an access service used by both MTS and WATS carriers, and it allows for the connection of MTS and WATS carriers at the end office. Feature group C is available for originating and receiving calls as well as for two-way service. Note that FGC is not readily available because it has been overridden by the use of feature group D.

Q. *What is feature group D?*

A. *Feature group D* (FGD) is a trunk-side switch connection, and it is also called *equal access*. FGD allows access to the following services:

- 1+ dialing for equal access
- International calling
- Automatic number identification

Q. *What is a CIC?*

A. A *caller identification code* (CIC) identifies a mobile unit. An example of a caller identification code is the 1787 portion of the 987-1787 number, that is, the *XXXX* portion.

Q. *What is a router?*

A. A *router* is a device used in *local-area network/wide-area network* (LAN/WAN) *applications* to direct data traffic to designated addresses. This component works by looking at the destination address in the data packet. A router is used, for example, to connect a wireless network to the Internet for TCP/IP applications. A CDPD is an example of a system that utilizes a router for connecting the wireless portion of the CDPD network to the Internet.

Q. *What is digit-by-digit dialing?*

A. *Digit-by-digit dialing* is the process of treating calls in such a way that the dialed digits are analyzed on a digit-by-digit basis to obtain specific routing and feature information.

Q. *What is a software load?*

A. A *software load* is software that is loaded into a wireless system's switching platform. Switch loads are often upgraded to resolve problems as well as to enhance a wireless network.

Q. *What is a CN?*

A. A *change notice* (CN) is a notification of adjustments in a customer's service profile. A CN is usually associated with Lucent equipment. Depending on whether or not a CN is a class A change determines whether or not the customer pays for the retrofit.

Q. *What is a CNA?*

A. A *CNA* is a patch associated with an Ericsson switch software load. Often there will be numerous CNAs employed between major software load upgrades.

Q. *What is a line side?*

A. A *line side* is an end office connection.

Q. *What is a wire center?*

A. A *wire center* is a building or structure in which one or more central offices are located for the purpose of providing telephony service.

Q. *What is a loop-around test?*

A. A *loop-around test* is a manual two-way transmission test of a T-1 or other facility. Typically, a loop-around test is performed as part of the acceptance procedure for a T-1 to a cell site or as part of the diagnostic procedure for a malfunctioning T-1.

Q. *What is equal access?*

A. *Equal access* is the process by which wireless customers can select a long-distance *primary interexchange carrier* (PIC) such as AT&T, MCI, or Sprint.

Q. *What is CAT1 wire?*

A. *CAT1 wire,* also called *category 1 wire,* refers to a type of wire used in telephony. Specifically, it is any wire other than phone wire that is used for transmissions. CAT1 wire, however, is not coaxial cable. Rather, CAT1 wire can be any untwisted wire of any American Wire Gauge (AWG) size. CAT1 wire is used for control and alarm wiring in wireless applications.

Q. *What is CAT2 wire?*

A. *CAT2 wire,* also called *category 2 wire,* refers to twisted-pair wire having an AWG size between 22 and 26. CAT2 wiring can be twisted or untwisted, and it is used for traditional telephone applications.

Q. *What is CAT3 wire?*

A. *CAT3 wire,* also called *category 3 wire,* refers to twisted-pair wire having an AWG size between 22 and 24. CAT3 wire can be twisted or untwisted. It is used for data communication up to 16 MHz, and it is used for T-1 and 10-base-T connections as well as standard telephone applications.

Q. *What is CAT4 wire?*

A. *CAT4 wire,* also called *category 4 wire,* refers to twisted-pair wire having an AWG size between 22 and 24. CAT4 wire can be twisted or untwisted. It is used for data communication up to 20 MHz, and it is used for T-1 and 10-base-T connections as well as standard telephone applications.

Q. *What is CAT5 wire?*

A. *CAT5 wire,* also called *category 5 wire,* refers to twisted-pair wire having an AWG size between 22 and 24. CAT5 wire can be twisted or untwisted. It is used for data communication up to 100 MHz, and it is used for T-1, 10-base-T, and 100-base-T connections as well as standard telephone applications.

Q. *What is an OSI?*

A. An *open-system interconnection* (OSI) is designed to allow full interoperability between network platforms. There are seven OSI layers, as listed in Table 3.6. OSI layer 7 is the highest while OSI layer 1 is the lowest. Layer 1 is also the physical layer that is the starting point for the other OSI layers.

Q. *What is a frame relay?*

A. A *frame relay* is a packet switching technology that is used for transporting data for a WAN. Frame relay is used in many applications for wireless communications systems, but it is used primarily for data control and messaging among different network nodes. Frame relay is a connection-oriented service that also affords different levels of *quality of service* (QOS) to ensure that information is passed without latency issues and reliably.

TABLE 3.6 THE SEVEN OSI LAYERS

OSI layer	Layer name	Comments
7	Application	Used for connecting the application program or file to a communication protocol.
6	Presentation	Performs the encoding and decoding functions.
5	Session	Establishes and maintains the connection for the communication processes in the lower OSI layers.
4	Transport	Performs error correction and transport. (Both Tx and Rx are performed here.)
3	Network	Performs switching and routing functions for the MSC.
2	Data link	Receives and sends data over the physical layer.
1	Physical	Supports actual media used for sending and receiving the communications. (Radio or fiber-optic wires are two examples.)

Q. *What is a class 4 (or toll) switch?*

A. A *class 4* (or *toll*) *switch* is used to connect several class 5 switches to the rest of the telephony network for call delivery and reception. There are numerous types of connection circuits that can be used to connect an end office switch to a tandem switch. For wireless applications, the MSC connects to a class 4 switch for delivery of all its calls to the PSTN. Often a tandem or class 4 switch will be associated with a specific NPA code, such as 914, for routing calls to all the local exchanges.

Q. *What is a toll center?*

A. A *toll center* is a class 4 switch used in the switching hierarchy. A toll center, or toll switch, is used to connect class 5 switches to other class 5 switches via the same toll switch or another toll switch connected to it. Often a toll center will be associated with a specific NPA code, such as 914, for routing calls to all the local exchanges.

Q. *What is a tandem switch?*

A. A *tandem switch* is used to connect several end offices together without connecting to an intervening toll switch. Thus a tandem switch is more efficient than a toll switch. Tandem switches are often used when different end offices are controlled by the same operator and a high volume of traffic is handled between those end offices, negating the need to transport the calls through a class 4 switch. A wireless switch is really a tandem switch. Often a tandem or class 4 switch will be associated with a specific NPA code, such as 914, for routing calls to all the local exchanges.

Q. *What is a circuit switch?*

A. A *circuit switch* is a wireless modem service offered to subscribers. This particular switch can handle data rates of from 9.6 kbytes/s up to 14.4 kbytes/s and higher. Circuit switches are offered for mobile unit subscribers to have modem dial-up capability. Circuit switch services are used for customers who typically use *file transfer protocol* (FTP) to transmit data to and from a mobile unit environment. If packet data are used, say, for e-mail, then a packet service like CDPD may be more cost effective and efficient than a circuit switch.

Q. *What is call delivery?*

A. *Call delivery* is the process used for connecting the originating customer to the terminating customer or to some other service. The term *call delivery* is used to describe the general procedure that is used by a wireless operator to process wireless calls to and from the subscriber, and it includes the treatment of the call and the features used by the subscriber.

Q. *What is international routing?*

A. *International routing* is the process used to deliver calls from one country to another. Note that calling restrictions often vary from country to country.

Q. *What is a SIM card?*

A. A *SIM* (subscriber identification module) *card* is a card that identifies the user of a wireless system. SIM cards are currently used with GSMs. The SIM card contains subscriber information that enables a subscriber to utilize different handsets when they travel from market to market or country to country. The SIM card comes in several sizes ranging from a small chip to a credit card, both of which fit into the subscriber unit itself.

Q. *What is a SIM2 card?*

A. A *SIM2 card* is the next generation of SIM cards, and it will have enhanced functions and features. The SIM2 card will enable the operator to have true over-the-air activation, instead of over-the-air notification.

Q. *What is a prepaid service?*

A. A *prepaid service* allows the subscriber to purchase a certain amount of phone usage in advance of his or her actual use of the phone. To offer prepaid service requires that the operator have real-time billing capability. There are numerous applications for prepaid wireless service, and many operators are offering it.

Q. *What is a SONET?*

A. A *synchronous optical network* (SONET) is a fiber-optic–based system or service that is used to provide alternative routing, self-healing, in the event of a fiber-optic line cut. A SONET employs synchronous transport signaling, which is the electrical version of OC-1.

Q. *What is a SONET ring?*

A. A *SONET ring* refers to the route that the signal within a SONET traverses for self-healing in the event of a fiber cut. A SONET ring is called a "ring" because the SONET starts at one point and returns to the same point after it has passed through numerous locations. The idea is that if a section of the ring is broken, traffic can still be transported on the ring but in the opposite direction.

Q. *What is authentication?*

A. *Authentication* is the process for exchanging information between a mobile station, subscriber unit, and an AUC, or authentication center, of a wireless system. The AUC is normally part of the HLR in many systems. Authentication is used to prevent fraud in a wireless network.

4

Radio Systems

Radio system projects always generate a host of questions ranging from what a particular phrase means to how to calculate a particular item that is to be part of the cell site or system design. The following is a list of the more commonly asked questions that befall the RF engineering department and managers who oversee these groups.

Q. *What is a Fresnel zone?*

A. The *Fresnel zone* is the area in a radio system where the primary energy for a radio link, propagation, is contained (Fig. 4.1). The calculation of the first Fresnel zone is essential for radio communication systems because the areas within the first Fresnel zone determine the change the radio wave will undergo. It is important to ensure that there are no obstructions within the first Fresnel zone for a communication system. Note that the Fresnel zone is three dimensional and that its path is not only vertical but also horizontal.

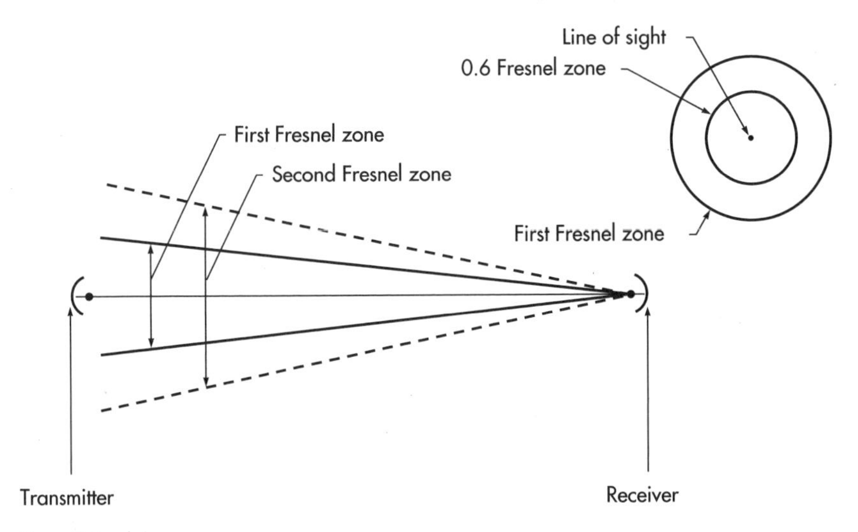

Figure 4.1 Fresnel zone.

Q. *What is a 401 contour?*

A. A *401 contour* defines the radio boundary for a wireless system using a 39-dBμ value. A *401-contour extension* must be coordinated with neighbor markets and filed with the FCC if a particular wireless operator's radio boundary crosses FCC–defined boundaries around rural service areas (RSAs), metropolitan service areas (MSAs), basic trading areas (BTAs), metropolitan trading areas (MTAs), or cellular geographic service areas (CGSAs).

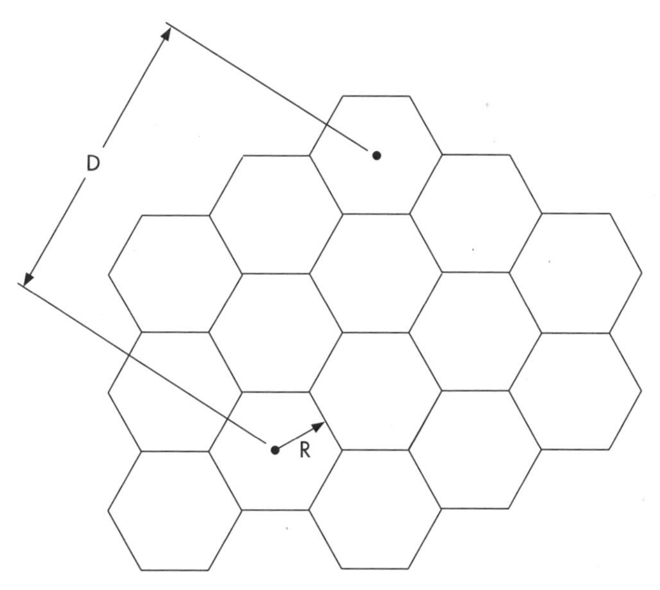

Figure 4.2 The D/R ratio. D/R = $\sqrt{3}$. N = reuse factor.

Q. *What is a D/R ratio?*

A. A *distance-to-radius* (*D/R*) *ratio* is a parameter used to describe the reuse factor for a wireless system. The D/R ratio for any wireless system determines the reuse factor as well as the distance between the reusing cell sites and of course the radii of the serving cell sites. Figure 4.2 and the table below illustrate standard D/R ratios for different frequency reuse patterns, *N*.

D/R	*N* (reuse pattern)
3.46R	4
4.6R	7
6R	12
7.55R	19

Q. *What is ducting?*

A. *Ducting* is an atmospheric thermal inversion whereby radio signals can propagate over distances greater than normal. Ducting is an advantage in amateur and high-frequency (HF) communications.

Q. *What is interference?*

A. *Interference* is any impairment of a radio signal. Usually the term *interference* refers to cochannel interference whereby the signal impairment is caused by another carrier or radio that is using the same frequency. There are many causes of interference such as signals from adjacent channels and intermodulation products.

Q. *What is clutter?*

A. *Clutter* is a numerical quantity in decibels that is used to describe the morphology in a particular radio transmission area. Accounting for clutter is essential in estimating the radio-wave propagation.

Q. *What is building clutter?*

A. *Building clutter* is a numerical quantity in decibels that describes the morphology specifically associated with urban and dense urban environments within a wireless system operating area. Accounting for building clutter is essential in estimating a radio-wave's propagation characteristics.

Q. *What is an RSA?*

A. A *rural service* (or *statistical*) *area* (RSA) is a geographic boundary defined by the FCC for the purpose of issuing a cellular license. RSA cellular systems are usually small and are not located in densely populated areas.

Q. *What is an MSA?*

A. A *metropolitan service* (or *statistical*) *area* (MSA) is a geographic boundary defined by the FCC for the purpose of issuing a cellular license. MSA cellular systems are usually large and are located in densely populated areas.

Q. *What is a BTA?*

A. A *basic trading area* (BTA) is a geographic boundary defined by the FCC for the purpose of issuing a PCS license in an area that is not densely populated.

Q. *What is an MTA?*

A. A *metropolitan trading area* (MTA) is a geographic boundary defined by the FCC for the purpose of issuing a PCS license in a densely populated area.

Q. *What is a CGSA?*

A. A *cellular geographic service* (or *statistical*) *area* (CGSA) is a defined geographic boundary for a cellular system.

Q. *What is a line of sight?*

A. A *line of sight* (LOS) is the straight path between a radio signal transmitting antenna and a radio signal receiving antenna when unobstructed by the horizon or objects in the radio transmission path (Fig. 4.3).

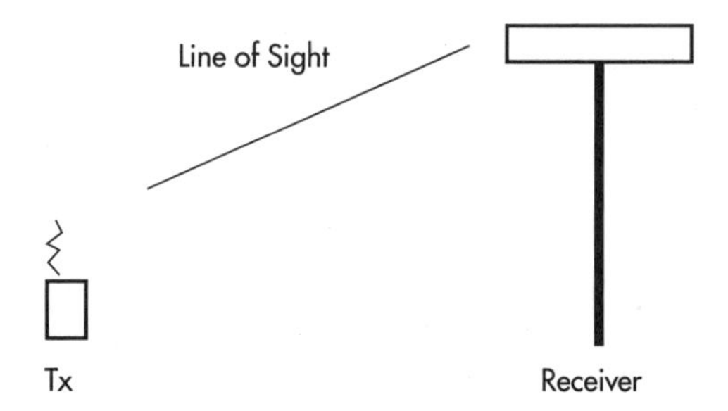

Figure 4.3 Line of sight.

Q. *What does the term Rayleigh fading mean?*

A. *Rayleigh fading,* also called *fast fading,* is commonly used to describe multipath fading in a two-way radio communication system that occurs when there is not a clear path between the transmitter and the receiver. *Rayleigh fading* describes the statistical distribution of the radio signal's power as received by a radio receiver. Rayleigh fading occurs when two or more waves from the source transmitter are reflected and form standing-wave pairs in space. When the standing-wave pairs occur, the signals are summed in amplitude, which causes irregular signal strength variations, which then usually results in a reduction in signal strength.

Rayleigh fading occurs when the receiving antenna moves through constructive and destructive wavefronts. The receiver's susceptibility to fading is a function of the frequency of operation and the receiver bandwidth. The higher the frequency, the shorter the distance is between wave crests. The wider the bandwidth, the less susceptible the receiver is to fading. A value of 5 dB is typically used in accounting for Rayleigh fading in the link budget.

Q. *What does the term Rician fading mean?*

A. *Rician fading* describes the statistical energy distribution of a direct wave path from a transmitter to a receiver. This is also referred to as the *line-of-sight path,* and it represents the variation in signal strength that occurs when the path from a transmitter to a receiver is not obstructed. *Rician fading* describes a condition that occurs when one dominant signal arrives at the receiver with several other weaker multipath signals. Rician fading is not that common in two-way communication because buildings or other objects usually obstruct a line of sight to the source.

Q. *What does the term spread-spectrum technology mean?*

A. *Spread-spectrum technology* is used to describe a variety of different technology platforms in wireless communications. A *spread-spectrum system* is any communication system that deliberately occupies more channel bandwidth than the minimum required for data transfer. The rationale behind utilizing spread-spectrum technology is to gain an improvement in the signal-to-noise ratio of the communication system itself. Three basic types of spread-spectrum formats are utilized with many perturbations: *code division multiple access* (CDMA), *time division multiple access* (TDMA), and *frequency division multiple access* (FDMA).

Q. *What is microwave path clearance?*

A. *Microwave path clearance* is the process of ensuring that there will be no interference with a microwave link either when installing a new microwave link for back-hauling network data or when installing a cell site in the PCS band where existing microwave users in the same band need to be cleared—that is, relocated—prior to the commencement of operation for that cell site or cell sites.

Q. *What is path clearance?*

A. *Path clearance,* an essential component for point-to-point communication systems, involves ensuring that there are no obstructions between the transmitting and receiving antennas or within the first Fresnel zone.

Q. *How is path clearance calculated?*

A. Path clearance is calculated as demonstrated in Fig. 4.4.

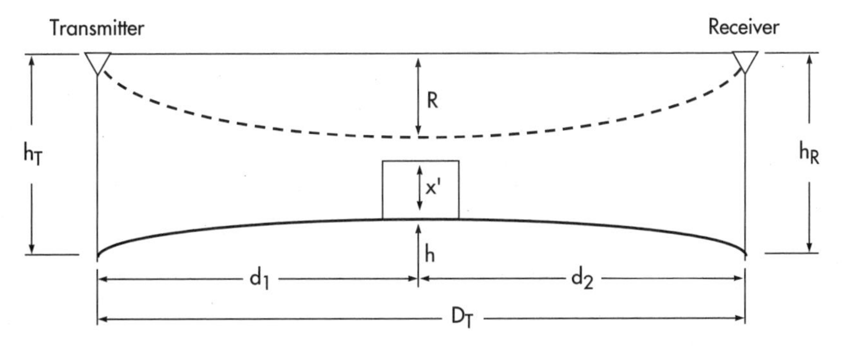

Figure 4.4 Path clearance.

Example 4.1

To determine the path clearance required for a point-to-point communication site refer to Fig. 4.4.

First Fresnel zone:

$$R = \sqrt{72 \frac{d_1 d_2}{D_T f}}$$

Earth curvature:

$$h = \frac{d_1\, d_2}{1.5k}$$

with f = frequency in GHz; d_1, d_2, D_T in miles; x, h, h_T, h_R in feet; and where $d_1 = 1.6$ mi, $d_2 = 2.1$ mi, $D_T = 3.7$ mi, and $f = 0.88$ GHz (or 880 MHz).

$$R = 72 \sqrt{\frac{(1.6)(2.1)}{(3.7)(.88)}} = \sqrt{\frac{3.36}{3.256}} = 73.14 \text{ ft}$$

$$R' = (0.6)R = 43.884 \qquad h = \frac{(1.6)(2.1)}{(1.5)(4/3)} = 1.68 \text{ ft} \qquad k = 4/3$$

Assume that the transmitter, receiver, and obstruction have the same ASML.

Earth curvature	1.68
0.6 Fresnel zone	43.88
Obstruction height	100
	145.56 ft

The minimum Tx and Rx heights for the system are $h_T = h_R =$ 145.56 ft.

Q. *How is two-branch diversity achieved?*

A. To achieve two-branch diversity, two separate receiver paths are used for receiving the desired signal (Fig. 4.5). Under ideal conditions the maximum allowable gain from a two-branch diversity system is 3 dB.

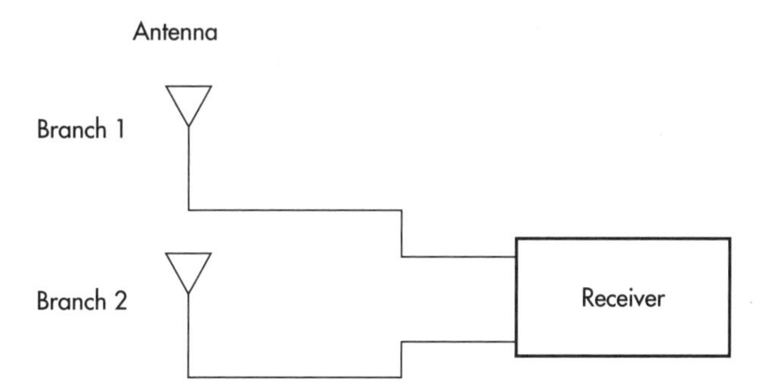

Figure 4.5 Two-branch diversity receive.

Q. *How is three-branch diversity achieved?*

A. To achieve three-branch diversity, three separate receiver paths are used for receiving the desired signal (Fig. 4.6). Under ideal conditions the maximum allowable gain from a three-branch diversity system is 4.7 dB.

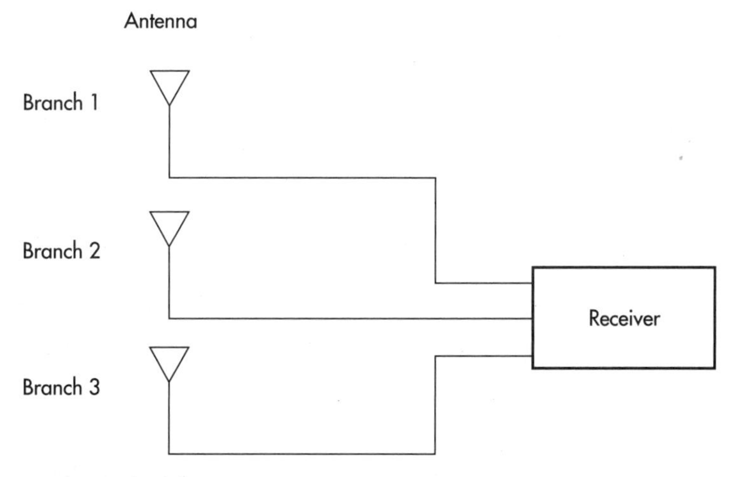

Figure 4.6 Three-branch diversity receive.

Q. *What items are included in a link budget?*

A. The items that make up a link budget are shown in Fig. 4.7.

Link Budget—Downlink

Frequency range, MHz	Tx	
Base station parameters		
Tx PA output power		
Tx combiner loss		
Tx duplexer loss		
Tx lightning arrester loss		
Tx cable loss		
Tx jumper loss		
Tx tower top amp gain		
Tx antenna gain		
	———	
		(Cell ERP)
Environmental margins		
TX diversity gain		
Fading margin		
Environmental attenuation		
Cell overlap		
	———	
Mobile parameters		
Antenna gain		
Rx diversity gain		
Antenna cable loss		
Coding gain		
Rx sensitivity		
	———	
		Downlink budget, dB

Figure 4.7 Link budget outline.

Q. *What are propagation models?*

A. *Propagation models* are computer simulations that attempt to estimate radio coverage and also interference for the purpose of facilitating the RF design process. There are numerous propagation models that are used now and that continue to be developed for use in wireless systems. However, most of the propagation models are derived from Okumura's field data, which have been modified to reflect more refined path-loss characteristics.

Q. *What is the quick propagation model?*

A. The *quick propagation model* is a fast estimating process that can be used in a field experiment to produce general propagation expectations for an area. The quick method shown below can be used to make a rough-draft cell design:

$$880\text{ MHz:} \quad PL = 121\text{ dB} + 36\log\text{ dB }(X\text{ km})$$

$$1900\text{ MHz:} \quad PL = 130\text{ dB} + 40\log\text{ dB }(X\text{ km})$$

Q. *What is the Carey propagation model?*

A. The *Carey propagation model* is used for submitting information to the FCC for cell site applications. The Carey curves were initially specified for 450 to 460 MHz, but they are effective from 450 to 1000 MHz. There are a few assumptions, or rather design constraints, that are placed on the use of this model, as is the case with any other propagation model.

The Carey model is defined as follows:

$$L_1 = 110.7 - 19.1 \cdot \log_{10} h_b + 55 \log_{10} R \qquad 8 \le R \le 48 \text{ km}$$

$$L_2 = 91.8 - 18 \cdot \log_{10} h_b + 66 \cdot \log_{10} R \qquad 48 \le R \le 96 \text{ km}$$

where h_b = 30 to 1500 m.

Q. *What does the term morphology mean?*

A. *Morphology* refers to the unique terrain characteristics for a given area. The unique terrain characteristics could be dense buildings or open flat terrain or rolling hills, to mention a few. Morphology is used to help predict the expected radio coverage and performance for a given area. There are several types of morphology commonly factored into radio wave propagation estimates; however, the actual values that are assigned to each are often open to debate and much manipulation for the better or worse by RF engineers. The types of morphology are dense urban, urban, suburban, and rural, and there is a possibility that more morphology classes will be added.

Q. *What does the term dense urban morphology mean?*

A. *Dense urban morphology* refers to a dense business district of a metropolitan area. The buildings in this type of area generally are 10 to 20 stories high or taller, consisting of skyscrapers and high-rise apartment buildings.

Q. *What does the term urban morphology mean?*

A. *Urban morphology* refers to an area where building structures are normally from 5 to 10 stories high.

Q. *What does the term suburban morphology mean?*

A. *Suburban morphology* refers to an area where there is a mix of residential and business buildings ranging from 1 to 5 stories high but mainly consisting of 1- to 2-story structures.

Q. *What does the term rural morphology mean?*

A. *Rural morphology* refers to an area that consists of generally sparsely populated, open areas with structures that do not exceed 2 stories.

Q. *What is the Hata model?*

A. The *Hata model* is an empirical model derived from a technical report made by Okumura so that the results could be used in a computational model. The *Okumura Report* [1] is a series of charts that are instrumental in radio communication system modeling.

The Hata model is shown below:

$$L_H = 69.55 + 26.16 \cdot \log_{10} f_c - 13.87 \log_{10} h_b - a(h_m) + (44.9 - 6.55 \log_{10} h_b) \cdot \log_{10} R$$

The range for which the Hata model is valid is listed below:

$$F_c = 150 \text{ to } 1500 \text{ MHz}$$
$$h_b = 30 \text{ to } 200 \text{ m}$$
$$h_m = 1 \text{ to } 10 \text{ m}$$
$$R = 1 \text{ to } 20 \text{ km}$$

The Hata model should not be employed if the path loss is less than 1 km from the cell site or if the site is less than 30 m in height. (This is an interesting point to note because sites are typically less than 1 km apart and often below the 30-m height.)

In the Hata model, the value h_m is used to correct for the mobile unit antenna height. (It is interesting to note that if a height of 1.5 m is assumed for the mobile unit, that value nulls out of the equation.)

The Hata model employs three correction factors based on the environmental conditions for which the path-loss prediction is evaluated: urban, suburban, and open. The environmental correction values are easily calculated but vary for different values of mobile unit height. For the values listed below, a mobile unit height of 1.5 m is assumed:

Urban: 0 dB
Suburban: −9.88 dB
Open: −28.41 dB

Q. *What is the Cost231 propagation model?*

A. The *Cost231 propagation model*, also known as the *COST231 Walfish/Ikegami propagation model* [1], is used for estimating the path loss in an urban environment for cellular communication. This model was first developed for use for GSM in Europe, but it has since been adapted for many other types of wireless systems used throughout the world. The Cost231 model combines empirical and deterministic methods over the frequency range of 800 to 2000 MHz. The Cost231 model is made up of three basic components:

1. Free-space path loss
2. Roof-to-street diffraction loss and scatter loss
3. Multiscreen loss

$$L_c = L_f + L_{RTS} + L_{ms}$$

where $L_{RTS} + L_{ms} \leq 0$, L_f = free-space loss, L_{RTS} = roof-to-street diffraction loss and scatter loss, L_{ms} = multiscreen loss

$$L_f = 32.4 + 20 \log_{10} R + 20 \log_{10} f_c$$

where R = km, f_c = MHz

$$L_{RTS} = -16.9 - 10 \log_{10} w + 10 \log f_c + 20 \log \Delta h_m + L_0$$

where w = street width in meters, $\Delta h_m = h_r - h_m$

$$L_0 = \begin{cases} -10 + 0.354\phi & 0 \leq \phi \leq 35 \\ 2.5 + 0.075\,(\phi - 35) & 35 \leq \phi \leq 55 \\ 4.0 - 0.114\,(\phi - 55) & 55 \leq \phi \leq 90 \end{cases}$$

where ϕ = incident angle relative to street

$$L_{ms} = L_{bsh} + k_a + k_d \log_{10} R + k_f \log_{10} f_c - 9 \log b$$

where b = distance between buildings along radio path

$$L_{bsh} = \begin{cases} -18 \log_{10}(1 + \Delta h_b) & h_b > h_r \\ \phi & h_b < h_r \end{cases}$$

$$k_a = \begin{cases} 54 & h_b > h_r \\ 54 - 0.8\, h_b & d \geq 500 \text{ m}, h_b \leq h_r \\ 54 - 1.6\, \Delta h_b \cdot R & d < 500 \text{ m}, h_b \leq h_r \end{cases}$$

NOTE: Both L_{bsh} and k_a increase the path loss with lower base station antenna heights

$$k_d = \begin{cases} 18 & h_b > h_r \\ 18 - 15 \dfrac{\Delta h_b}{\Delta h_r} & h_b \leq h_r \end{cases}$$

$$k_f = \begin{cases} 4 + 0.7 \cdot \left(\dfrac{f_c}{925} - 1\right) & \text{for midsized city and suburban area with moderate tree density} \\ 4 + 1.5 \cdot \left(\dfrac{f_c}{925} - 1\right) & \text{for metropolitan center} \end{cases}$$

In the above equations, the following items bound the equations' useful range (it is always important to know the valid ranges for a particular model):

$$f_c = 800 \text{ to } 2000 \text{ MHz}$$
$$h_b = 4 \text{ to } 50 \text{ m}$$
$$h_m = 1 \text{ to } 3 \text{ m}$$
$$R = 0.02 \text{ to } 5 \text{ km}$$

Some additional default values that apply to the Cost231 model when specific values are not known are listed below (the default values can and will significantly alter the path-loss values):

b = distance between buildings (20 to 50 m)
w = width of street ($b/2$)
h_r = height of roof, m [3 · (number of floors) + height of roof, m]
Height of pitched roof = 3 m
Height of flat roof = 0 m
θ = incident angle = 90°

Q. *What is an antenna?*

A. An *antenna* is a device used to couple electromagnetic energy between the air and a radio system. Specifically, when an antenna is transmitting, it converts electric current into an electromagnetic wave, and conversely, when it is receiving, it converts an electromagnetic wave into electric current.

Q. *What types of antennas are available?*

A. There are many types of antennas available, and all of them perform specific functions depending on the application at hand. The type of antenna used by a system operator can be a *collinear, log periodic, folded dipole,* or *yagi,* to mention a few. In addition, an antenna may be either active or passive. An *active antenna* usually has some type of electronic device associated with it to enhance its performance. A *passive antenna* is the more classical type, and no electronic devices are associated with it to enhance its performance.

Q. *What is the difference between a dBi antenna and a dBd antenna?*

A. The difference between a dBi antenna and a dBd antenna is 2.14 dB. Specifically, a *dBi antenna* is a theoretical *isotropic antenna* that radiates equally in all directions and does not exist. A *dBd antenna* is a *dipole antenna* that has 2.14 dB more gain than an isotropic antenna. (See Fig. 4.8.)

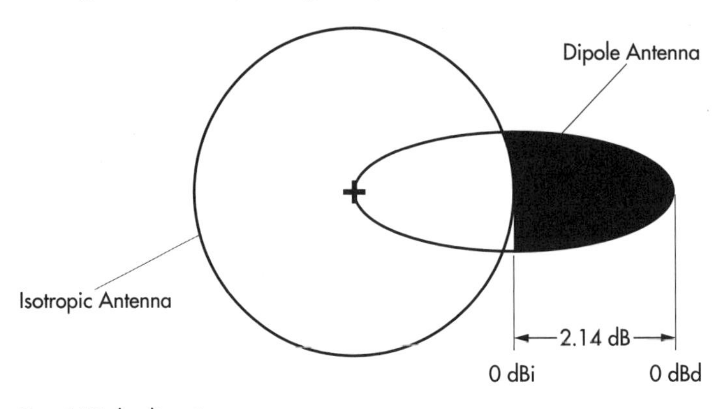

Figure 4.8 Dipole and isotropic antennas.

Q. *How do you convert the gain for a dBi antenna to the gain for a dBd antenna?*

A. To convert the gain value for a dBi antenna to the equivalent gain value for a dBd antenna, the following equation is used:

$$\text{Gain for dBd antenna} = \text{gain for dBi antenna} - 2.14 \text{ dB}$$

Table 4.1 shows the calculated values along with their nearest approximate values.

TABLE 4.1 CONVERSION FROM DBI ANTENNA GAIN TO DBD ANTENNA GAIN

dBi antenna gain, dB	dBd antenna gain, dB (rounded-off value)
5	2.86 (3 dBd)
10	7.86 (8 dBd)
12	9.86 (10 dBd)
14	11.86 (12 dBd)
18	15.86 (16 dBd)
21	18.86 (19 dBd)

Q. *How do you convert the gain for a dBd antenna to the gain for a dBi antenna?*

A. To convert the gain value for a dBd antenna to the equivalent gain value for a dBi antenna, the following equation is used:

$$\text{Gain for dBi antenna} = \text{gain for dBd antenna} + 2.14 \text{ dB}$$

Table 4.2 shows the calculated values along with their nearest approximate values.

TABLE 4.2 CONVERSION FROM DBD ANTENNA GAIN TO DBI ANTENNA GAIN

dBd antenna gain, dB	dBi antenna gain, dB (rounded-off value)
3	5.14 (5 dBi)
10	12.14 (12 dBi)
12	14.14 (14 dBi)
14	16.14 (16 dBi)
18	20.14 (20 dBi)
21	23.14 (23 dBi)

Q. *What does the term diversity mean?*

A. *Diversity* as it applies to an antenna system used in wireless communications is a method for comparing radio signal fading in a particular environment. *Diversity gain* is calculated as the gain that would have occurred had a diversity technique not been used. In the case of a two-branch diversity system, if the received signal into both antennas is not of equal signal strength, then there can be no diversity gain. This is an interesting point considering that most link budget calculations incorporate diversity gain as a positive attribute. The only way diversity gain can be incorporated into a link budget is if a fade margin is included in the link budget and the diversity scheme chosen attempts to improve—that is, reduce—the fade margin that is included there.

Q. *What types of diversity are there?*

A. There are numerous types of diversity that can be used in a wireless system, all of which fall into one of the following categories.

The type of antenna diversity used can be, and often is, augmented with another type of diversity that is accomplished at the radio level.

- Spacial
- Horizontal
- Vertical
- Polarization
- Frequency
- Time
- Angle

Q. *What is a leaky feeder, or slotted coax?*

A. A *leaky feeder,* also called a *slotted coax,* is used for providing radio coverage inside solid structures such as buildings, tunnels, and elevator shafts. The leaky feeder acts as an antenna, delivering the RF energy along its length. The leaky feeder is a coax cable with slots located along its ridges. The size and spacing of the ridges and slots determine the frequency of operation best matched to the leaky feeder.

Q. *What is polarization?*

A. The *polarization* that is used for cellular, PCS, and enhanced specialized mobile radio (ESMR) systems is primarily vertical. There are a few, rare instances in which the polarization is not vertical, for example, a polarization diversity antenna and the horizontal polarization antenna used for increased isolation for a control station with ESMR. However, the polarization is determined by the E field used, which, in almost all wireless communication systems, means that the polarization is vertical.

Q. *What is diversity spacing?*

A. Diversity spacing is associated with the antenna spacing, and it is a design requirement that is stipulated by RF engineers. *Diversity spacing* is a physical separation between the receive antenna, which is needed to ensure that the proper fade margin protection is designed into the system. Horizontal space diversity is the most common type of diversity scheme that is used in wireless communication systems. The following is a brief rule of thumb used to determine the horizontal diversity requirements for a cell site.

$$\text{Diversity spacing } (f) = \frac{\text{AGL of antenna (ft)}}{11} \times \left(\frac{835}{f_0}\right)$$

Q. *What is the physical spacing required for wall mounting antennas?*

A. For many building installations, it may not be possible to install the antennas above the penthouse or other structures on the building. Often it is necessary to install the antennas onto the penthouse or water tank of an existing building. Rarely has a building architect factored into the building design the potential installation of antennas. Therefore, as shown in Fig. 4.9, the building walls and structures may meet one orientation needed for the system but rarely all three for a three-sector configuration. Consequently, it is necessary to determine what the offset from the wall of the building structure needs to be. Figure 4.10 illustrates the wall mounting offset that is required to ensure proper orientation for each sector.

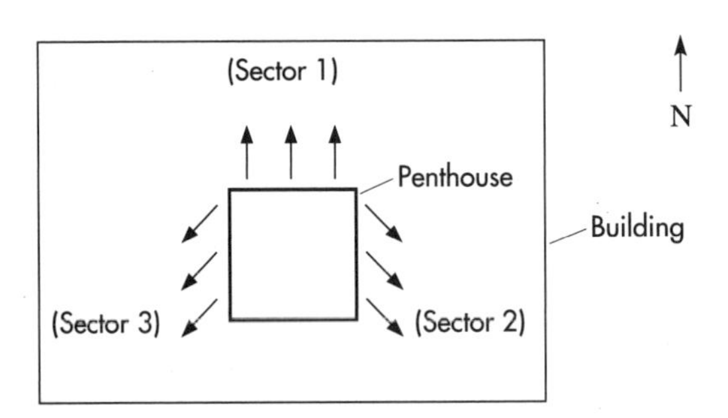

Figure 4.9 Three-sector building configuration.

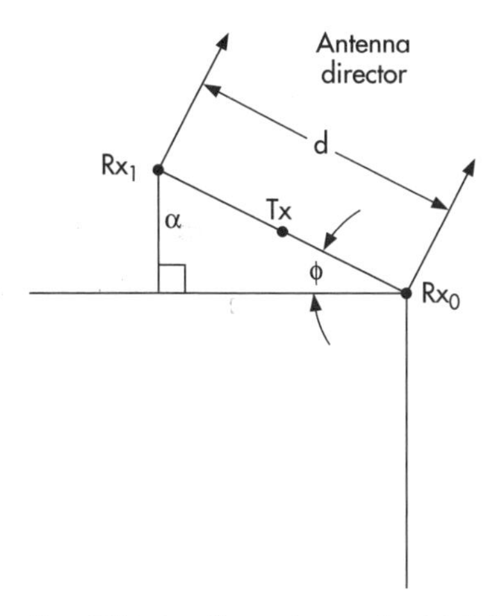

Figure 4.10 Antenna offset mounting. $\alpha = d \times \sin(\phi)$, where d = diversity separation (ft), α = distance from wall (ft), and ϕ = angle from wall (°).

Q. *What is the antenna installation tolerance?*

A. The antenna installation tolerances apply directly to the physical orientation and plumbness of the antenna installation itself. There are usually two separate considerations: how accurate the antenna orientation needs to be and how plumb the antenna installation needs to be. The obvious issues here are not only the engineering design requirements but also the practical implementation of the antennas. Therefore, the following guidelines should be used:

Type	Tolerance
Orientation	±5% or antenna's horizontal pattern
Plumbness	±1° (critical)

Q. *What is an obstruction?*

A. An *obstruction* is a physical impedance to a radio signal. There are several types of obstructions, and they can exist at either the cell site or near the cell site. Obstructions must be considered during the installation of the antenna system. For example, a tall building next to a cell site would be an obstruction if the building obscured the radio path. Another form of obstruction would be a mountain if it obscured a radio path. Other obstructions that normally occur around buildings are HVAC systems, window washing apparatus, or another competitor's radio or antenna equipment.

Q. *What does the unit of measure dBc describe?*

A. The unit dBc (decibels, carrier) defines the number of either positive or negative decibels gained in terms of the radio signals received by a *carrier* antenna, the c component of dBc. For example, if a carrier had an ERP of 100 W, +50 dBm, and at some distance away there was a path loss of 50 dB, then the value at that location would be −50 dBc, which would equate to 0 dBm. If the attenuation was 40 dB, then the value would be −40 dBc with an RSSI of 10 dBm.

Q. *What is EIRP?*

A. *Effective isotropic radiated power* (EIRP) is the gain in decibels of an isotropic antenna, dBi.

Q. *How is the EIRP calculated?*

A. Calculating EIRP for a cell site or any wireless system is straightforward. The antenna gain is supplied in either dBi or dBd. The best way to calculate EIRP is to convert the antenna gain to dBi if it is given with respect to dBd, as shown in the following example.

Antenna gain = 10 dBd = 12.14 dBi
(To simplify the example, 12 dBi will be used.)

If the following conditions apply (the parentheses around numbers mean that the number has been rounded off):

Power amplifier, 40 W	= +46 dBm
Combined losses	= −3.5 dB
Tx filter loss	= −0.5 dB
Feed-line and jumper loss	= −4 dB
Cell site antenna gain	= +12 dBi

Then, Power out = +50 dBm = 100 W EIRP

Q. *What is path loss?*

A. *Path loss* is the attenuation that a radio signal undergoes when it travels from a source antenna to a target antenna. In a downlink direction, the *source antenna* is the cell site, and the *target antenna* is the subscriber unit. The path loss is measured in decibels of attenuation, and it is reciprocal in that the attenuation that is experienced in the downlink is fundamentally the same in the uplink. Therefore, the same path-loss value can be used for either direction. For example, if the following conditions apply:

Cell site ERP (100 W) = +50 dBm
Subscriber unit, RSSI = −85 dBm

Then, Path loss = −135 dB

Q. *What is terrain?*

A. In wireless communications terms, *terrain* describes the various land-form obstacles that lie in the radio signal's path between the cell site and the subscriber unit. *Terrain* typically refers to the contours of the land that the signal traverses—that is, mountains, bodies of water, marshes, and forest. An attenuation value expressed in decibels is associated with each type of terrain and is used in the estimation of the signal level for radio signal coverage.

Q. *What is terrain, or environmental, attenuation?*

A. *Environmental,* or *terrain, attenuation* is the attenuation in decibels that a radio signal undergoes from a source to a target. Table 4.3 lists some general attenuation estimates that are used in propagation models. Glancing at the table, it is immediately apparent that the difference between the urban and suburban environments is more than 3 dB. Obviously, each area has its own unique propagation characteristics. Thus it is possible to make a best-guess estimate of the potential attenuation for that particular area.

Table 4.3 Environmental Effects on Path Loss, dB

Foliage	
Sparse	6
Light	10
Medium	15
Dense	20
Very dense	25
Buildings	
Water/open	0
Rural	5
Suburban	8
Urban	22
Dense urban	27
Vehicle	10–14

Q. *What is shadowing?*

A. Shadowing, also called *slow fading,* is caused largely by partial blockage from a building or by environmental absorption from obstructions such as trees.

Q. *What is a fade margin?*

A. A *fade margin* is a value in decibels that is used in the link budget for the purposes of estimating the amount of attenuation, or rather fading, that will take place in a wireless communication path. A fade margin that is used for wireless systems ranges from 5 to 20 dB depending on whether the application is in a building or on a street. Incorporating a fade margin is meant to ensure that on average a particular fade in signal strength is expected and that it is accounted for in the link budget.

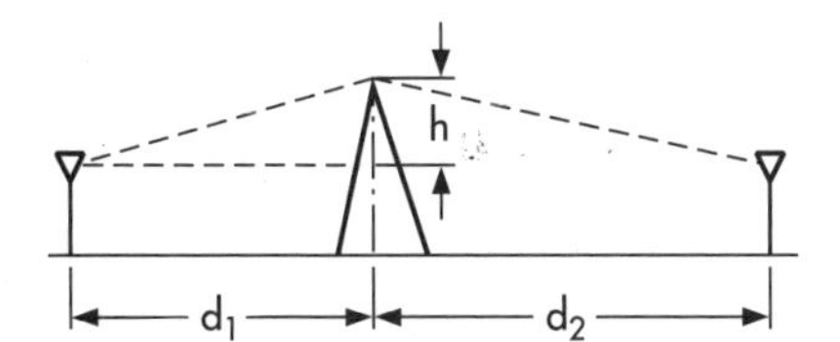

Figure 4.11 Relationship between distance from an obstruction and height of the transmitter and receiver, $d_1 \leq d_2$.

Q. *What is diffraction?*

A. *Diffraction* describes the amount that a radio signal propagates around and over an obstruction. This propagation is measured in decibels. Depending on the severity of the diffraction, the radio signal may be impaired. There are several types of diffraction methods modeled in RF, and they are smooth and knife edge. As a general rule of thumb, the larger the obstruction, the greater the diffraction, or rather attenuation.

Q. *How is the amount of diffraction loss predicted?*

A. The method used to calculate the amount of diffraction loss expected is shown below, for a single obstacle in the path. (See also Fig. 4.11.)

$$v = h\sqrt{\frac{2(d_1 + d_2)}{\lambda d_1 \cdot d_2}}$$

$$G_d(\text{dB}) = 20 \log_{10} |F_v|$$

$$\therefore G_a(\text{dB}) = 0 \qquad v \leq -1$$

$$= 20 \log_{10} (0.5 - 0.62v) \qquad -1 \leq v \leq 0$$

$$20 \log_{10} (0.5 - 0.95v) \qquad 0 \leq v \leq 1$$

$$20 \log_{10} (0.4) - \sqrt{0.1184 - (0.38 - 0.1v)^2} \qquad 1 \leq v \leq 2.4$$

$$20 \log_{10} \left(\frac{0.225}{v}\right) \qquad v > 2.4$$

For 880 MHz

$$d_1 = 1 \text{ km} \qquad d_2 = 1.5 \qquad h = 200 \text{ m} \qquad \lambda_{880} = 0.34 \text{ m}$$

$$v_{880} = 200\sqrt{\frac{2(2500)}{(0.34)(1.5 \times 10^6)}} = 19.8$$

$$G_d = 20 \log_{10}\left(\frac{0.225}{v}\right) = 20 \log_{10}\left(\frac{0.225}{19.8}\right) = -38 \text{ dB}$$

or a diffraction loss of 38 dB. For 1900 MHz

$$v_{1900} = 200\sqrt{\frac{2(2500)}{(0.157)(1.5 \times 10^6)}} = 29.14$$

$$G_d = 20 \log_{10}\left(\frac{0.225}{29.14}\right) = -42.2 \text{ dB}$$

or a diffraction loss of 42.2 dB

Q. *What types of impairments interrupt the radio path?*

A. Impairments to a radio path are any obstructions that degrade the signal. The four basic types of impairments that occur in a communication path are path loss, shadowing, multipath, and doppler shift.

Q. *What is multipath impairment?*

A. *Multipath signal impairment* most frequently occurs in an urban environment where more than one reflection in a transmission path is normal. (See Fig. 4.12.) The issues of multipath problems in communication systems show up as delay spread and Rayleigh fading.

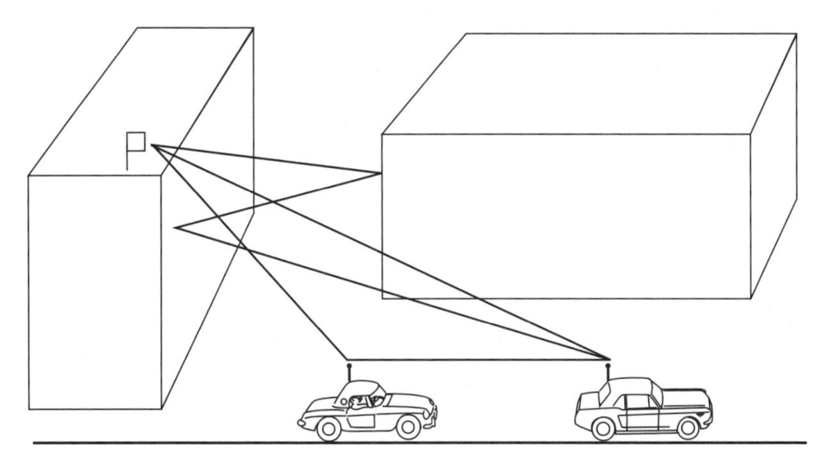

Figure 4.12 Multipath signals in an urban environment.

Q. *What is delay spread?*

A. *Delay spread,* also called *time dispersion,* occurs in a digital wireless system when multiple signals arrive at different times to the receiver and the difference in time between the signals' arriving are on the order of a bit period. The multiple signals' arriving in the receiver within a bit period may cause a distortion in the representation of the bit desired or even the wrong bit to be decoded because the delayed, or reflected, signal is stronger than the direct signal itself. When the time delay spread is no longer negligible with respect to the modulation bandwidth, the received information will be distorted due to the different paths transferred with the multipath incoming waves. The delay spread is more pronounced with higher data rates since the effect can cause symbols to overlap, producing inter-symbol interference.

Q. *What is time dispersion?*

A. *Time dispersion* is another name for *delay spread* in a digital wireless system.

Q. *What is a link budget?*

A. A *link budget* is a power budget and it is used to determine for modeling purposes the amount of loss and gain that can be expected by an RF signal as it traverses from the transmitter of the cell to the mobile unit or vice versa.

Q. *What is included in a link budget?*

A. There are many factors and parameters that are considered in establishing a link budget. However, the brief example given in Fig. 4.7 can and should be used as a basic boilerplate when designing a wireless system. There are some issues that are covered in the link budget that may or may not pertain to the particular potential cell site or technology platform utilized.

Q. *What method is used to determine if a system is uplink or downlink limited?*

A. The determination of whether a system is uplink or downlink limited is made by examining the link budget. Note that it is more common that a system is uplink limited. The table below illustrates uplink and downlink calculations:

Uplink limited	Downlink path loss, dB > uplink path loss, dB
Downlink limited	Uplink path loss, dB > downlink path loss, dB

Q. *What are some commonly used types of cell sites?*

A. There are many different types of cell sites, and often a wireless system will deploy multiple types of cell sites in order to meet the system design requirements. The list that follows is not all inclusive, and there are many perturbations of those that are shown.

1. Omni
2. Directional
3. Three sector
4. Six sector
5. Micro-cell
6. Pico-cell

Q. *How can blocking be reduced at a cell site?*

A. There are numerous ways to reduce the blocking at a cell site. The simplest means is to turn the cell site off, but this method is not effective for improving capacity for a system or region. The following is a brief list of possible methods that can be used, independent of the technology platform used:

- Radio additions
- Parameter adjustments
- Antenna system alterations
- New cells

Q. *What is offloading?*

A. *Offloading* is a process engineers use to shift traffic from one cell to another. Offloading is typically associated with the introduction of a new cell site to a given area where the new cell site removes traffic that was handled by the original cell site to the new cell. Offloading can also be achieved through traffic balancing in that some of the existing traffic may be offloaded to an adjacent cell site through use of antenna or parameter adjustments.

Q. *What are FCC guidelines?*

A. The *FCC guidelines* have been compiled by the FCC to regulate radio service. The guidelines should be known by the design engineer and the technical management of the company. All the FCC guidelines affecting the radio community are contained in various parts of the *Code of Federal Regulations Title 47,* CFR 47. Which section of CFR 47 applies to the particular system at hand depends on the license that the operator is utilizing to provide service. Each different part of CFR 47 dictates what can and cannot be used for a particular service. However, because there is a continuous flow of regulation changes, it is strongly advisable that operators obtain the most recent CFR 47 and check the FCC's Web page for dockets. For example, CFR 47, Part 22, applies to cellular phones, Part 24 applies to PCS devices, and Part 90, to SMR operators.

Q. *What FCC guidelines apply to cellular systems?*

A. The FCC guidelines that apply to cellular systems are CFR 47, Chapter 22. Of course, Chapter 22 refers to other chapters in CFR 47, but Chapter 22 is the starting point.

Q. *What FCC guidelines apply to PCS systems?*

A. The FCC guidelines that apply to PCS systems are CFR 47, Chapter 24. Of course, Chapter 24 refers to other chapters in CFR 47, but Chapter 24 is the starting point. It is important to note that the abbreviation *PCS* may have a different meaning than that used by marketing departments for a wireless company. The term may refer to a particular frequency band.

Q. *What is FAA compliance?*

A. FAA compliance is mandatory for all sites within a system. Although you do not need to file every site with the FAA, it is necessary to ensure that every site is within compliance. The files for ensuring compliance are listed in CFR 47, Part 17.

The compliance can be checked using a glide slope calculation as a starting point, and this procedure should be part of all site acceptances from engineering.

Q. *What are the key elements that need to be followed for FAA compliance?*

A. The key elements that need to be followed for FAA compliance are height, glide slope, alarming, and marking and lighting.

Q. *What is glide slope?*

A. *Glide slope* is a value that approximates the approach of an airplane to a runway depending on its distance from the runway. An example of a glide slope calculation is shown in Fig. 4.13.

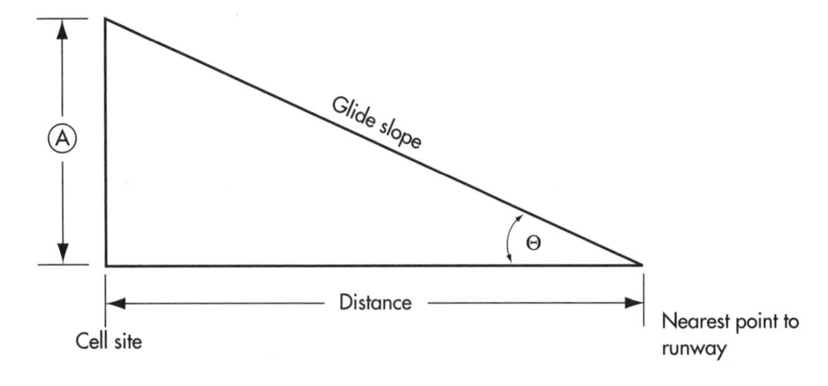

NOTE:

A = tallest point on structure

$$\text{Glide slope} = \frac{\text{distance (feet)}}{\text{factor}}$$

Factor = 100 if runway is > 3200 ft and within 20,000 ft of cell site
50 if runway is < 3200 ft and within 10,000 ft of cell site
25 if runway is a heliport and within 5000 ft of cell site

Figure 4.13 Glide slope calculation.

Q. *When does the FAA need to be notified about the site?*

A. Figures 4.14 and 4.15 show when the FAA is required to be notified of an impending communication site. The figures are meant as a guideline to follow when installing antennas on an existing building. Please note that in Figs. 4.14 and 4.15, it is assumed that the sites do not violate glide slope requirements. If the structure is in the glide slope path, it is prudent to register the site. Finally, note that the communication facility, if it requires registration, must receive the proper clearance and/or requirements prior to its construction.

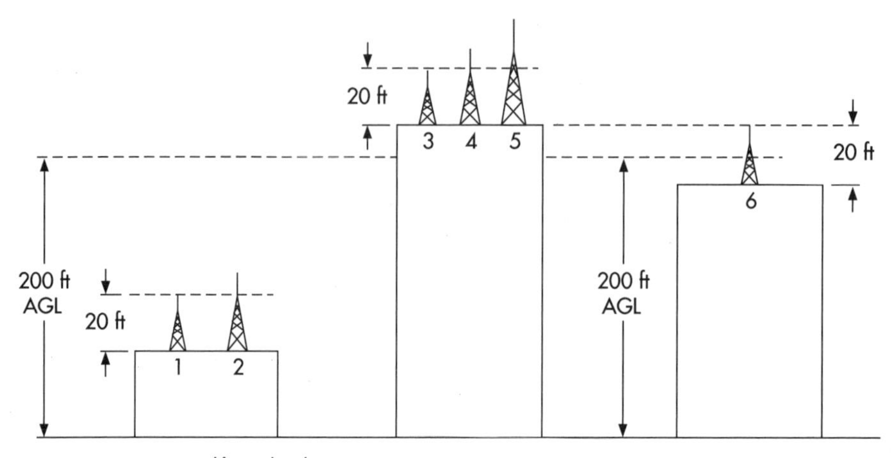

Figure 4.14 Registration required for 4 and 5 only.

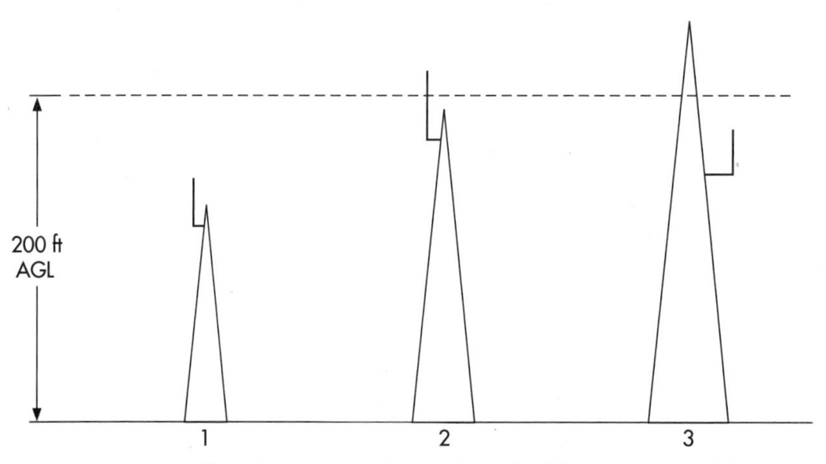

Figure 4.15 Registration required for 2 and 3. Note registration for 3 is needed regardless of where antenna(s) are installed.

Q. *What is EMF compliance?*

A. *Electromotive force* (EMF) *compliance* refers to FCC guidelines that stipulate allowable RF exposures within which all cell sites must operate.

Q. *What is an EMF power budget (or, simply, an EMF budget)?*

A. An *EMF power budget,* also called an *EMF budget,* is a power calculation that is used to determine if the site and its location are within RF exposure requirements set forth in the FCC guidelines.

A sample EMF power budget is shown in Fig. 4.16. The EMF power budget should be signed off by the manager for the department and shared with the operations department of the company. An EMF budget needs to be completed for every cell site in operation and also for those proposed.

Q. *How is an EMF power budget calculated?*

A. The method for calculating the EMF power budget is included in the IEEE C95.1—1991 specification with measurement techniques included in C95.3. Both cellular and PCS systems utilize the same C95.1—1991 standard. These currently, however, are different guidelines that apply to different wireless services. Recently OET-65 from the FCC was issued with revisions for EMF power calculations.

Q. *How can a copy of OET-65 be obtained?*

A. A copy of OET-65 is currently available via the FCC Web page http://www.fcc.gov.

Q. *What is RG58 wire?*

A. *RG58 wire* is the designation for a type of coaxial cable that is used primarily for LAN applications and that has a transmission impedance of 50 Ω. RG58 is used in wireless applications in the receive path between the radios and the receive multicoupler as well as in some transmitter applications where the distances are extremely short and flexibility is desired.

Q. *What is RG59 wire?*

A. *RG59 wire* is the designation for a type of coaxial cable that is used primarily for cable, television, and video applications due to its transmission impedance of 75 Ω.

Cell:
Date:

Sector 1:

Number of channels	19
ERP/channel	100 W
Total power	1900 W

Colocated transmitters:

Paging (931.875 MHz)	
ERP	1000 W

Data points:

Location 1		**Distance**	**Total power**	**Power density**	**Max for band**	**% budget**
Cell site		25 ft	1900 W	0.260527768	0.586666667	44%
Paging		20 ft	1000 W	0.214249809	0.62125	34%
					Total	**79%**
Location 2						
Cell site	100 ft	1900 W	0.016282986	0.586666667	3%	
Paging		110 ft	1000 W	0.007082638	0.62125	1%
					Total	**4%**

Figure 4.16 EMF power budget.

Q. *What is a colocation procedure?*

A. A *colocation procedure* is a process used to coordinate between wireless companies. The outline below lists the steps taken in a typical colocation process. Each of the steps described can be further expanded.

Real Estate

- The real estate department initiates the initial contact with the service provider at the colocation.
- Real estate contacts are exchanged between the two providers.
- Three business days can elapse before an escalation procedure is begun.

Engineering Review and Approval

- The engineering contacts are begun, and a verbal description of what is desired is used for discussion.
- The site-particular parameters are exchanged between both service providers.
- A sample lease exhibit drawing is generated and submitted to the existing operator for review and comment. (This exhibit may be in memo form and may be faxed.) (Five days may be needed for review.)
- The operator initials the drawing to signify approval.
- Six business days can elapse before an escalation procedure is begun.
- The company initiating the request generates a revised lease exhibit drawing and resubmits it to the colocation operators for their records.

Construction

- Two weeks prior to the commencement of construction, the company initiating the request contacts the existing operator's engineering department to notify them of the impending action (real estate contact).

Q. *Why are transmitters combined?*

A. The purpose of combining transmitters is to maximize the use of an antenna, which means that one antenna can be used for several transmitters. If transmitters were not combined, an antenna would be required for every transmitter used at a cell site.

Q. *How are decibels, milliwatts (dBm), converted to decibels, microvolts (dBμ)?*

A. Often sensitivity is quantified in terms of negative decibels, milliwatts (−dBm), or decibels, microvolts (dBμ). The following equation demonstrates how to convert from decibels, milliwatts, to decibels, microvolts:

$$\begin{aligned} 0\ \text{dBm} &= 1\ \text{mW} = 1 \times 10^{3}\ \mu\text{W} \\ 1\ \mu\text{V} &= (50\ \mu\text{W})^{1/2}\ \text{referenced to } 50\ \Omega \\ \therefore -116\ \text{dBm} &= 2.511\ \text{pW} \\ &= 0.224\ \mu\text{V} \end{aligned}$$

Q. *What is an SIG channel?*

A. An SIG channel is a control channel. Motorola used the name SIG to describe the control channel for a cell site, which is why a control channel is known by both names.

Q. *What is a directional coupler?*

A. A *directional coupler* is an electromagnetic device that provides separate outputs for forward and reflected power for the signal that is input into the device. The forward and reflected power outputs are usually about 20 dB lower in signal strength than the signal that is put through the device. The objective of the directional coupler is to provide a coupling point to which the cell site diagnostics may be connected without impeding the signal that is being measured.

Q. *What is a diplexer?*

A. A *diplexer,* also known as a *duplexer,* is a device that permits simultaneous transmission and reception with the same radio antenna.

Q. *What are the major components of a radio system?*

A. The major components of a radio system are listed below. It is important to note that many of the components listed are also used in every technology platform.

- Subscriber equipment (portable or mobile units)
- Cell site(s)
- Antenna system at cell site
- Cell site equipment room or area
- Telco interface
- Mobile switching center (MSC)
- Public switched telephone network (PSTN)

Figure 4.17 highlights the various components listed above.

Q. *How many cells are required for a given network?*

A. How many cell sites are required for any given network is a question frequently asked. The answer varies according to the number of variables that play into the decision. Some of the variables that need to be determined involve the following topics:

1. Time frame
2. Physical area to which the design applies
3. Subscriber and system usage amounts
4. The link budget
5. Coverage requirements or constraints
6. Spectrum available
7. Technology platform utilized
8. D/R, C/I, and/or reuse constraints
9. Cell site capacity constraints
10. Capital constraints
11. Equipment constraints

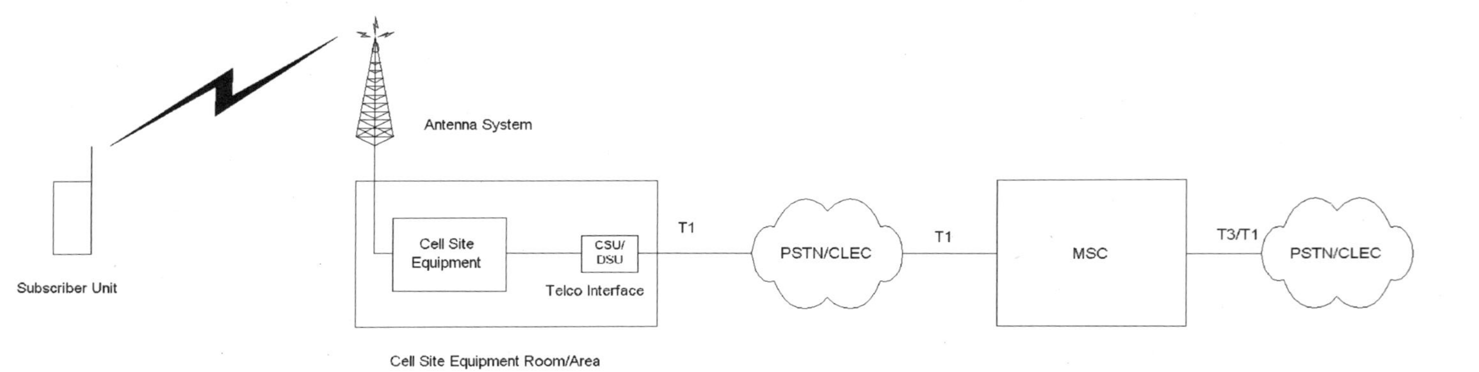

Figure 4.17 Major components of a radio system.

The preceding list is not all inclusive, and more items may need to be factored into the formation of the answer.

The following is a brief example of how the number of cell sites is determined for a simple system design. (See also Fig. 4.18 and Tables 4.4 to 4.6.)

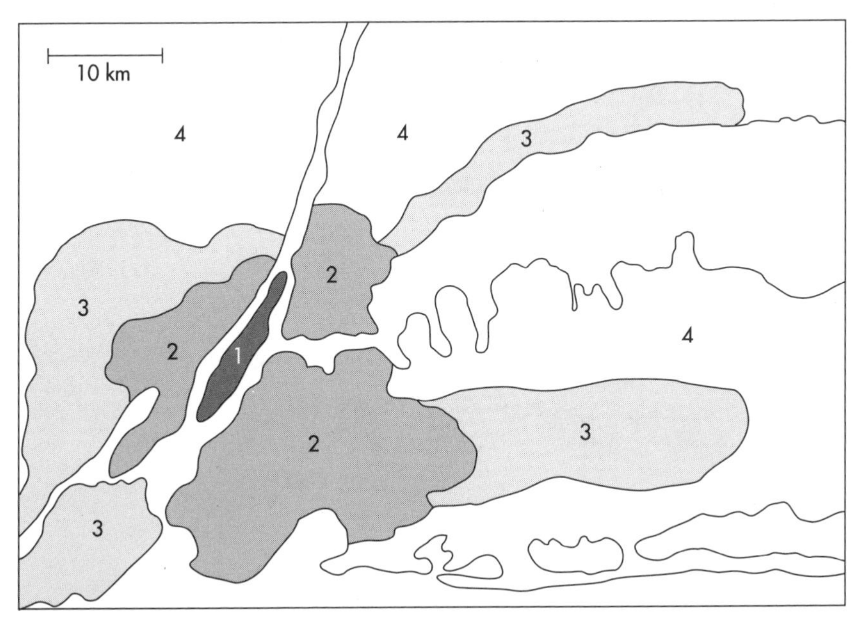

Figure 4.18 System design. Requirements for the geographic regions 1 through 4 are listed in Table 4.4.

TABLE 4.4 SYSTEM TRAFFIC DISTRIBUTION

Region	Area, km^2	Traffic, %	Erlangs	Erlangs/km^2
1	100	20	700	7.0
2	1,500	40	1,400	0.933
3	2,500	30	1,050	0.420
4	5,900	10	350	0.0593
Total	10,000	100	3,500	0.35

500,000 subscribers at 7 merlangs/subscriber = 3500 erlangs, system busy hour.

TABLE 4.5 SUBSCRIBER TRAFFIC

Region	Area, km^2	No. of subscribers	Erlangs	Erlangs/km^2
1	100	100,000	700	7.0
2	1,500	200,000	1,400	0.933
3	2,500	150,000	1,050	0.420
4	5,900	50,000	350	0.0593
Total	10,000	500,000	3,500	0.35

0.7 merlang/subscriber.

TABLE 4.6 CELL SITES NEEDED.

	Region			
	1	2	3	4
Coverage	7	49	43	54
Capacity				
Omni	57	114	86	29
Sector	22	43	33	11
Design	22 sector*	49 sector	43 sector	54 omni

*Because of the dense urban environment, 22 sector sites were chosen for region 1.

EXAMPLE 4.2

To determine capacity (cell erlang):

1. Omni = 12.3 erlangs/cell
2. Sector erlang × sector gain = 12.3 erlangs/sector × 2.64 = 32.47 erlangs/cell

Now, determine the number of capacity cells for region 1:

$$N_{capacity} = \frac{\text{capacity region 1}}{\text{capacity cell}} = \frac{700}{12.3} = 57 \text{ cells} \qquad \text{omni}$$

Or,

$$\frac{700}{32.47} = 22 \text{ cells} \qquad \text{sector}$$

Region 1:

$$\text{PL} = 121 + 36 \log_{10}(d) + \text{overlap}$$

$$137.8 = 121 + 36 \log_{10}(d) + 4 \text{ dB}$$

$$\therefore d = 2.2675 \text{ km}$$

$$A_{\text{cell region 1}} = \pi r^2 = 16.15 \text{ km}^2$$

Now, to determine the number of cells, this equates to region 1 for coverage only:

$$N_{cell} = \frac{\text{area region 1}}{\text{area cell}} = \frac{100 \text{ km}^2}{16.15} = 7 \text{ cells}$$

Therefore, 22 sector cells are needed because region 1 is the capacity driven.

Region 2:

$$\text{PL} = 121 + 36 \log_{10}(d) + \text{overlap} - 5 \text{ dB}$$

$$\therefore d = 3.12 \text{ km}$$

$$A_{\text{cell region 2}} = \pi r^2 = 30.62 \text{ km}^2$$

$$N_{cell} = \frac{\text{area region 2}}{\text{area cell}} = \frac{1500 \text{ km}^2}{30.62} = 49 \text{ cells}$$

$$N_{capacity} = \frac{\text{capacity region 2}}{\text{capacity cell}} = \frac{1400}{12.3} = 114 \text{ cells} \qquad \text{omni}$$

Or,

$$N = \frac{1400}{32.47} = 43 \text{ cells} \qquad \text{sector}$$

Region 3:

$$PL = 121 + 36 \log_{10}(d) + 4 - 10 \text{ dB}$$

$$\therefore d = 4.298 \text{ km}$$

$$A_{\text{cell region 3}} = 58.05 \text{ km}^2$$

$$N_{\text{cell}} = \frac{\text{area region 2}}{\text{area cell}} = \frac{2500}{58.05} = 43 \text{ cells}$$

$$N_{\text{capacity}} = \frac{1050}{12.3} = 86 \text{ cells} \qquad \text{omni}$$

Or,

$$N_{\text{capacity}} = \frac{1050}{32.47} = 33 \text{ cells} \qquad \text{sector}$$

Therefore, 43 sectors are needed for region 3.

Region 4:

$$PL = 121 + 36 \log_{10}(d) + 4 - 15 \text{ dB}$$

$$\therefore d = 5.918 \text{ km}$$

$$A_{\text{cell region 4}} = 110.05 \text{ km}^2$$

$$N_{\text{cell}} = \frac{5900}{110.05} = 54 \text{ cells}$$

$$N_{\text{capacity}} = \frac{350}{12.3} = 28.45 \qquad \text{omni}$$

Or,

$$N = \frac{350}{32.47} = 11 \qquad \text{sector}$$

Therefore, 54 omni sites are needed for region 4.

Q. *What is a link budget?*

A. A *link budget* is a power budget that is calculated for the talk-out or talk-back paths. A simplified talk-out link budget is shown below for reference. The list below helps describe the components of the talk-out link budget for an in-building coverage design using the path-loss equation shown. (Note that numbers enclosed in parentheses have been rounded off.) If the following conditions apply,

Power amplifier, 40 W	=	+46 dBm
Combined losses	=	−3.5 dB
Tx filter loss	=	−0.5 dB
Feed-line and jumper loss	=	−3 dB
Cell site antenna gain	=	+10 dBd
In-building attenuation	=	−20 dB
Subscriber unit receiver sensitivity	=	−102 dBm

then the maximum path loss allowed is 131 dB.

Taking this one step further and using the Quick 800 propagation model, the designer can estimate what the expected *received signal strength indication* (RSSI) will be at 1 km from the cell site:

$$\text{Path loss} = 121 + 36 \log (\text{km})$$

Thus the path loss at 1 km from the cell site is 121 dB.

In contrast, if the gains and losses to the subscriber unit—that is, the talk-back budget—are the following (please note the subtle difference):

Power amplifier, 40 W	=	+46 dBm
Combined losses	=	−3.5 dB
Tx filter loss	=	−0.5 dB
Feed-line and jumper loss	=	−3 dB
Cell site antenna gain	=	+10 dBd
In-building attenuation	=	−20 dB
Path loss at 1 km	=	−121 dB

then the RSSI at 1 km is −92 dBm.

There are several steps that must be taken in constructing a viable link budget. The first is to account for all the components and obstructions to the radio path from the transmitter to the receiver. The second is to keep in mind that many of the components between the cell site and the subscriber unit are estimates and that, since physically measuring every point is not possible, the value is only an estimate. The third is that it is necessary to verify that the system is a balanced path.

Q. *What is a talk-out path?*

A. The *talk-out path* is the RF path from the cell site antenna to the receiver at the subscriber unit, as shown in Fig. 4.19. It is also referred to as the *downlink path*. It should be noted that the talk-out path power budget is normally different from the talk-back path power budget.

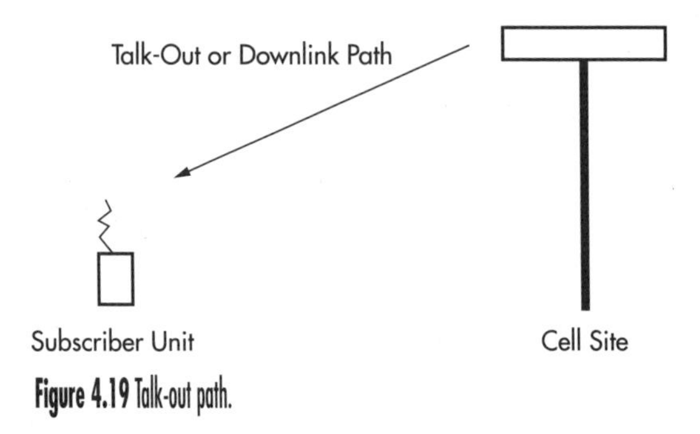

Figure 4.19 Talk-out path.

Q. *What is a talk-back path?*

A. The *talk-back path* is the RF path from the subscriber unit to the receiver at the cell site. This path is in the direction opposite from the talk-out path except the power values and the link budget result should be different. The reason for the difference lies in the simple fact that the components that make up the talk-back path are different from those in the talk-out path. An example of a talk-back path is shown in Fig. 4.20.

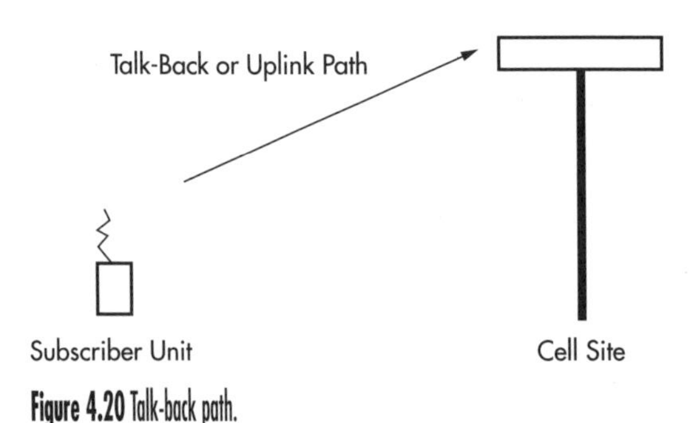

Figure 4.20 Talk-back path.

Q. *What is a balanced path?*

A. A *balanced path* is an RF path for which the power budgets for the talk-out and talk-back paths are equal. Thus, a *balanced path* is what the RF designer is striving for in establishing the link budget.

To achieve this balance, the designer determines the link budget for the talk-out path first and the talk-back path second. Both values give a maximum path loss (in decibels) that can be sustained in the paths. If there is a difference between them, then the talk-out or talk-back paths are individually increased or decreased until the paths are balanced.

An example of this balancing procedure follows. Note that the example is a simplified version of an actual situation and it does not contain all the components normally present. Note that numbers in parentheses have been rounded off.

For a talk-out path, if

Cell site transmit ERP	= (+50 dBm)	(100 W)
Subscriber receiver sensitivity	= −102 dBm	

then the maximum talk-out path loss allowed is

$$(+50 \text{ dBm}) + -102 \text{ dBm} = +152 \text{ dB}$$

The designer then figures out the path loss in the talk-back direction.

For a talk-back path, if

Subscriber maximum transmit ERP	= (+28 dBm)	(0.6 W)
Cell site receiver antenna gain	= +12 dBd	
Feed-line loss	= −2 dB	
Cell site receiver sensitivity	= −105 dBm	

then the maximum talk-back path loss allowed is

$$+28 \text{ dBm} + 12 \text{ dBd} + -2 \text{ dB} + -105 \text{ dBm} = 143 \text{ dB}$$

To determine the path balance, or rather imbalance if it exists, the designer subtracts the talk-out path value from the talk-back value:

Talk-out path loss − talk-back path loss = path-loss imbalance

or

$$152 \text{ dB} - 143 \text{ dB} = 9 \text{ dB}$$

There are many ways to balance the path losses, but two simple methods are as follows:

1. Reduce the cell Tx power by 9 dB, to equal 12.5 W.
2. Increase the gain of the cell site receiver antenna by 9 dBd, to equal 21 dBd.

The use of a 21-dBd antenna is not a practical solution, and based on the coverage requirements, it may not be practical to reduce the ERP by a full 9 dB. Therefore, lowering the ERP of the cell site while improving the receiver path is a more likely scenario. Reducing base Tx power by 6 dB to 0.25 W obtains a higher gain antenna of 15 dBd; this achieves the desired 9-dB improvement in link budget.

Q. *How is a desirable signal level calculated?*

A. A desirable signal level is commonly considered to be the RSSI in decibels, milliwatts, required for the subscriber unit to operate properly. Just what this value is depends on a multitude of variables, and it is related directly to the design objective and the technology platform chosen. For instance, the desirable signal level for on-street coverage would be different from the desirable signal level for a mobile unit or a portable unit to be used inside a building. Furthermore, some technology platforms require a 7-dB energy per bit per noise ratio (E/N) while others require a 22-dB *carrier to interferer* (C/I).

A simple way to establish a desirable signal level follows. If the following conditions apply,

Distance from the cell	=	1 km
Subscriber sensitivity	=	−102 dBm
C/N	=	+19 dB
In-building attenuation	=	−20 dB
Path loss = 121 + 36 log (km)	=	121 dB
Cell site ERP, 25 W	=	+43 dBm
In-building attenuation	=	−20 dB
Path loss, 1 km from cell	=	−121 dB
Subscriber sensitivity	=	−102 dBm
C/N	=	(+19)

then the RSSI at 1 km equals −79 dBm.

Therefore, using the above numbers, a signal level of −79 dBm should be maintained on the street level to ensure that the in-building attenuation is optimal. An obvious modification to the above table would be to change the signal requirement for an in-car unit, which would change the in-building attenuation from −20 dB to an in-car attenuation of −14 dB. Recalculating the math again yields a value of −85 dBm.

Therefore, for an in-building system, a signal level of −79 dBm is needed, but for an in-car system, a signal level of −85 dBm is needed, an obvious 6-dB difference, meaning there would be fewer cell sites for the in-car system.

Q. *What is gain?*

A. *Gain* is the ratio of the power (in decibels) produced by a device (its output) to the power (in decibels) consumed by the device (its input). If the device adds power to the signal, then the gain is a positive value, and if the device is passive like a combiner, feed line, or attenuator, the gain is negative, reflecting a loss in signal level. The basic equation that should be used when calculating gain is as follows:

$$\text{Gain, dB} = 10 \log (P_{out}/P_{in})$$

With the help of a few simple examples, the application of the above gain equation will become apparent.

Input, W	Output, W	Gain, dB
10	100	10
5	100	20
5	10	3
10	5	−3

Q. *What is a guardband?*

A. A *guardband* is a portion of an operator's spectrum that is set aside for nonuse for the purpose of either protecting the operator's own system or potentially a competitor's system. Figure 4.21 shows a typical location of a guardband within an operator's system. In the diagram, a portion of a B-band cellular operator's spectrum has been set aside to protect the new and existing services from each other.

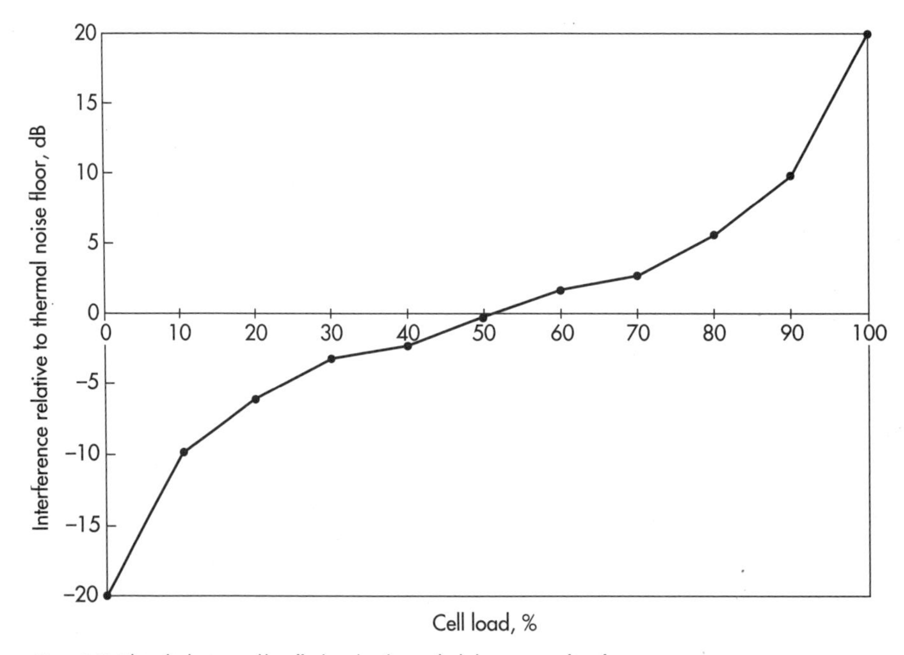

Figure 4.21 Relationship between usable traffic channels and system load, showing points of interference.

Q. *What is a guardzone?*

A. A *guardzone* is typically part of a CDMA system that has been superimposed upon an existing AMPS. To ensure good communication and maximum performance for the CDMA, an exclusion area, or guardzone, is set up where the AMPS channels would normally operate. The frequency spectrum that the CDMA system then occupies is thus designed to exclude all AMPS channels. A diagram of a guardzone is shown in Fig. 4.22. Of course, the larger the guardzone, the more restrictive the capacity and frequency reuse will be for an operator.

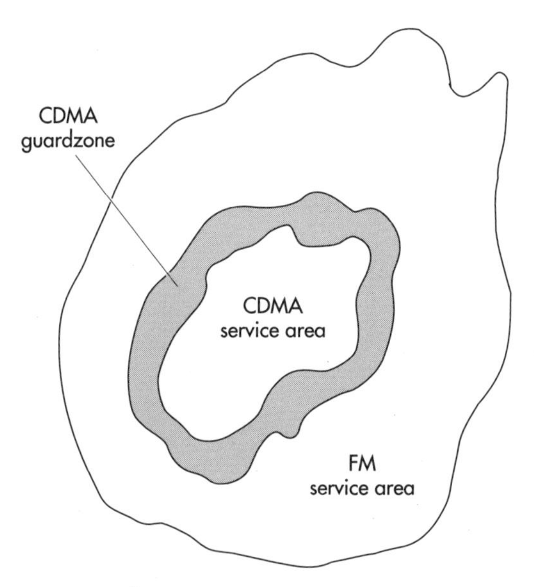

Figure 4.22 Guardzone.

Q. *How is ERP calculated?*

A. Calculating the *effective radiated power* (ERP) for a cell site or any transmitting device is a rather straightforward process, as shown below:

$$\text{Gain, dB} = 10 \log \left(\frac{P_{out}}{P_{in}}\right)$$

Manipulating the above equation, the ERP for a site is P_{out} in watts:

$$P_{out}\ (\text{W}) = P_{in}\ (\text{W}) \cdot \text{antilog}\left(\frac{\text{gain, dB}}{10}\right)$$

The gain, in decibels, is the net gain or loss from all the devices between the transmitter and the air, which will include the gain of the antenna. The P_{in}, in watts, is typically the power at either the output or the transmitter. Two examples follow; the first includes several items in the Tx path, while the second example has a known Tx power at the input of the feed line.

Example 4.3

If the following conditions apply,

Power amplifier, 40 W	= +46 dBm
Combined losses	= −3.5 dB
Tx filter loss	= −0.5 dB
Feed-line and jumper loss	= −3 dB
Cell site antenna gain	= +10 dBd

then the power out equals +49 dBm, or 79.5 W. Thus:

$$\text{ERP} = 39.8\ \text{W} \cdot \text{antilog}\left(\frac{10 - 3.5 - 0.5 - 3}{10}\right) = 79\ \text{W}$$

Example 4.4

If the following conditions apply,

Power into feed line	= +42 dBm
Feed-line and jumper loss	= −3 dB
Cell site antenna gain	= +10 dBd

then the power out equals +49 dBm, or 79.5 W. Thus:

$$\text{ERP} = 15.8\ \text{W} \cdot \text{antilog}\left(\frac{7}{10}\right) = 79\ \text{W}$$

Q. *At what decibel level should the ERP be set?*

A. Normally the ERP for a cell site should be set to maintain a balanced-path condition between the talk-out and talk-back paths provided that the talk-out path is greater than the talk-back path. However, in the following example, the talk-out path is greater than the talk-back path, leading to a system imbalance, and a procedure is described for establishing an optimal ERP. It is important to realize that the example that follows is a simplified situation.

If the following conditions apply for a talk-out path,

Cell site transmitter ERP = X
Subscriber receiver sensitivity = −102 dBm

then the maximum talk-out path loss allowed equals X + 102 dB.

If the following conditions apply for a talk-back path,

Subscriber maximum transmitter ERP = (+28 dBm) (0.6 W)
Cell site receiver antenna gain = +12 dBd
Feed-line loss = −2 dB
Cell site receiver sensitivity = −105 dBm

then the maximum path loss allowed in the talk-back direction is 143 dB.

To determine the path balance, or rather imbalance if it exists, the talk-back path loss is subtracted from the talk-out path loss:

Talk-out path loss − talk-back path loss = path-loss imbalance

or

$$X + 102\ \text{dB} = 143\ \text{dB}$$

Thus,

$$X = 41\ \text{dBm} = 12.5\ \text{W}$$

Q. *What is an SQE?*

A. A *signal quality estimate,* usually associated with an iDEN system, is a type of C/I value that is used to establish an acceptable BER for the system. An SQE within a nominal range of 12 through 25 can be directly equated to the C/I, but it does not maintain the same linearity outside of that range. Therefore, if the design for an iDEN system calls for a 20-dB C/I level, then the expected SQE would be 20.

Q. *What is a C/I ratio?*

A. A *C/I ratio,* expressed in decibels, represents the ratio of *carrier energy* (C) *to interference energy* (I). The C/I ratio is a normal design criterion that is used by the RF engineering department.

As an example, if the desired C/I ratio for a system is 17 dB, then the carrier C must have a signal greater than 17 dB over the interferer I. The following table shows additional calculations.

Carrier C, dBm	Interferer I, dBm	C/I, dB
−80	−100	20
−90	−100	10
−90	−80	−10*
−100	−118	−18

*Interferer is 10 dB stronger than the carrier.

Q. *What C/I level is optimal?*

A. The C/I level that is optimal for a particular system depends on a multitude of factors. One way to determine this information is to utilize the design criteria recommended by vendors for the technology involved. The other method is to conduct tests to determine if the C/I value can be relaxed, reduced, or increased relative to the values recommended.

Table 4.7 gives C/I levels for various systems. However, it should be noted that each system's requirements and service levels can be different.

TABLE 4.7 C/I LEVELS FOR VARIOUS TECHNOLOGY PLATFORMS

Technology	C/I, dB
GSM	13
AMPS	17
TACS	17
IS-136	19
IS-54	19
iDEN	20
IS-95	NA (This uses an E_b/N_o value)

The above C/I values should produce a BER that is acceptable for good voice communication.

Q. *What is a C/N ratio?*

A. A carrier-to-noise (C/N) ratio expressed in decibels, describes the relationship between the carrier and the noise floor for the system or cell. C/N values are sometimes interchanged with C/I values, but they shouldn't be because they are different. The C/N value is used for a *noise-limited system* while the C/I value is used for an *interference-limited system.*

As an example, if the desired C/N ratio for a particular system is 19 dB, then the carrier C must have a signal greater than 19 dB over the noise N. The following table provides sample C/N calculations:

Carrier C, dBm	Noise N, dBm	C/N, dB
−80	−100	20
−90	−110	20
−90	−93	3
−100	−118	18

Q. *What is a C/A ratio?*

A. The *carrier-to-adjacent-channel-interference* (C/A) *ratio,* expressed in decibels, is not a common criterion included in the design or expansion phases for a wireless network.

As an example, if the desired C/A ratio for a particular system is 5 dB, then the carrier C must have a signal greater than 5 dB over the adjacent channel A. The following table shows similar calculations:

Carrier C, dBm	Adjacent channel A, dBm	C/A, dB
−80	−100	20
−90	−95	5
−90	−93	3
−100	−90	−10

Q. *What is dBm?*

A. *Decibels, milliwatts* (dBm) are radio system units of measure. Typically, a decibel, milliwatt, value is referenced to a 50-Ω system. The decibel, milliwatt, equals 1 milliwatt (mW) of power or 0.001 watt (W) (see Table 4.8).

Table 4.8 Relationships between Watts and Decibels, Milliwatts

W	dBm
0.001	0
100	50
1	30
0.000001	−30
0.000000001	−90

Q. *What is 0 dB?*

A. The term *0 dB* means "unity gain," or rather, no gain. Since a decibel is used to express a ratio, if there is no change between the input and the output of a device, that is, there is no gain—then the net gain for the device is 0 dB.

Q. *How are watts converted to decibels, milliwatts?*

A. The conversion of watts (W) to decibels, milliwatts (dBm) is rather straightforward and can be accomplished by several methods. A fairly simple method follows:

$$X \text{ W} = Y \text{ dBm}$$

Process:

1. Convert watts to milliwatts.
2. Enter value into equation Y dBm = 10 log (X) where X is in milliwatts.
3. Solve for Y.

$$100 \text{ W} = Y \text{ dBm}$$

Process:

1. 100 W = 100,000 mW
2. Y dBm = 10 log (100,000)
3. Y dBm = 50 dBm

Q. *How are decibels, milliwatts, converted to watts?*

A. The conversion of decibels, milliwatts (dBm) to watts (W) is rather straightforward and can be accomplished by several methods. A fairly simple method follows:

$$Y \text{ dBm} = X \text{ W}$$

Process:

1. Enter value into equation Y dBm = 10 log (X/0.001) where X is in watts.
2. Solve for X.

$$36 \text{ dBm} = X \text{ W}$$

Process:

1. 36 dBm = 10 log (X/0.001)
2. X = 3981 · 0.001 = 3.98 W, or approximately 4 W

Q. *What does the term 20 decibels per decade refer to?*

A. The term *20 decibels per decade* usually refers to free-space path loss. It is a measure of attenuation in the signal that says the signal will be attenuated by 20 dB for every decade of distance that is traversed. The formula that yields 20 decibels per decade is 20 log R where R is the distance from the source. The difference between 40 decibels per decade and 20 decibels per decade is obviously 20 dB. Note that the expression for 40 decibels per decade is now 40 log R, which is derived as follows:

$$\text{40 decibels per decade} = 40 \times \frac{1}{R^2}$$

Using the log of the value,

$$\text{Path loss} = 10 \log \left(\frac{1}{R_n}\right)$$

Therefore,

$$\text{40 decibels per decade} = \frac{1}{R_4} \text{ PL} = 40 \log R$$

or

$$\text{PL} = 40 \log R$$

Q. *What is free-space path loss?*

A. *Free-space path loss* (PL) is usually the reference point for all the path-loss models commonly employed. The term *free-space path loss* refers to the type of path loss, or attenuation, that a radio signal undergoes when it traverses between a transmitting antenna and a receiving antenna. The value used to compute free space is dependent only on the distance that it is from the source transmitter. There is a frequency-of-operation component in the free-space equation, but it is a one-time value.

Free space is assigned the value of 20 decibels per decade, which means that for every decade, factor of 10, that is traversed, 20 dB of attenuation is added to the signal. In other words, for every decade of distance that the signal goes through the air, free-space attenuation means the signal decreases by 100 times.

The equation used to calculate free space follows:

$$\text{Free-space path loss, dB} = 32.4 + 20 \log R + 20 \log f$$

where R equals the distance from the source in kilometers and f equals the frequency in megahertz. The 20-decibel-per-decade comment comes from the part of the equation that changes in value, which is the distance from the source, 20 log R.

The free-space equation is usually the reference point for all the path-loss models employed. Each propagation model predicts more accurately than free space the attenuation experienced by the signal over that of free space.

TABLE 4.9 FREE-SPACE CALCULATIONS

	Free space/path loss, dB	
Distance, km	**880 MHz**	**1900 MHz**
1.0	91.29	97.97
2.0	97.31	103.99
3.0	100.83	107.51
4.0	103.33	110.01
5.0	105.27	111.95

The base-line assumptions used to construct Table 4.9 are that distances are expressed in kilometers and the frequencies are 880 and 1900 MHz. Table 4.9 shows that the only change that occurs between 880 and 1900 MHz is a constant value of 6.68 dB, with more attenuation for the 1900-MHz band than the 880-MHz band.

5

Implementation

Evidence of the rapid growth of the wireless communications industry can be seen not only in the increasing number of subscribers to cellular systems but also in the increasing number of cell sites in operation. The higher volume of wireless communications traffic raises many system design issues.

The implementation of a cellular system design, which includes the construction of the cell site, must be accomplished carefully to ensure high-quality system performance.

This chapter addresses many commonly asked questions pertaining to the implementation of a wireless system design.

Q. *In wireless cellular communications technology, what does the term construction refer to?*

A. In wireless cellular communications terminology, the term *construction* refers specifically to the building of a cell site. Although some operators use the terms *construction* and *implementation* interchangeably, in fact cell site construction is only one part of the design implementation. *Construction* refers narrowly to the actual modification of a location so that the wireless equipment can be installed and operated properly. Thus the construction phase of the system design implementation may involve a tenant improvement (TI) process or the erection of a tower with the associated shelter.

Q. *What is a cell site, and what does it consist of?*

A. A *cell site,* or rather a wireless communication site, consists of the building and equipment used to receive and transmit cellular communication signals within a particular geographical area. The components of a typical cell site are shown in Fig. 5.1.

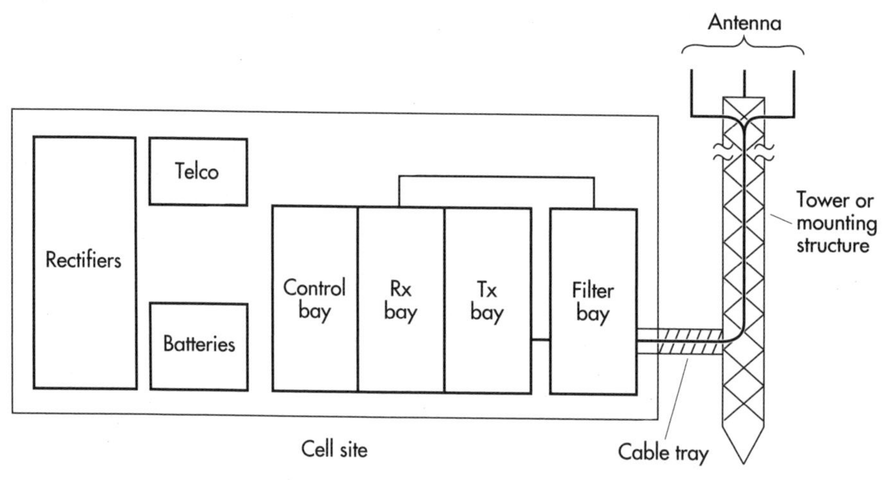

Figure 5.1 Communication site.

Q. *What permits are required for constructing a cell site?*

A. The permits required for constructing a cell site vary from municipality to municipality. Many times, to obtain a permit, a variance of the local zoning codes must be applied for if the proposed construction and/or use is not granted by right. It is the local building inspector who determines the specific requirements that must be met to obtain the building and ancillary permits.

It is strongly suggested that building permit requirements be thoroughly researched and considered during the initial project feasibility study.

Q. *What is a macro cell site?*

A. A *macro cell site* is either a large geographical area or a large equipment installation. Thus a defining characteristic of a macrocell is its service area, or rather coverage area. However, in terms of system design implementation, a macrocell is defined as such by the size of its physical footprint—that is, the amount of space and power it requires. Presently, sites considered to be macrocells require 200 ft^2 of space to house their equipment.

With the advent of the PCSs, the need for macrocells was initially considered a thing of the past. However, as the PCSs have evolved technologically, their cell sites often have the look, feel, and size of macrocells.

There are many cell site component configurations that work well with each type of technology platform that can be used for a particular communication system. For instance, the AMPS, TACS, GSM, CDMA, and NADC systems can be configured successfully as omni, bidirectional, or three-sector cell sites.

The type of antenna installation chosen for a cell site depends on the physical location of the site and the number of antennas needed, as well as whether the site is an omni or sector cell site.

Q. *What is an omni cell site?*

A. An *omni cell site* has no sectors and a coverage area that radiates 360 degrees from the cell site base. Usually, an omni cell site has one transmitter antenna and two receiver antennas arranged in a two-branch diversity scheme, as shown in Fig. 5.2. A typical coverage pattern for an omni cell site is shown in Fig. 5.3.

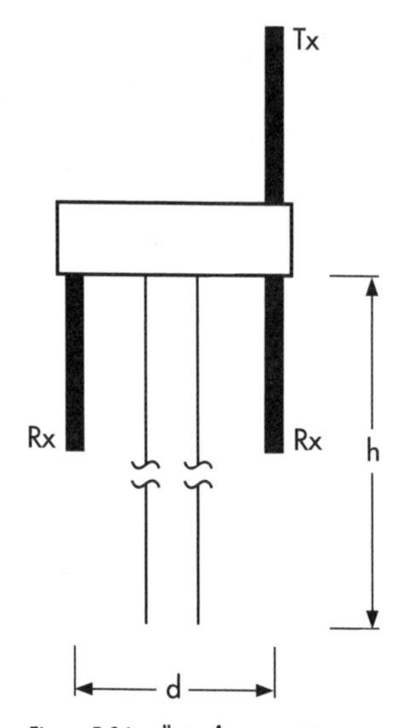

Figure 5.2 Installation for an omni site on a monopole. $d/h = 13$ or $d = h \times 13$ (ft).

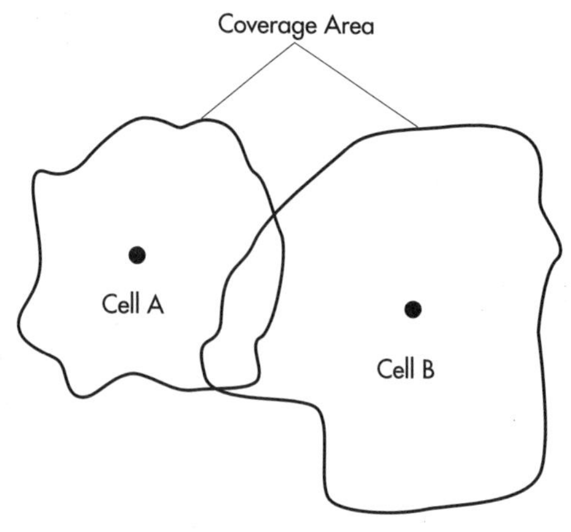

Figure 5.3 Omni cell coverage.

Omni cell sites are commonly found in rural areas or in new cellular system market areas because, although they are not as spectrally efficient as sectored cell sites, they are more cost effective in their use of equipment and capital. Often an omni cell site in later years will be converted to a sectored cell site once the traffic increases to warrant the configuration change.

Q. *What is a single-sector cell site?*

A. A *single-sector cell site* is a directional cell site that utilizes only one sector to provide coverage for a given area. The single-sector cell site utilizes directional antennas, which means that it cannot cover a full 360-degree area. However, in its switch database and construction aspects, this type of cell site will take on the characteristics of an omni cell site. The chief difference between these two types of cell sites is that a single-sector cell site uses directional antennas whereas an omni cell site uses omni antennas. Figure 5.4 illustrates the difference between single-sector and omni cell sites.

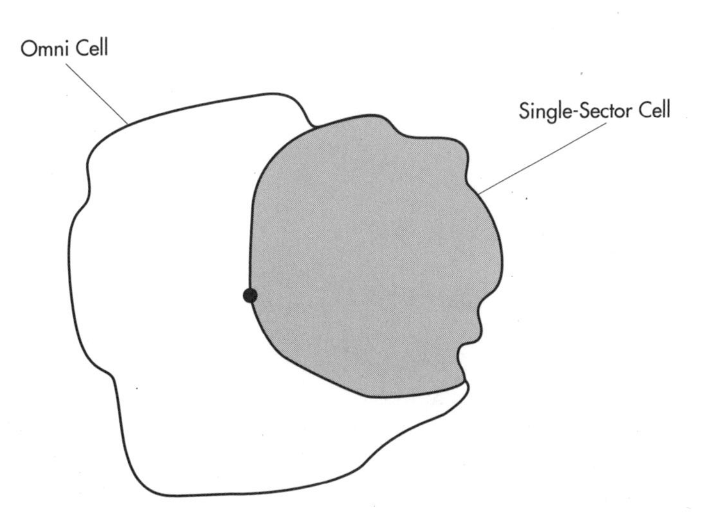

Figure 5.4 Single-sector cell versus omni cell.

Q. *What is a monopole?*

A. A *monopole* is a type of tower structure that is used in wireless communication systems for the installation of antenna equipment. The monopole can be used for either an omni cell site or a sectored cell site. It is possible to have more than one carrier occupy part of a practice monopole, known as *colocation.* Figure 5.5 is a diagram of a monopole.

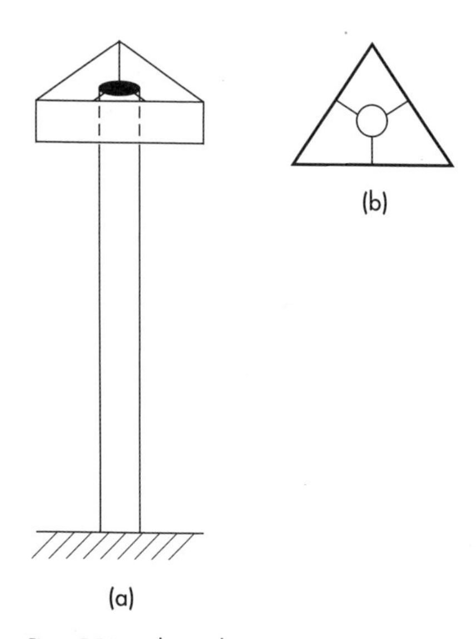

Figure 5.5 Monopole. (*a*) Side view. (*b*) Top view.

Q. *What installation issues commonly exist at cell sites?*

A. The nature and extent of issues that arise during the development of a particular cell site will vary tremendously from one location to another. Some issues arise at almost every site. For example, the various contractors used for an installation must be carefully orchestrated because very often project B cannot begin until project A has been completed. Other types of issues arise sporadically.

The following is a brief list of some of the more common installation issues that are encountered when a cell site is being developed:

1. Access to the site
2. Hours allowed for construction (usually a tenant improvement issue)
3. Cell site equipment delivery
4. Telco acceptance
5. Building permits
6. AE drawing approvals from internal groups as well as landlord
7. Antenna system installation
8. Landlord claims of property damage as a result of installation work
9. Power system upgrade
10. Floor loading (primarily for batteries)
11. HVAC venting and installation
12. Noise abatement for the cell site and HVAC system
13. Alternative power requirements (generator hookup)
14. Parking and bathroom facilities

Q. *What are the antenna mounting requirements for an omni cell site?*

A. There are several methods that can be used to mount antennas on omni cell sites. The first is to simply install the antenna on a monopole, as shown in Fig. 5.6. The transmit antenna is highest on the structure, and the receive antennas are located under the platform. The distance allowed between the receive antennas should adhere to the separation requirements for the network, which most likely will match the results of the equation given in the figure itself.

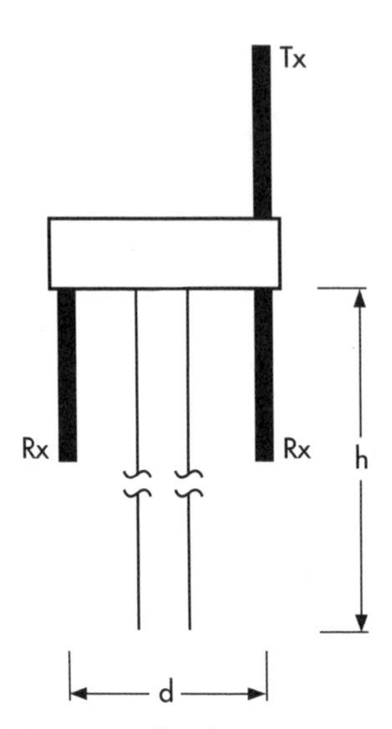

Figure 5.6 Installation for an omni site on a monopole. $d/h = 13$ or $d = h \times 13$ (ft).

Q. *What is antenna setback?*

A. The term *antenna setback* refers to the physical distance an antenna must be from the edge of the building or other structure to which it is attached. The antenna setback is a requirement that is dictated by the local municipality and/or the landlord or engineering department. An antenna setback is established usually for asthethic reasons. Figure 5.7 shows an antenna setup with a setback requirement.

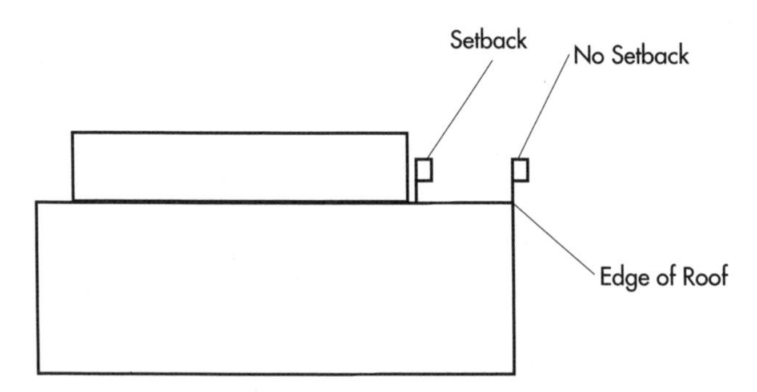

Figure 5.7 Antenna setback.

Q. *How should the antennas for an omni cell site be installed?*

A. The way in which omni antennas should be installed is determined in consideration of a variety of factors including setback and structural issues. However, focusing primarily on the implementation aspect, there are a few variables that control omni installations such as architectural features of buildings and the number of antennas to be installed. Figure 5.8 depicts an antenna installation that occurs on a building. Please note that the location of antennas needs to meet the required setback rules. If the setback rules cannot be adhered to, then it is possible to install the antennas near the edge of the roof. However, in that location, they may become visible to the public. In addition, the landlord may object to this type of installation or the local ordinances may prohibit it.

The placement of the receive antennas should be such that, if there is only one major road in the area for the cell to cover, then the horizontal diversity placement for the antennas should maximize in the direction toward the road.

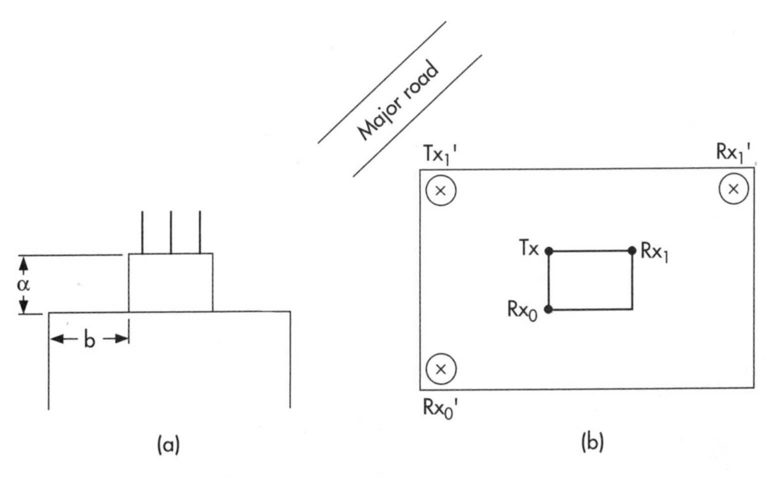

Figure 5.8 Existing rooftop. (*a*) Side view. $a = 5 \times b$. (*b*) Top view.

Q. *What is a directional antenna cell site?*

A. A *directional antenna cell site* receives or transmits radio signals only in one or more directions. The term is also used to mean a single-sector cell site. Depending on the terminology used within a particular operator's organization, a *directional antenna cell site* can mean either a single-sector cell site or a sectorized cell site consisting of from two to six or more sectors.

Figure 5.9 is a single-sector cell site that is meant to cover only a portion of a highway. The antenna structure for this application is located on a billboard, and it is positioned so that coverage follows the roadway through the use of a highly directional antenna.

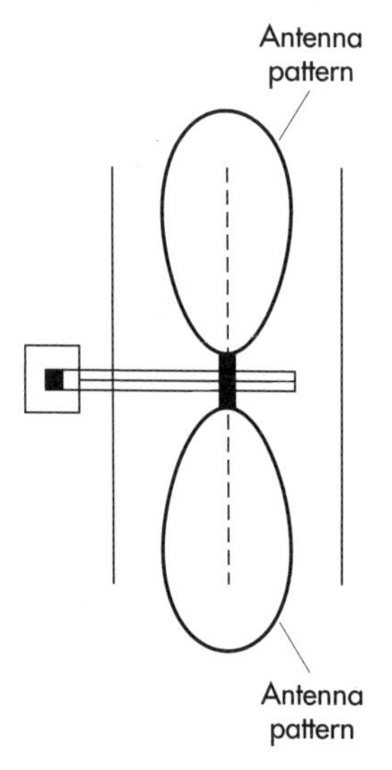

Figure 5.9 Directional site.

Q. *What is a three-sector cell site?*

A. A *three-sector cell site* has three antenna sectors that span 120 degrees each, thus covering a full-circle, or 360-degree area. The three-sector cell site is one of the most popular cell site configurations utilized in the wireless industry, next to the omni cell.

There are a multitude of transmit-receive antenna combinations that can be used for establishing a three-sector cell site. However, in the examples given in Figs. 5.10 and 5.11, only a few of the more basic configurations are shown. Note that additional configurations can be derived easily from the material presented.

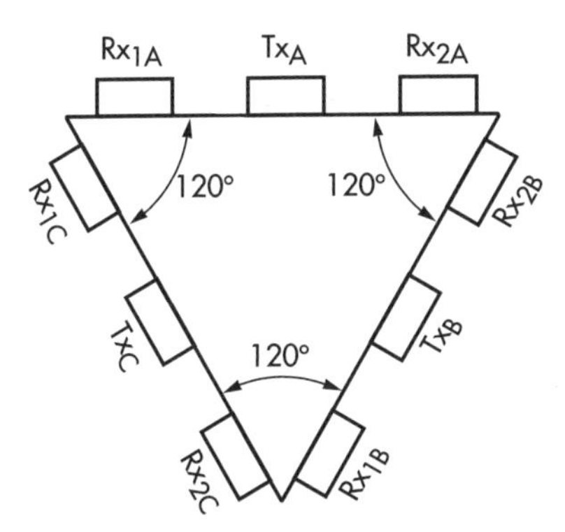

Figure 5.10 Three-sector antenna array.

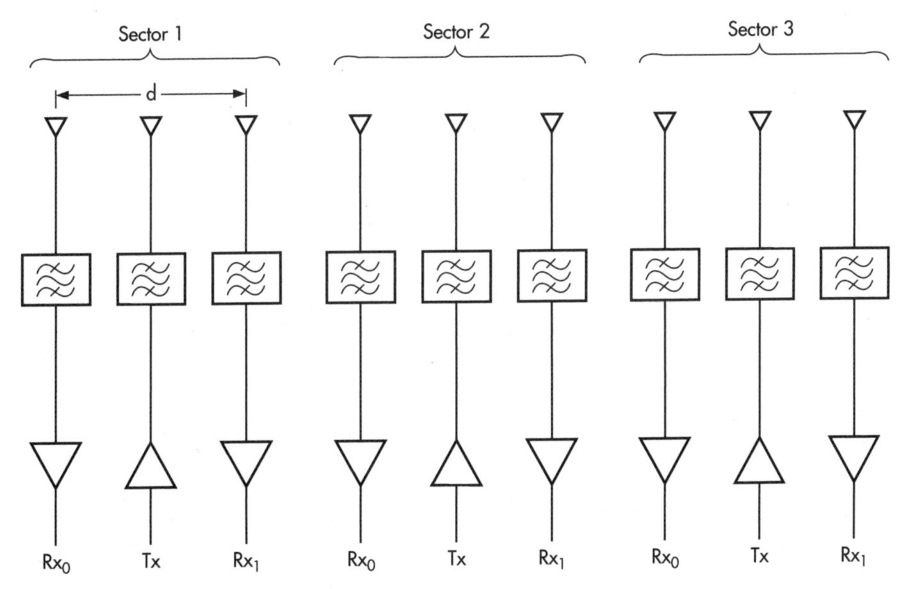

Figure 5.11 Three-sector Tx and Rx plumbing.

Figure 5.10 shows a typical antenna arrangement for a monopole application. The configuration shown in the figure has a separate antenna for each transmit and receive leg of the cell site. The general transmit and receive configuration for a three-sector cell site is shown in Fig. 5.11. Note that if an operator anticipates having to install an additional transmit antenna, it should be incorporated in the initial design for the site so as to avoid configuration problems in the future.

Q. *What is a six-sector cell site?*

A. A *six-sector cell site* has six antenna sectors, which could be laid out as shown in Fig. 5.12. The use of only one transmit antenna is shown in the drawing, but a second one could be added depending on the configuration requirements.

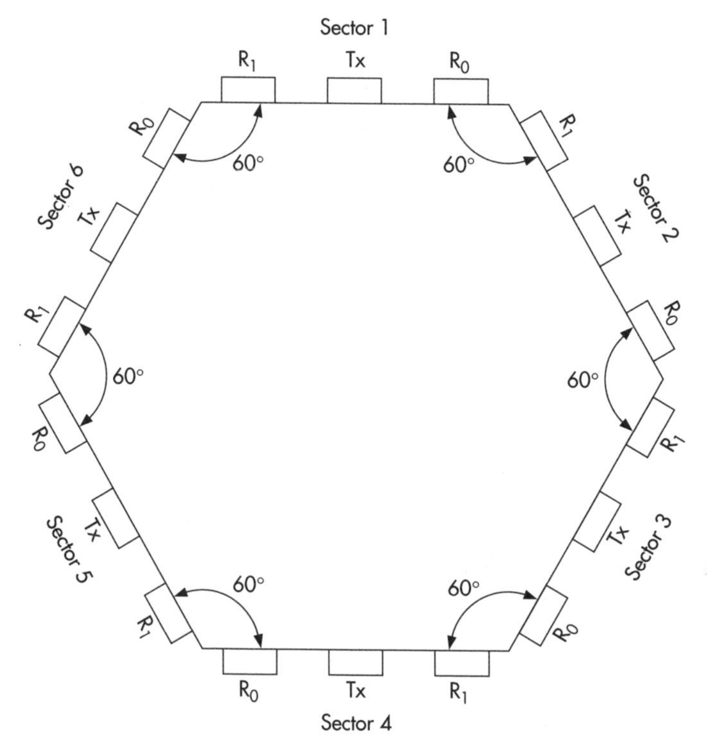

Figure 5.12 Six-sector configuration, top view.

Q. *What is a microcell?*

A. There is no true definition of what constitutes a microcell versus a macrocell versus a picocell, and various manufacturers define their equipment differently. However, for general installation and construction purposes, a *microcell* is a site that takes up less than 50 ft^2 of space. Microcells are becoming prevalent in wireless systems as operators strive to reduce the physical size of the operating sites. Normally, reducing the physical size of the equipment correspondingly reduces cell site coverage area.

Note that in engineering terms, *microcell* refers to the site's coverage area while in construction terms, *microcell* refers to the equipment's size and power.

Q. *Where are microcells used?*

A. Microcells are deployed to provide coverage within buildings, subway systems, and tunnels and to resolve unusual coverage problems.

Q. *What types of microcells are available?*

A. There are currently several types of technology platforms that fall into the general categorization called *microcells:*

1. Fiber-fed microcell
2. T-1 microcell
3. Microwave microcell
4. High-power rerad
5. Low-power rerad
6. Bidirectional amplifier

Q. *What cable or coaxial cable issues need to be considered for installation?*

A. Some of the physical installation issues that need to be factored into the design involve cable runs from the antenna system, each leg, to the base station equipment. Often there are situations where the desired routing of the cables is not practical, making the real installation length much longer than would otherwise be necessary. Having to add cable length for installation may make the site undesirable. If this situation occurs too far down the stream of the construction process, it will be too late to reject the site or to make the appropriate design alterations to correct the situation.

Q. *What antenna mounting issues need to be considered for installation?*

A. The antenna mounting is critical in the success or failure of a cell site in its ability to deliver coverage, and therefore this phase of system implementation must be accomplished with extreme care. The following brief checklist can be a starting point to ensure that antenna mounting concerns are addressed prior to acceptance of a cell site:

1. The number and type of antennas to be installed
2. Maximum cable run allowed
3. Obstructions that would alter the desired coverage
4. Rx antenna spacing and diversity requirements
5. Isolation from other services
6. Antenna above ground-level (AGL) requirements
7. Antenna mounting parameters
8. Intermodulation requirements
9. Path clearance requirements

In addition, when installing an antenna on a tower, the physical spacing (or offset from the tower) must be calculated to ensure that the tower structure either enhances or does not alter the antenna coverage pattern desired.

Q. *What is diversity spacing?*

A. The term *diversity spacing* refers to the physical separation between the receive antennas, which is needed to ensure that the proper fade margin, as designed for the system, is maintained. *Horizontal space diversity* is the most common type of diversity scheme used in wireless communication systems. Usually the RF engineering department specifies the distance between the antennas. However, the following equation can serve as a rule of thumb in determining the optimal horizontal diversity spacing for a particular cell site:

$$\text{Horizontal spacing, ft} = \frac{\text{height antenna above ground, ft}}{11}$$

Q. *How should an antenna be mounted on a roof?*

A. When installing an antenna on an existing roof or penthouse, consideration must be given to how high the antenna must be with respect to the roof surface. The ideal location would be at the roof edge. However, if that location is not a viable option, then alternative placements must be considered. Note that for any placement other than the roof edge, the height of the antenna must be in proportion to the distance between the edge of the roof and the antenna installation, as depicted in Fig. 5.13. In the figure, it is assumed that there are no additional obstructions between the proposed antenna location and the roof edge, like HVAC units and window-cleaning apparatus.

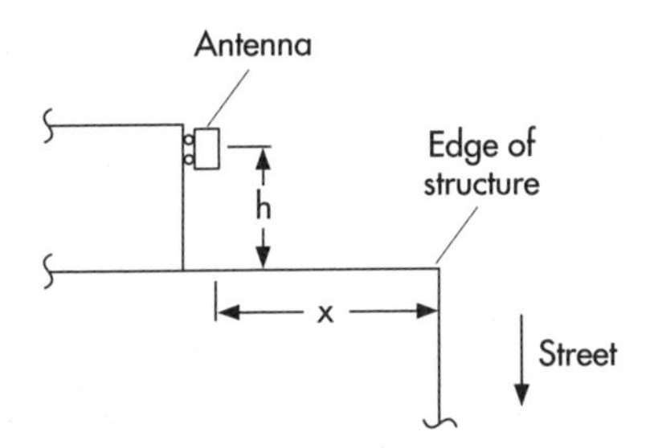

Figure 5.13 Roof mounting. $h = x/5$, where h = height of antenna above roof (ft) and x = distance from edge of roof (ft).

Q. *What types of towers are suitable for antenna installations?*

A. There are numerous types of towers that can be used for wireless communication equipment installations. These types of towers can be grouped into three general classes: monopole, guy wire, and self-supporting.

Q. *What is a self-supporting tower?*

A. A *self-supporting tower* requires no additional support like a building or guy wires. The self-supporting tower is usually the most expensive type of tower, but it has probably the most flexibility in terms of configuration possibilities. An example of a self-supporting tower is shown in Fig. 5.14.

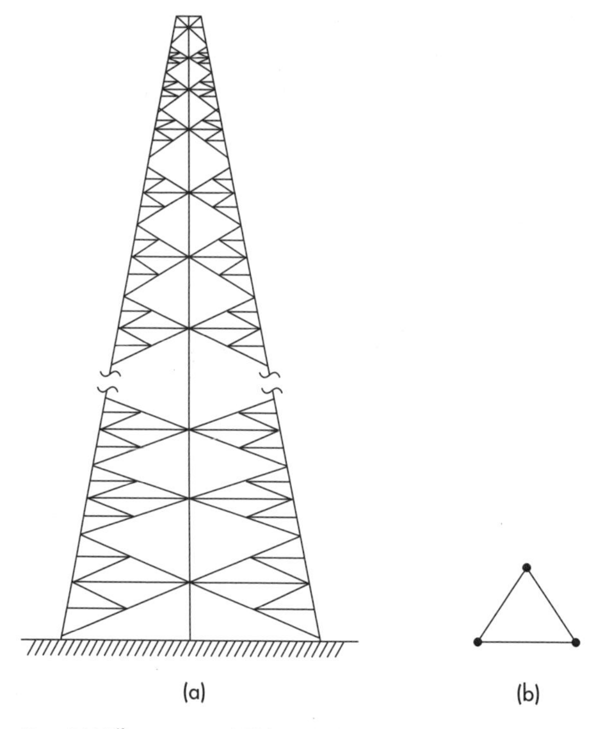

Figure 5.14 Self-supporting tower. (*a*) Side view. (*b*) Top view.

Q. *What is a guy-wire tower?*

A. A *guy-wire tower* requires a series of cables to keep it erect and help distribute the structural load. The guy-wire tower is usually the cheapest to construct, but it requires more physical space due to the placement of the support wires. The diagram in Fig. 5.15 shows an example of a guy-wire tower.

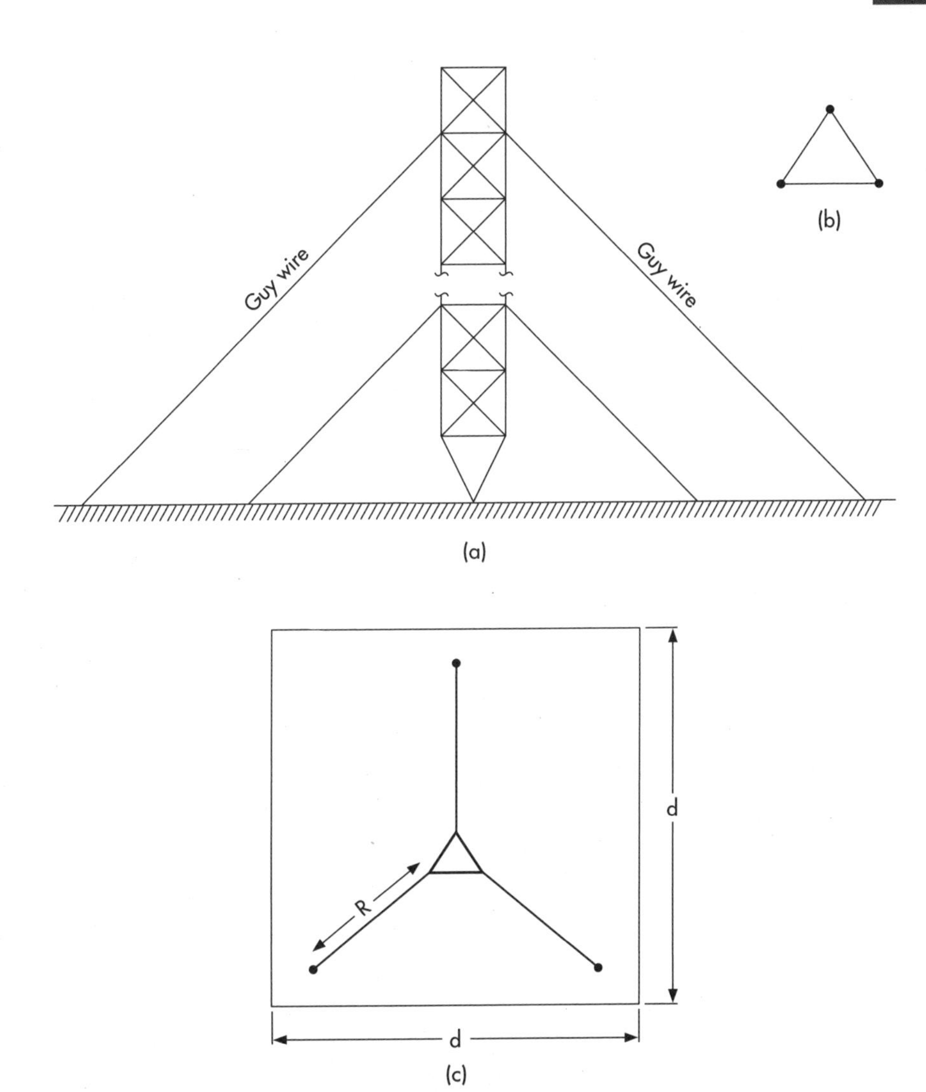

Figure 5.15 Guy-wire tower. (*a*) Side view. (*b*) Top view. (*c*) R = height of tower × guy radius (%), $d = 2 \times R + 30$ (ft).

Q. *What is a monopole tower?*

A. A *monopole* is a single pole, hence the name, which then has a "top hat" on the top to which the antennas are mounted. The monopole's top hat comes in a variety of sizes, but a typical top hat is either 10 or 15 ft from point to point on any edge. A monopole can also have several top hats associated with it to facilitate multiple colocation possibilities. A monopole tower is the tower type that is normally depicted in cellular and wireless communication design guidelines. A diagram of a monopole is shown in Fig. 5.16.

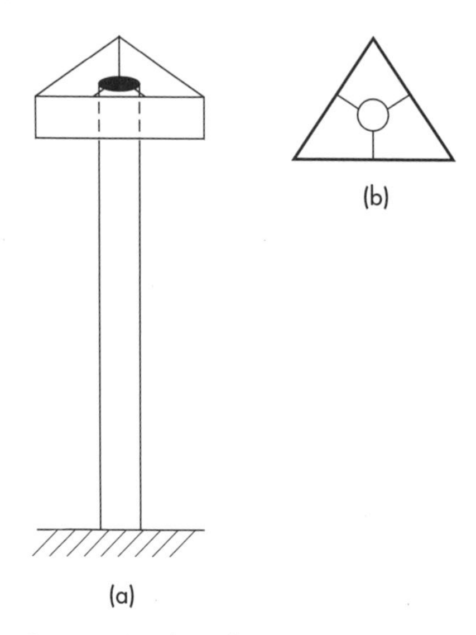

Figure 5.16 Monopole. (*a*) Side view. (*b*) Top view.

Q. *How should an antenna be mounted on a wall?*

A. For many building installations, it may not be possible to install the antennas above the penthouse or other structures. Instead, it is often necessary to install the antennas onto the penthouse or water tank for the building, as shown in Fig. 5.17.

Problems frequently arise when installing antennas onto an existing structure because the building architect has not factored the antennas into the original building design. Therefore, for a three-sector configuration as shown in Fig. 5.17, the building walls may meet one orientation needed for the system but rarely all three.

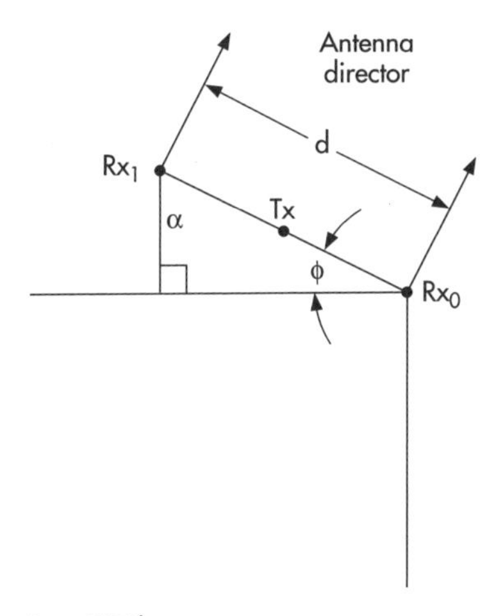

Figure 5.17 Three-sector antenna array.

Q. *What is the antenna installation tolerance?*

A. There are usually two separate tolerance requirements—one concerning how accurate the antenna orientation should be and the other concerning how plumb the antenna installation should be. Tolerances must be considered not only in relation to design but also in relation to cost effectiveness. The following tolerance guidelines should be used:

Type of tolerance	Amount of tolerance
Orientation	±5% or antenna's horizontal pattern
Plumbness	±1 degree (critical)

Q. *What is the antenna orientation tolerance?*

A. The *antenna orientation tolerance* is a function of the antenna pattern, and it therefore can be unique for each type of cell site. Obviously, for an omni cell site there are no orientation requirements since the site is meant to cover 360 degrees. However, for a sectored directional cell site, the orientation tolerance is a critical issue.

To have no error in the orientation of an antenna is desirable but rather impractical. Thus, an orientation tolerance should be specified by the RF engineering department. However, in the absence of this, the guideline is that the tolerance should be within 5 percent of the antenna's horizontal pattern. The following chart will help illustrate this issue using some of the more standard types of antenna patterns.

Antenna horizontal pattern, degrees	Tolerance from bore site, degrees
110	±5.5
92	±4.6
90	±4.5
60	±3.0
40	±2.0

As the antenna pattern becomes tighter, the tolerance for the orientation error is reduced. The objective defined here is ±5 percent, but that number can be either relaxed or tightened depending on the particular system requirements. A 5 percent tolerance should also be factored into any potential building sway, which can and does occur, although sway is usually a nonissue due to the height of the buildings used for wireless installations.

Q. *What is the vertical tolerance for antennas?*

A. The *vertical tolerance* for an antenna installation involved with wireless communications is a tight and rigid requirement that is often poorly documented. The tolerance needs to be tight due to the direct impact of the tolerance amount on the coverage of the cell site. A vertical tolerance that is too lax could have the same impact as the downtilting of the cell site. The vertical antenna tolerance, plumbness, is ±1 degree from true vertical.

Q. *What procedure should be followed in the commissioning of a wireless communication cell site?*

A. The items in Table 5.1 should be evaluated prior to or during the commissioning of a wireless communication cell site. This list is generic and should be tailored to a particular application, with

TABLE 5.1 COMMUNICATION SITE CHECKLIST

Topic	Received	Open
Site Location Issues		
1. 24-hour access		
2. Parking		
3. Directions to site		
4. Keys issued		
5. Entry and access restrictions		
6. Elevator operation hours		
7. Copy of lease		
8. Copy of building permits		
9. Lien releases obtained		
10. Certificate of occupancy		
Utilities		
1. Separate meter installed		
2. Auxiliary power (generator)		
3. Rectifiers installed and balanced		
4. Batteries installed		
5. Batteries charged		
6. Safety gear installed		
7. Fan and venting supplied		
Facilities		
1. Copper or fiber		
2. Power for fiber hookup (if applicable)		
3. POTS lines for operations		
4. Number of facilities identified by engineering		
5. Spans shaked and baked		
HVAC		
1. Installation completed		
2. HVAC tested		
3. HVAC system accepted		

Table 5.1 Communication Site Checklist (*Continued*)

Topic	Received	Open
Antenna System		
1. FAA requirements met		
2. Antennas mounted correctly		
3. Antenna azimuth checked		
4. Antenna plumbness checked		
5. Antenna inclination verified		
6. SWR check of antenna system		
7. SWR record given to operations and engineering		
8. Feed-line connections sealed		
9. Feed-line grounds completed		
Operations		
User alarms defined		
Engineering		
1. Site parameters defined		
2. Interference check completed		
3. Installation MOP generated		
4. FCC requirements documentation completed		
5. Drive test completed		
6. Optimization completed		
7. Performance package completed		
Radio Infrastructure		
1. Bays installed		
2. Equipment installed according to plans		
3. Rx and Tx filters tested		
4. Radio equipment ATP'd		
5. Tx output measured and corrected		
6. Grounding completed		
7. Equipment bar coded		

items added or removed as appropriate. This list can serve as an excellent first step in ensuring the operability of the cell site.

Q. *What is a tower-top amplifier?*

A. A *tower-top amplifier* (TTA) is an active piece of electronics equipment requiring power that is installed either in the antenna or right at the outside of the antenna. The implementation issues for tower-top amplifiers relate directly to proving the power supply and physically installing the unit correctly.

There are many applications for tower-top amplifiers in wireless communications systems. The primary purpose of the tower-top amplifier is to improve the receiver's sensitivity by eliminating the feed-line loss component.

The location of the TTA is shown in Fig. 5.18.

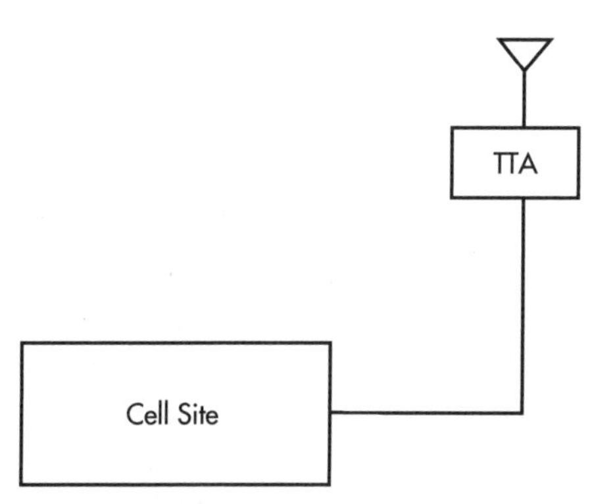

Figure 5.18 TTA location.

Q. *What does the term power factor mean?*

A. The term *power factor* (PF) describes the match between the source and the load for a power system. Under ideal conditions, the power factor should be 1.0, indicating maximum power transfer from the source to the load. The PF rating is important because a poorly matched power system will be inefficient and costly to operate. In many cases the rectifier plant at the cell site has a power factor close to 1.0. However, the actual power factor value should be measured and where required corrected so that it is as close to 1.0 as possible.

Q. *What is a British thermal unit?*

A. A *British thermal unit* (Btu) is a measure of heat:

$$1 \text{ Btu} = 2.930 \times 10^{-4} \text{ kWh}$$

British thermal units are used in the dimensioning of the cell site HVAC system whether it is in an interior or exterior location. For cabinet installations, if the equipment comes preinstalled from the manufacturer's factory, the heat dissipation requirements have already been factored into the cabinet design. Some of the manufacturers who consider heat dissipation requirements are Nortel, Motorola, Lucent, and Ericsson.

Q. *What does the term phase mean?*

A. The term *phase* refers to the electric power system of a cell site or an MSC. The two power phases that are most commonly encountered in wireless systems are the single phase (1-O) and the three phase (3-O). The single-phase system is more common, but the three-phase system allows the site to operate more economically because it is more efficient.

Single-phase systems are more common in buildings. The single-phase system, which consists of three wires, is supplied from one leg of a delta or wye transformer.

The three-phase system is supplied by either a delta or wye transformer. This power system consists of four wires, one for each of the three phases plus a fourth one for a neutral phase. Each phase in a three-phase electrical system is 120 degrees out of phase from the other phases.

Q. *What is a delta transformer?*

A. A *delta transformer* is one of two transformer types, the other being a wye transformer, that can be used to provide either three- or single-phase power to a cell site or an MSC. The delta transformer configuration is shown in Fig. 5.19.

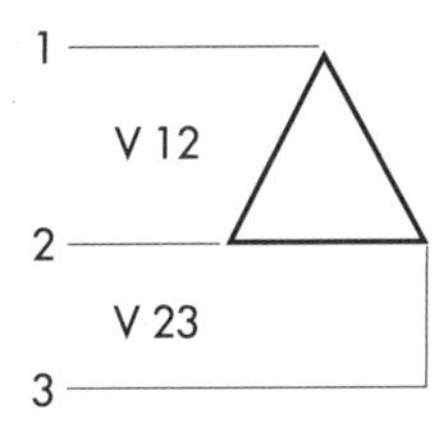

Figure 5.19 Delta transformer.

Q. *What is a wye transformer?*

A. A *wye transformer,* also called a *Y transformer,* is one of two transformer types, the other being a delta transformer, that can be used to provide either three- or single-phase power to a cell site or an MSC. The wye transformer configuration is shown in Fig. 5.20.

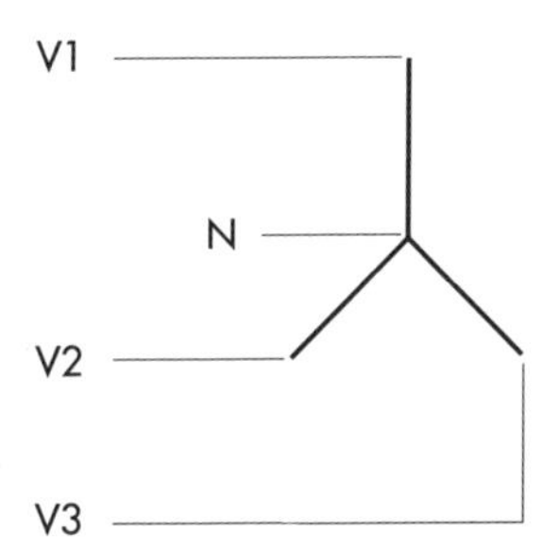

Figure 5.20 Wye transformer.

Q. *How are batteries used in wireless communication systems?*

A. Batteries are an integral part of a wireless communication system, and they are used at both the cell sites and the MSCs. Often other ancillary installation equipment also relies on batteries. The batteries are a backup power supply that can provide −48 Vdc continuous electric power to the system when there is an interruption in the commercial power supply.

The two types of batteries most commonly used in wireless communication systems are (a) *wet cell* and (b) *sealed.*

Of the two types, wet cells are more often used at MSCs because they have an exceptionally long life. Wet cells are not used as readily elsewhere in the system installation because they require regular maintenance over and above keeping them charged and their terminals clean.

Sealed batteries, in contrast to wet cells, are used in more cell site locations because they allow for more possible installation configurations, including many inside buildings. Sealed batteries will not vent gases, and therefore they do not require a hydrogen gas-venting system. Sealed batteries do not have as long a shelf life as do wet cells; however, they require far less maintenance.

Q. *What is a battery string?*

A. A *battery string* is a series of batteries connected together for the purpose of supplying 48 Vdc to a communication system. The number of battery strings will vary depending on the number of ampere-hours required from the communication site. For example, if it is determined that two strings of batteries are required for 4 hours of battery backup power, then to provide 8 hours of backup power would require an additional two battery strings—that is, four battery strings.

Q. *What is ampacity?*

A. *Ampacity* is a unit of measure that is used to describe the ampere capacity of a battery string or system. Ampacity ratings are given in terms of amperes per hour, which describes how many amps the battery system can supply the load over a given period of time. For example, if a battery string were able to deliver 500 A/h of power, then that would mean that the battery system could deliver 500 A for 1 hour until it was completely drained. However, if the equipment required only 100 A of dc power per hour, then a 500-A/h rating would allow 5 hours of continuous usage.

Q. *What is galvanized pipe?*

A. *Galvanized pipe,* also referred to as *hot-dipped galvanized pipe,* is pipe that has been electroplated with steel so that it will resist corrosion. The use of galvanized pipe is prevalent in antenna installations for all cell sites to protect the antenna structure from rusting and possibly becoming a source of intermodulation products.

Q. *What is a feed line?*

A. A *feed line* is a coaxial cable that connects the radio equipment of a cell site to the antennas.

Q. *What are coaxial cables?*

A. A *coaxial cable* is a cable that connects the RF equipment to the antenna system. There are many locations within a cell site where coaxial cables are sometimes used. The most common use for coaxial cables is to connect the radio to the antenna system, which is a setup often referred to as a *feed line.* However, coaxial cables are also used to connect the feed line to the antennas, the radios to the filters, and other ancillary equipment to the radio.

Q. *What is an eyewash station?*

A. An *eyewash station* is a place where people can rinse foreign objects or substances out of their eyes. Eyewash facilities are included in a cell site for the safety of personnel who may be performing battery maintenance at the site. An eyewash station is wall mountable. It is rarely used, but it is essential to have one on the premises to ensure personnel safety. An eyewash station should be a regular piece of cell site equipment, just as is the first-aid kit.

Q. *What is Halon?*

A. *Halon* is a fire-suppressing gas. Although it is no longer being deployed in new installations, there are many legacy cell sites that still utilize Halon for their fire suppression system. Halon was and is very effective in putting out electrical and other types of fires; however, it is not environmentally safe, and it has therefore been removed from the market. When fire suppression systems are upgraded, they use an FM200 system for extinguishing fires.

Q. *Should a desk be included in cell site designs?*

A. Yes. An onsite desk will facilitate equipment troubleshooting and maintenance. For some unknown reason, including a desk as part of the cell site has always been controversial.

Q. *What is a fire suppression system?*

A. A fire suppression system is a system that is designed to put out, or rather suppress, a fire through automatic or manual activation. The types of fire suppression systems that are found in cell sites involve water sprinklers (very bad), Halon, and FM200. For an MSC, the more commonly used system is FM200. Halon is considered detrimental to the environment, and it is not used in new systems for that reason. However, Halon systems may be found in older cell sites that predate the designation of Halon as too hazardous. Water sprinklers are sometimes found in cell sites, but these systems are so damaging to electronic equipment that they should be avoided.

Q. *What are sprinklers?*

A. *Sprinklers* are used in water-based fire suppression systems that deliver water to a fire through the use of sprinklers normally attached to the ceiling of a room. A sprinkler system can be either dry or wet, and it is activated through heat detection.

Q. *What is an FM200?*

A. An *FM200* is a fire suppression system that is commonly used in both cell sites and MSCs. The FM200 system has replaced the older fire suppression systems that utilized Halon gas. The equipment used to store and deliver FM200 is so similar to Halon equipment that it is almost a direct replacement.

Q. *What is a wet system?*

A. A *wet system* is a fire-suppressing sprinkler system that has standing water in its pipes to feed the sprinklers so that when there is a fire problem, the source of water is ready for delivery. The problem in using a wet system in a building is that when installing a cell site, sprinklers are not needed to protect the equipment and the water system, should it be accidentally activated, can actually cause damage to the equipment. Therefore, it is common to type and disable the sprinkler system for the cell site room during installation. Sometimes, however, the system is already charged with water and it is logistically impractical to shut off the system. Thus an existing sprinkler system may be left in place.

Q. *What is a dry system?*

A. A *dry system* is a type of sprinkler system that does not have standing water in the pipes that feed the sprinklers. The water is provided only when a problem is detected so that the pipes remain empty of water when the water is not needed.

Q. *What access issues are likely to arise during cell site construction?*

A. Access issues that affect cell site construction are different for every location. The most pressing access issue is gaining permission to perform whatever work needs to be done at the operator's convenience rather than that of the landlord or tenants. Obviously, there are compromises to unlimited access. Therefore, how and when the installation team enters or leaves the equipment room and antennas needs to be reviewed with the real estate team and the landlord to prevent potential problems.

Q. *What is battery life?*

A. *Battery life* is the length of time the battery or battery string will function according to the specifications for that particular battery. For example, the battery life for the battery strings at a cell site is defined as 5 years at a given load of, say, 500 A/h. Thus, for 5 years those batteries should be able to deliver 500 A/h of power when called upon if the proper charge and maintenance have been performed.

Q. *What is an ampere-hour?*

A. An *ampere-hour* (Ah) is a unit of measure that describes the amount of electricity that passes a particular point in 1 hour. Thus it is a measure of the ability of a battery or battery string to deliver a specified amount of power. For example, if a battery string were rated at 500 A/h, then for a period of 1 hour, the battery string could deliver 500 A of dc power at the voltage specified for the batteries. Also, if the system required only 100 A of dc power, then a 500-Ah battery string would provide 5 hours of power at 100 A.

Q. *What is an in-building wireless communication system?*

A. An *in-building wireless communication system* provides coverage for the interior of the building in which it is installed. An in-building system is not designed to provide exterior coverage to the same level as that of a macro cell site. However, in-building systems typically provide some level of exterior coverage by which to facilitate handoffs to and from the system.

Q. *How are elevator shafts used in cell site installations?*

A. *Elevator shafts,* as their name implies, are the shafts or passageways that house the elevators and provide access between various floors of a building through the use of elevators. The elevator shafts are important because cables can be run through them when installing an in-building system, which then provides coverage within the elevators.

Q. *How are risers used in cell site installations?*

A. A *riser* is a passageway between floors in a building. A riser could be a dumbwaiter shafter or an actual cable riser that was built into a building for the express purpose of passing cables from floor to floor. A riser is used for routing coaxial cables from the cell site equipment room to the antenna structure, which is not complicated by special construction or installation issues.

Q. *How are telco closets used in cell site installations?*

A. A *telco closet* is the location within a building that houses the telecommunication equipment for that building. The telco closet is normally a demarcation point for the telco for the building. The physical size of the telco closet can range from a full-size, 10- by 20-ft room to a small broom closet. The location of the telco closet is unique to each building.

Q. *What is trenching?*

A. *Trenching* is the digging process that is sometimes used for delivering either power or telco facilities to the wireless communication site. Trenching can involve digging up a parking lot, a street, or simply the lawn of a building.

Q. *What is grounding?*

A. *Grounding* is the process of connecting an electric circuit to a particular object that makes an electrical connection to the earth, thereby incapacitating an electric charge. Electrical circuits must be grounded to ensure personnel safety as well as equipment integrity. There are several types of grounding systems that can be deployed at a cell site or an MSC. The grounding system used for a cell site or an MSC should be either a single-point or multipoint ground system, and it should meet all the specifications from the equipment manufacturer and local building codes. In general, an effective grounding system can conduct 5 Ω of electric current.

A ground system usually consists of grounding rods that have been driven into the earth near the site, which are then connected to the building ground system. Sometimes water pipes or chemical bath solutions are also used with the necessary precautions having been taken to ensure good ground integrity.

Q. *What are cable trays?*

A. *Cable trays* are raceways upon which cables are placed to facilitate their being connected from point *A* to point *B* of the MSC or cell site. Cable trays are used in almost all wireless communication facilities.

The most common installation question that arises with the installation of cable trays concerns how to orient them. Figure 5.21 shows the correct orientation for cable trays. As can be seen in the diagram, the trays must be positioned in such a way that when the cables egress from the trays, a stress point is not created leading to future maintenance problems.

Figure 5.21 Cable tray.

Q. *What types of architectural and engineering drawings are made for an installation project?*

A. Many types of *architectural and engineering* (A&E) *drawings* are made to depict the locations and installation issues associated with a cell site or an MSC. Drawings are made at various stages of the installation project: the initial concept, no drawing, 30 percent, 60 percent, 90 percent, 100 percent, and of course, the as-built. Depending on the operator's organization structure, different groups review and sign off on the various drawings at different checkpoints throughout the project's life. Typically the 60 and 90 percent drawings are given the widest review and thus require the most inter-department coordination. The as-builts are often never recorded, which leads to many problems later when increasing the system's capacity or renegotiating the lease for the system site.

Q. *What is a 60 percent A&E drawing?*

A. A *60 percent A&E drawing* is a drawing that is a work in progress. The drawing is referred to as "60 percent" because it is supposed to be 60 percent complete. However, what constitutes 60 versus 50 or 70 percent is rather subjective. The purpose of the 60 percent drawing is to allow construction, real estate, and engineering the opportunity to review the work in progress and either agree with the methodologies used or object to them and propose alternatives. Some of the issues brought up in the 60 percent drawings involve antenna and equipment placements. Often the 60 percent drawings are used for obtaining a building permit or applying for a zoning variance.

Q. *What is a 90 percent A&E drawing?*

A. A *90 percent A&E drawing* is a drawing that is a work in progress. The drawing is referred to as "90 percent" because it is supposed to be 90 percent complete. The purpose of the 90 percent drawing is to allow construction, real estate, and engineering the opportunity to review the work and ensure that any changes proposed in the 60 percent phase were in fact executed. The 90 percent drawings are very close to 100 percent drawings.

Q. *What is an as-built A&E drawing?*

A. An *as-built A&E drawing* reflects how the cell site or MSC was actually constructed. The as-built drawings are important documents that are used by engineering for later expansion of the cell site or MSC. The as-builts are also used in the lease negotiations, and they should therefore be stored in the lease folder.

Q. *When, during the implementation process, is FAA approval necessary?*

A. The Federal Aviation Administration (FAA) must approve certain cell site and/or antenna structures. The agency's approval is given in the form of a *final-determination letter* detailing the agency's particular requirements for marking and/or lighting an antenna or other site structure to ensure that it does not become a navigation problem. A final determination is not needed for every cell site or antenna structure—only those that violate the glide slope requirements. However, the glide slope should be calculated for every cell site and antenna structure that is either proposed or used by a wireless system operator.

Q. *What is an installation interval?*

A. An *installation interval* is the time it takes to install the fixed network equipment either in a cell site or in an MSC.

Q. *How are cranes used at cell site installations?*

A. Cranes are used in wireless communication systems either in the testing of a potential cell site to see if it meets coverage requirements or in the delivery of radio and ancillary equipment to a cell site. For example, a crane might be used to bring the equipment cabinets to the roof of a building. Cranes are not used too often because of the expenses they entail and the logistical problems they sometimes create.

Q. *What is datafill?*

A. In the context of wireless system construction and implementation, *datafill* is the process of programming the necessary parameters into the switch that controls and populates the parameters established for a cell site. The datafill is needed to optimize the cell site.

Q. *What is cell site acceptance?*

A. *Cell site acceptance* is the point during the construction and implementation process at which operations accepts responsibility for the cell site or MSC from the construction and implementation department. Although the acceptance is often accompanied by a punchlist of open issues that need to be resolved, the site is near enough to completion that it can be brought into service for the purpose of providing revenue for the system.

Q. *What purpose do site visits serve?*

A. A site visit provides engineering, construction, and operations the opportunity to review a potential cell site for use in the network and to identify any potential problems prior to negotiating or signing the lease. In general, site visits are conducted early in the implementation process.

Q. *What is a punchlist?*

A. A *punchlist* is a list of unresolved issues. A punchlist associated with the construction of a cell site or an MSC involves issues related to the building of the cell site or the lack of particular equipment such as a battery string or an eyewash station. The punchlist is generated by either operations, construction, and/or engineering, and it is used to ensure that all the issues are eventually closed. If the punchlist is associated with a subcontractor, either the entire payment or only part of it is withheld until the issues are resolved.

Q. *How is an MSC laid out?*

A. A typical MSC layout is shown in Fig. 5.22. There are numerous perturbations of MSC layouts depending on the vendors used, the ancillary equipment installed, and the future expansion plans for the location.

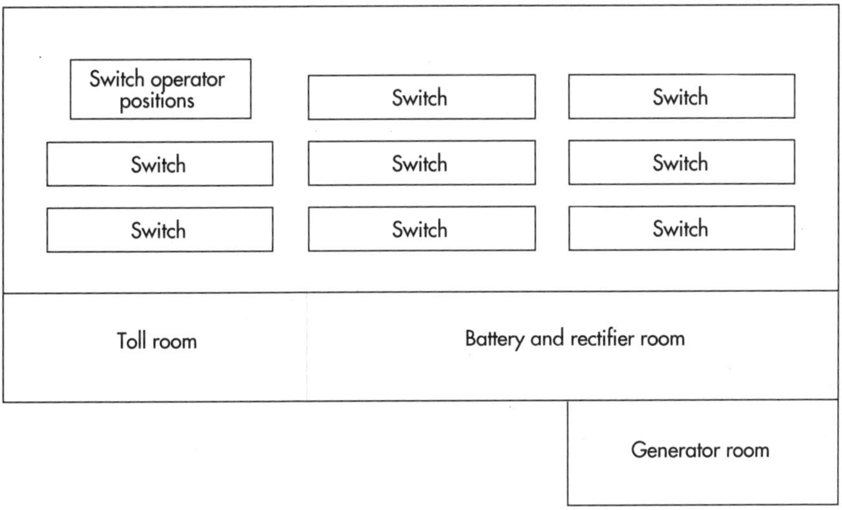

Switch operator positions
Switch
Switch
Switch
Switch
Switch
Switch
Switch
Switch
Toll room
Battery and rectifier room
Generator room

Figure 5.22 Generic MSC configuration.

Q. *What is an interior cell site?*

A. An *interior cell site,* or *tenant improvement* (TI) *cell site,* is a particular space within a building to which the owner or landlord has granted certain access privileges to a tenant. A sample interior cell site layout is shown in Fig. 5.23. Needless to say, the layout shown can be modified to accommodate the equipment and the growth strategy for the operator. However, the layout used for the system should be standardized to prevent the design *à la mode* process. Of course, the standardization needs to have a certain flexibility inherent to it since each interior TI room has its own set of issues.

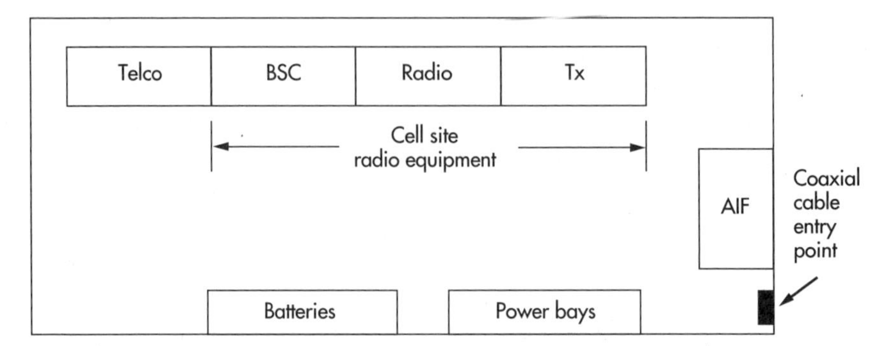

Figure 5.23 Generic cell site configuration. BSC = base site controller, Radio = cellular radios, Tx = amplifier, AIF = antenna interface, Telco = T-1 microwave interconnect.

Q. *What is an exterior cell site?*

A. An *exterior cell site* is usually a cell site that is located in a shelter or hut or that is contained within its own cabinet. The term *exterior cell site* is used in many ways, but the primary meaning is a cell site that is not located within an existing building. Exterior cell sites are most common in rural applications and in some suburban installations where a monopole is erected with the accompanying shelter.

Q. *What is a hut?*

A. A *hut* is another name for an equipment cabinet that is used for exterior installations.

Q. *What is an emergency generator?*

A. An *emergency generator* is a generator that is used with either a cell site or an MSC to provide ac power, via either a single- or three-phase power supply, during times that commercial ac power is not available. The generator for a cell site is smaller in terms of kilowatts than the generator for an MSC. The number and location of generators throughout the wireless communication network are determined from the operator's disaster recovery program.

Q. *What is a generator hookup?*

A. A *generator hookup* is a connector that is located on the outside of a building or shelter. The generator and its hookup are designed to supply ac power temporarily to the cell site. The generator hookup for a cell site should be located in such a way so as to facilitate delivering the generator with a small truck.

Q. *What is floor loading?*

A. The term *floor loading* refers to the ability of a building or room to withstand a particular weight per square foot. For example, the battery strings for a cell site can weigh 250 lb/ft^2 or more, and the room that houses this equipment needs to be able to support that load. The weight that a floor can support depends on the structural integrity of the room or building. The equipment load, or weight per square foot, is usually provided in the equipment literature.

Q. *What is roof loading?*

A. *Roof loading* is similar to floor loading except it applies to the roof of a structure. In the northern United States, usually a roof is designed to withstand a particular snow load, or rather roof load. The roof load limits need to be checked when installing equipment on the roof to ensure that the equipment will remain on the roof and not crash through to the room or rooms below.

Q. *What is equipment loading?*

A. The term *equipment loading* refers to the actual power drain, in amperes, required to run particular equipment. For example, if the equipment load for a radio was 5 A and there were five radios, then the radio equipment load would be 25 A.

Q. *What type of HVAC system is required at cell sites?*

A. Some level of *heating, ventilation, and air-conditioning* (HVAC) *system* is required at all cell sites and MSCs. The HVAC specifications are given in British thermal units as measures of heat that must be removed, and that figure is then converted to a particular tonnage for air-conditioning. Normally the primary use of the HVAC system is to remove heat from the equipment room.

However, in some cases heat may also be required since all the equipment, including batteries, are designed to operate within a given temperature range. Often the HVAC requirements may necessitate an upgrade to the power plant for the cell site or MSC.

Q. *What is heat dissipation?*

A. *Heat dissipation* is the releasing of heat; in wireless communication systems, it is the heat released by the equipment in a cell site or an MSC in the course of being on and processing calls. The rate of heat dissipation for each piece of equipment used either at the cell site or the MSC is obtained either from on-site physical measurements, or it is provided by the manufacturers. The heat dissipation is usually defined in terms of British thermal units, which in turn defines the HVAC requirements for the location.

Q. *What is hydrogen venting?*

A. *Hydrogen venting* is the process of proving venting for battery gases in a cell site or an MSC.

Q. *What is a tower installation?*

A. A *tower installation* is a cell site installation at which the antennas are located on a tower. The radio equipment, however, can be located in an interior building or in a separate shelter controlled by the wireless operator.

Q. *What is a roof-top installation?*

A. The term *roof-top installation* usually refers to a cell site that has been installed on the roof of a building. The mere location of the antennas on a roof top is not referred to as a "roof-top installation" by itself.

Q. *What is a cell on wheels?*

A. A *cell on wheels* (COW) is a temporary installation where the cell site is brought into a location in a trailer or van and operates until the permanent cell site is installed. The purpose of the COW is to expedite the operation of the cell site to provide coverage or capacity relief for an area.

Q. *What is a switch on wheels?*

A. A *switch on wheels* (SOW) is a temporary installation where the switch is brought into a location in a trailer or van and operates until the permanent switch is installed. The purpose of the SOW is to expedite the operation of the switch to provide capacity relief.

Q. *What is an interior installation?*

A. An *interior installation* (also called a *tenant improvement*) is the process of constructing the room used to house the cell site equipment in the interior of an existing building.

Q. *What is a colocation installation?*

A. A *colocation installation* is a location, sometimes a roof or tower, that is hosting more than one operator's equipment. *Colocation* can also refer to the installation of switching equipment in another telco facility.

Q. *What is a car installation?*

A. A *car installation* is the process of installing a cellular or wireless phone into a car.

Q. *What is a repeater installation?*

A. In the wireless industry, a *repeater installation* is usually the installation of a rerad or bidirectional amplifier within the system.

Q. *What is a microwave system?*

A. In the wireless cellular and PCS communication industry, a *microwave system* is a point-to-point system that is used to provide telco facilities, usually a T-1/E-1 equivalent.

Q. *What are some standard power conversions?*

A. The power conversions listed in Table 5.2 should prove to be helpful.

TABLE 5.2 SOME STANDARD POWER MEASURE CONVERSIONS

From	To	Multiply by
Horsepower	British thermal units per minute	42.418
Horsepower	Kilowatts	0.746
Kilowatts	Horsepower	1.341
Kilowatthours	British thermal units	3413
British thermal units	Kilowatthours	0.000293
Watts	British thermal units per hour	3.413
British thermal units per hour	Watts	0.293

Q. *What are some standard distance measurement conversions?*

A. Table 5.3 contains many of the standard conversions needed for various measurements encountered in a wireless communication system. Often the more commonly used conversions deal with converting from metric units to standard units and vice versa.

TABLE 5.3 SOME STANDARD DISTANCE MEASURE CONVERSIONS

From	To	Multiply by
Meters	Feet	3.28
Feet	Meters	0.3048
Miles	Kilometers	1.609
Kilometers	Miles	0.6214
Kilometers	Feet	3281
Feet	Kilometers	0.0003408
Liters	Gallons	0.2642
Gallons	Liters	3.785
Rods	Feet	0.06061
Yards	Feet	3
Yards	Meters	1.094
Inches	Centimeters	2.54
Centimeters	Inches	0.3937
Feet	Centimeters	30.48
Centimeters	Feet	0.03281

Q. *What are the standard temperature conversions for wireless communication systems?*

A. The following table contains the standard temperature conversions needed for a wireless system:

From	To	Multiply by
Fahrenheit	Kelvin	(F + 459.67)/1.8
Celsius	Fahrenheit	(C · 9/5) + 32
Fahrenheit	Celsius	(F − 32) · 5/9
Celsius	Kelvin	C + 273.1

Q. *What are some of the environmental requirements that must be met by the MSC?*

A. The following are some of the environmental issues that need to be addressed for a typical MSC:

- Air cooling
- Grounding
- Over-voltage protection
- EMI
- Floor load
- Ceiling height
- Climatic state
- Dust filtering
- Relative humidity of air
- Corrosion
- Floor covering
- Sound level
- Vibration

Q. *What is a single-point ground system?*

A. In a *single-point ground system,* all the cell site equipment is connected to the grounding system for the site at a single point. The communication equipment, cabinet frames, radios, and power plant are connected to the single point so that all the equipment samples the ground potential at one and only one point.

Q. *What is a halo grounding ring (or halo ground)?*

A. A *halo grounding ring,* or *halo ground,* is a grounding wire that encircles the interior of a cell site, and all the equipment cabinets, cable trays, and so on are connected to it. The halo grounding ring usually consists of a bare no. 2 AWG wire, which is tied to the master ground bus bar at the cell site. The ring usually is installed about 1/4 to 1 ft below the finished ceiling with the appropriate standoffs from the wall. Depending on the manufacturer of the cell site equipment and the vintage of the grounding practices used, either all the equipment in the cell site is connected to the halo grounding ring or only equipment like the cable tray and auxiliary racks are connected to it.

Q. *What is an external ground system?*

A. An *external ground system* for a communication site consists of a series of ground rods that are planted at each corner of the communication shelter, the commercial ac power entrance, and of course the wave-guide entrance. The ground rods are not only connected via the ground but also through a solid no. 2 AWG tinned copper wire that encircles the perimeter of the communication site. The connections from the ground rods to the exterior ground ring should be cad welded. The exterior ground ring needs to be buried at least 30 in below the ground and also needs to be bonded to the tower system, if applicable, at two different locations.

Q. *What is a ground ring?*

A. A *ground ring* is an external ground system. The ground ring is the physical wire that encircles the communication site and consists of a solid no. 2 AWG tinned copper wire that encircles the perimeter of the communication site. Ground rods that are planted at each corner of the communication shelter, the commercial ac power entrance, and of course the wave-guide entrance are then connected to the ground ring. The connections from the ground rods to the exterior ground ring should be cad welded. The exterior ground ring needs to be buried at least 30 in below the ground and also needs to be bonded to the tower system, if applicable, at two different locations.

Q. *What ground resistance is needed for a communication site?*

A. The ground resistance needed for a communication site is typically less than 5 Ω. Ideally it should be 0 Ω but this is not practical.

Q. *What does the term deadmen mean?*

A. The term *deadmen* describes the foundations located at each of the guy-wire anchor heads on a guy-wire tower. The deadmen consist of buried concrete blocks that have an anchor rod or bolt that is exposed above the finished concrete so that the guy wire can be connected to it.

Q. *How are the wind load requirements determined for an antenna system?*

A. The minimum wind load requirements for an antenna system should be 80 mi/h, but the actual requirements, which are determined by the geographic area as well as by the cross-sectional area of the installation, are determined from the EIA/TIA 222-E standards.

6

Real Estate

Land-use acquisition, commonly called real estate in the wireless industry, plays a critical role in both switched and wireless network applications. All too often the design and future expansion options a network provider has at his or her disposal are driven not by technological issues but by real estate issues, or rather land-use acquisition issues.

What follows is a brief listing of many frequently asked questions related to the real estate aspects of a wireless operation. Within the FAQs are some terms and their definitions that might appear in leases or other real property negotiations and transactions. Real estate issues have a strong and profound impact on network design.

Q. *What is site acquisition?*

A. *Site acquisition* is the process that is utilized by an operator or potential operator to obtain property for a cell site, an MSC, an office, or any other facilities. Site acquisition activities include identifying a prospective property that meets specified conditions for potential use, obtaining a lease, obtaining a right of way, or actually purchasing a property.

Q. *When is an environmental impact study needed?*

A. An environmental impact study is required for any parcel (that is, property) that has been ruled to be environmentally sensitive such as wetlands, flood plains, and steep slopes. The environmental impact study is conducted in three parts—Phase 1, Phase 2, and Phase 3—to which a wetlands survey is added if necessary.

Q. *What is an environmental impact statement?*

A. An *environmental impact statement* is the written report that describes the findings of the environmental impact study. The statement is the basis of the site review conducted by the relevant regulatory agencies. These agencies may require that certain remedial actions take place before a property is altered.

Q. *What is a Phase 1 Environmental Audit?*

A. A *Phase 1 Environmental Audit,* also often referred to as a *Phase 1 Environmental Study,* is a visual analysis of the site, which involves analysis and research of the site and surrounding sites to determine if a potential environmental risk exists and if additional testing is required.

For wireless communication operators, Phase 1 consists of a visual inspection of the property, with particular emphasis on the intended location for the communication facility, which may possibly be an intrusive action. Phase 1 also involves a historical review of what the property was used for to see if there were any potential environmental issues that may be at the site. In addition, an *agency file check* is done, which is simply a check to see if there are any regulatory compliance problems with the property. The physical inspection is either just a brief visual inspection or a more thorough inspection in direct response to the issues raised in the historical review or agency file check.

Last, a report is written that summarizes the findings and draws a conclusion. The conclusion can be that no problems were found or that additional analysis is necessary, which would be a Phase 2 Environmental Audit.

An example of a Phase 1 Environmental Audit is an EMF–exposure study.

Q. *What is a Phase 2 Environmental Audit?*

A. A *Phase 2 Environmental Audit* occurs when a Phase 1 analysis deems that further investigation and analysis are warranted. In the Phase 2 Environmental Audit, site material suspected to be contaminated is tested. If there is reason to suspect that contamination may have spread from an adjacent property such as a landfill, then samples may be taken and tested from that property as well. In addition to possibly physical testing, more research is also conducted to track down and possibly resolve any issues raised by the regulatory authorities during their review of Phase 1.

The objective of Phase 2 testing and research is to establish whether any environmental suspicions about the property are correct and the relationship of the issue to the property. If an environmental issue is raised, then this process moves to Phase 3 Remediation. It should be stressed that not all Phase 2 audits move automatically to Phase 3 Remediation; sometimes issues discovered are also resolved during Phase 1 or Phase 2.

An example of a Phase 2 Environmental Audit could be a power density study at a building adjacent to the communication site. Another example could be the investigation of a particular tenant improvement site for the presence of asbestos or lead paint.

Q. *What is Phase 3 Remediation?*

A. *Phase 3 Remediation* is the cleanup and abatement of any unacceptable environmental contaminants that were uncovered in the Phase 2 Environmental Audit.

Phase 3 work may consist of such activities as the removal of asbestos or lead paint or the realignment of the antennas at the site of their power output.

Q. *What is a Wetlands Survey?*

A. What the *Wetlands Survey* consists of and when it must be done depend upon Environmental Protection Agency (EPA) and DEC statutes. The Wetland Survey is an examination of all the types of plants, animals, water flow, and soil composition on the site. The findings of the study may either affirm or negate the property's status as a wetland. If the site is deemed to be in a wetlands area, then there usually are severe restrictions placed on its development. In general, wireless facilities avoid any wetlands areas due to the excessive restrictions usually placed on its development.

Q. *What is a search ring?*

A. A *search ring,* also referred to as a *candidate ring,* is a document prepared by the RF engineering department stipulating what the requirements are for a potential cell site they are interested in establishing for the network. A search ring can also be used in the procurement process for locations for other types of facilities such as an MSC or even a hub. However, most often a search ring is commonly associated with the cell site build program. Some of the information commonly included in a search ring are the following items. (Of course, the requirements for the potential location can and do change based on the vendor's equipment.)

Search ring area
Date location required
Space required
AMSL and potential AGL
Number of antennas to be installed

Q. *What is a balloon test?*

A. A *balloon test* consists of elevating a balloon, of a predetermined size, over a potential cell site so that residents of the community and board members can assess the visual impact the site may have on the horizon and the local area.

Q. *What are site candidates?*

A. *Site candidates* are potential cell site locations that by either an internal or external real estate department for the company has identified as meeting the search area requirements specified in the search ring. Typically there will be two or three site candidates identified per search ring depending on the requirements or constraints placed upon the search. The reason for identifying more than one site candidate per search ring is that there is a multitude of potential issues that might disqualify one or the other site.

The steps taken by the real estate department to narrow the field of possible sites to three viable candidates are the following:

Obtain search ring criteria.
Analyze zoning data for culling possible suitable locations.
Analyze affiliate or bulk sites for matching search criteria.
Obtain candidate site lists.
Generate reports on all possible candidates for comparing against search ring criteria.
Coordinate site visits and possible transmitter tests.

Q. *How is a site finally chosen from among the candidates?*

A. The final candidate selection is made through a process that involves identifying and subsequently approving a potential cell site that meets the real estate, engineering, and construction requirements. Often the candidate selection will involve deciding which of potentially three properties best meets the requirements from a leasing, RF coverage, and construction point of view.

Q. *What is a search ring form?*

A. A *search ring form* is a document that stipulates the requirements that a cell site must meet in order to be approved by engineering. The form needs to have specific material in it detailing the requirements of the design objective. Figure 6.1 is an example of a search ring form that can be used as a general boilerplate.

Q. *What is a lease hold improvement?*

A. A *lease hold improvement* is the improving of a location leased to an operator, usually involving some type of structural change. A lease hold improvement is typically the interior renovation that takes place for a leased facility to accommodate the wireless equipment whether it is radio or switched network equipment.

RF Engineering

Search area code:__________ Capital funding code:__________

Target-on-air date:__________

Search area type: (capacity, coverage, frequency plan, competitive, new technology)

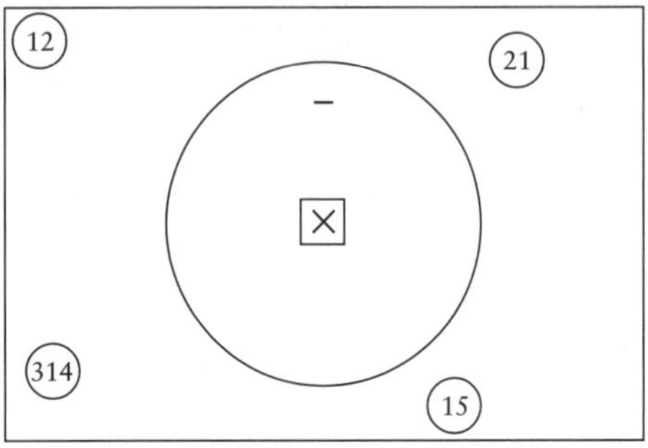

12

21

314

15

Search area map

Cell site configuration: (omni, 3 sector, 6 sector, other)

Type of infrastructure:__________

Physical size of equipment room:__________(ft^2)

Antenna info:

1. Number of antennas:__________
2. Type of antennas: (attach manufacturer's specifications)
3. Antenna height

   AGL:__________

   AMSL:__________

Maximum cable length:__________

Comments: __________

Search area request

Document:__________ Date:__________

Design engineer:__________

Review by:__________

Revision

Figure 6.1 Search area request.

Q. *What is a tenant improvement?*

A. A *tenant improvement* (TI) is an alteration made within an existing building for the purpose of housing cell site equipment. Usually the alteration is made during the interior fit-up process and is the result of preparing the room that is leased to house the radio equipment. Some issues to be resolved in a tenant improvement situation involve the need for a separate power feed, an HVAC system, and antenna cable runs.

Q. *What is an architectural review board?*

A. An *architectural review board* (AE board) is one of the boards that a wireless operator may have to interface with and obtain approval from in order to obtain a building permit to construct a cell site or erect an antenna. Often the AE board is involved with the site-building process in order to ensure that certain aesthetic qualities are maintained or preserved with the construction of the proposed cell site. The actual requirements and function of the AE board are specific to each municipality.

Q. *What is the role of the architectural engineer during site development?*

A. An *architectural engineer* (AE) is usually employed by the wireless operator to design and place a stamp of approval upon the building or tenant improvement design that is put forth by the wireless operator for the potential cell site. The AE might also be involved with zoning or planning board testimony in the course of securing the proper governmental approvals for the cell site.

Q. *What are local ordinances?*

A. Local ordinances are laws that have been passed by towns or cities to govern development within their geographic boundaries. Ordinances may be passed to regulate a wide range of activities from erecting signs to altering existing buildings. For wireless communications issues, the local ordinances define whether a wireless facility is allowed to operate within a given area by right or by variance. Variance applications must be submitted to local boards for their approval.

Q. *Can local zoning authorities deny a request for a cell site?*

A. Local zoning authorities do have the right to deny a request for a tower or any wireless facility as a nonconforming use for a particular zone. However, the denial must be rational and it cannot be made because of RF emissions—that is, provided that there are no emissions violations. In addition, it cannot summarily ban the construction, placement, or modification of wireless facilities on the sole basis that it relates to a wireless facility.

In general, the initial variance an operator seeks specifies an installation plan and a certain configuration or number of antennas that the facility will utilize. If the operator veers from the original specifications for the antennas, or the overall site plan, he or she may have to reapply for another variance.

Q. *What is a building permit?*

A. A *building permit* is the formal written permission given by the local government, usually the building inspector, for the operator to construct the proposed cell site. In some cases the building permit is secured through right—that is, it is authorized within a particular building code. However, in most cases the building permit can be obtained only after the planning board has approved the project. The particular issues and sequence associated with obtaining a building permit differ from municipality to municipality. However, in all cases the requirements for obtaining a building permit are available from the local building department.

Q. *What is an expeditor?*

A. An *expeditor* is a service that ushers building permit applications through the various boards and officials who must approve building projects. It is used in some large municipalities. There is a fee associated with this effort that is not trivial. However, there are situations in which the use of an expeditor is part of the process to obtain a building permit and cannot be avoided.

Q. *What is testimony?*

A. *Testimony* is used to prove information to various municipal boards for the purpose of obtaining the necessary governmental approvals to obtain a building permit to construct the cell site. By tradition, the following groups are involved with providing testimony to the local boards:

Real estate broker or his or her representative
AE firm or operator's construction department
RF engineering department
Expert witnesses as required

An interesting issue is that the use of the operator's RF engineering department to provide testimony is a tradition, but that testimony could be presented just as well by the real estate representative.

Q. *How is a cost-of-living adjustment factored into a lease?*

A. A *cost-of-living adjustment* (COLA) is inserted in a lease to indicate the amount of rent increases expected to be incurred by the wireless operator. A COLA is either a fixed amount, say, 5 percent per year, or it varies based on the Consumer Price Index (CPI). The value and methodology used for a COLA should be examined in detail since a 5 percent yearly increase over the course of the life of the cell site may seriously increase the operating costs.

Q. *How do local zoning laws affect wireless communications operators?*

A. A great many zoning ordinances stipulated by a municipality directly influence how a wireless operator can construct a wireless facility in that community. The largest effect of zoning ordinances is felt in procuring cell site locations because the local zoning ordinances specify whether a cell site can be constructed in a given area. Usually the ordinance will not specify a wireless facility explicitly, and in general, it is understood that if the use for a cell site is not permitted by the ordinance, then it is by default excluded and requires a variance. The ordinance stipulated in the zoning guidelines will describe several types of zones, usually industrial, commercial, or residential. There are, of course, additional classifications to the three zones listed; however, all municipalities utilize the three general classifications.

Q. *What is a residential zone?*

A. A *residential zone* is a geographic area that is restricted by local ordinances to residential use only. Many types of housing qualify as "residences," and these are usually referenced in terms of R1,

R2, and so on, and their definitions are contained in a local ordinance book. If an operator suggests locating a cell site within a residential zone, the opposition from the local community will usually be great. Consequently, it is generally preferable to locate a cell either on an existing tower or in a commercial or industrial zone for the community.

Q. *What is a commercial zone?*

A. A *commercial zone* is a geographic area that is restricted by local ordinances to commercial use. Many endeavors qualify as "commercial," and each municipality relies on its own definitions, which are spelled out in the local ordinance book. An example of commercial use is a convenience store.

Q. *What is an industrial zone?*

A. An *industrial zone* is a geographic area in which industrial activities are permitted. Each municipality creates its own definitions of "industrial activities," and those definitions are spelled out in their local ordinances. All municipalities, however, recognize that there is a range of activities from light industrial, such as a warehouse, to heavy industrial, such as a textile factory. An area zoned for industrial use is a good candidate for a cell site because its construction is usually met with minimal community resistance.

Q. *What is a nonconforming use?*

A. A *nonconforming use* is an activity prohibited in a certain location by local zoning laws. An example of a nonconforming use could be the establishment of a commercial business in a residential zone.

Q. *What is a variance?*

A. A *variance* is written permission obtained from the local zoning authority to build a structure or to use an existing structure in a way that is expressly prohibited by the current zoning laws that apply to an area. The variance is basically an exception from the zoning ordinance. Wireless operators must often obtain a variance to operate a wireless facility since doing so is not granted by right. In addition, in almost all cases, operators must obtain a variance to erect an antenna. The height restrictions most commonly applied are that no structures may be taller than a two-story home or a 65-foot silo for a farm. Most wireless facilities exceed those heights. To put the height issue in perspective, note that the average tree line is 70 feet in height.

Another situation in which a variance must be obtained occurs when the proposed building site is in a residential zone. Sometimes the permitting process can be bypassed by simply moving the cell site from a residential zone into a commercial or light industrial zone.

Q. *What is a landmark?*

A. A *landmark* is either a defined area or a building within a community that is protected for historical preservation. Landmarks have unique aesthetic requirements that must be considered during cell site development. Note that the term "landmarks" is also used in references to a local landmarks preservation board within a community.

Q. *What is a real estate license?*

A. A *real estate license* is issued by a state to an individual to permit him or her to purchase, sell, buy, rent, or lease real properties. A state grants a license to a person after he or she has attended a course and passed a written exam. In addition, a state requires the license applicant to obtain a sponsor for holding the license once it is awarded. Note that a real estate license is not required to obtain leases for potential cell sites since there is no commission being paid by the landlord to the company representative, or agent.

Q. *What is a sublease?*

A. A *sublease* is a lease obtained from a lessor who is himself or herself already leasing the location or part of it from someone else. It is important for wireless operators to identify the actual owner of a particular property to ensure that the wireless facility complies with the owner's rules. Note that many roof-top management firms are the master lease holders. Thus a sublease must live up to the agreement in the master lease. For example, if the master lease is to expire in three years, signing a five-year lease with several renewable options may not be a good idea.

Q. *What is access?*

A. Simply put, the term *access* refers to the ability, freedom, and permission to enter and leave a facility owned by someone else. Usually leases define *access* for a wireless facility as unlimited—24-hour access, seven days a week, all year round. Ensuring that a lease grants unlimited access is important so that operators are able to perform preventive maintenance and take care of emergencies such as a malfunctioning radio or antenna system. Note that access rights may be slightly different for roof tops than for interior locations. Usually access to a roof is not as critical because the only equipment on a roof is most often the antenna system, which probably won't need frequent repairs. However, if the equipment located on a roof is in a cabinet enclosure, then unrestricted access to the roof becomes a paramount issue.

Q. *What is space?*

A. *Space* is the area that the cell site equipment will occupy. Some typical examples of space requirements could be a 10- by 20-ft room, or rather 200 ft^2.

Q. *What is roof-top management?*

A. *Roof-top management* is a situation in which a firm has obtained the roof rights from a landlord. The roof-top management firm then oversees the process of obtaining new tenants for the roof top or tower, which maximizes the landlord's revenue for the structure. The roof-top management firm usually obtains their fee as a function of the rent charged, an amount typically equal to one month's rent every year. The particular functions a roof-top management firm performs on a site vary widely and are usually stipulated in the lease terms.

Q. *How do commissions factor in to a real estate transaction?*

A. A *commission* is a fee paid to a broker in return for his or her obtaining a sale for the seller. Usually a commission is a percentage of the overall price of a property or the rent to be paid over the life of a lease. For wireless operations, usually there is no commission paid to any of their defined agents. If a broker is utilized to secure potential cell sites, the commission may be paid by the landlord and is a function of the lease cost.

Q. *How much electric ac power does a cell site normally require?*

A. A cell site usually requires 100 to 200 A from either a single-phase or a three-phase power supply. This amount of power may not be available at an existing building, necessitating an upgrade to the power plant to accommodate a new cell site. If an operator is considering an existing building, he or she should take extra care in evaluating the power supply so that the extra funds for an upgrade are factored into the budget and the extra time to complete an upgrade is factored into the schedule. In addition, if the proposed cell site is to be located on undeveloped property (a *green field*), the time and expense to install utility poles must be factored into the plans for the site.

Q. *What is a submeter?*

A. A *submeter* is a separate electric service that provides electricity to the cell site. The advantage in using a submeter is that, because it measures only the electricity used by the cell site, the bill is sent directly to the wireless operator. Care must be exercised to ensure that the monthly bills actually make it to the operator and do not sit unpaid in the corner of a room while the utility company contemplates disconnecting the service due to nonpayment.

Q. *What is a meter check?*

A. A *meter check,* sometimes also called a "check meter," is a situation in which an operator or landlord checks the usage tracked by the electric meter associated with the cell site. It is important to note that the meter check is not conducted by the utility company. It involves visually checking the meter in conjunction with working out a payment process from the operator to the landlord.

Q. *What does the term rental rates mean?*

A. For wireless operators, the term *rental rates* refers to the lease costs associated with the cell site.

Q. *What issues disqualify cell site candidates?*

A. Some of the issues that tend to disqualify a cell site candidate are insufficient space to house the equipment, insufficient space to meet system expansion needs, inadequate feed-line length, and inadequate power supply. The most serious issues are inadequate locations for antennas, obstructions in the radio frequency path, and an antenna height that would have to be above the average terrain.

Q. *Which benchmarks need to be tracked in cell site real estate procurement?*

A. The benchmarks that should be tracked in the course of developing a cell site location include the identification of cell site candidates, the review of the candidate cell sites, the final selection of a site, and the negotiation of a lease or the purchase of a site. The benchmarks in Table 6.1 can form the basis of a tracking system for real estate procurement.

Q. *What are the steps and process for finding a cell site?*

A. The steps and process for finding a cell site are different from market to market and also from wireless company to company. However, 90 percent of the actions listed in the outline below should be accomplished in any real estate procurement project including PCS, cellular, SMR, WLL, paging, and LMDS. The actions listed are executed primarily by the real estate and RF engineering departments within a wireless company. The time durations for completing each of the steps are not included because they depend directly on the size of the system as well as the time-to-market requirements. The plan below can be adapted for both new systems and expansions of existing systems. Note that it is assumed that the RF design has already been prepared, and thus, the outline begins with the issuance of the search ring itself.

Identification of Cell Site Candidates

- Company identifies search areas.
- Company identifies search area requirements.
- Real estate department reviews database of existing cell sites and friendly sites.
- Real estate department identifies three candidate sites.

Table 6.1 Site Acquisition Tracking Form

	Planned	Actual
Site identification number	N/A	
Search ring issued		
Candidate sites identified		
Initial site visit		
SQT scheduled		
SQT results due		
RF approval/rejection		
Lease process begins		
Implementation site visit		
Site approved		
Lease exhibits generated		
Internal lease review		
Fully executed lease		
Building permit application submitted		
Variance required (yes/no)		
Variance application submitted		
Zoning meetings begin		
Zoning/variance approval		
Building permit obtained		
Construction begins		
Construction completed		
Equipment installed		
Cell site ATP completed		
Site integrated		
Site on-air date		

- Real estate department contacts landlord if site candidate is to be leased or real estate broker if site is to be bought.
- Real estate department sets up and coordinates site visits.
- Real estate and RF engineering departments visit sites.
- Real estate and RF engineering departments make preliminary recommendations on whether to accept or reject site candidates.
- Real estate department reviews local ordinances for properties the operator intends to purchase.
- RF engineering and real estate generate report based on site visit findings.

Evaluation of Real Estate Issues

- RF engineering determines if pretest is needed.
- Real estate informs landlord or property owner of test.
- Real estate secures insurance certificate and HH.
- Real estate coordinates pretest.
- RF engineering informs real estate of test results.
- Real estate informs landlord or property owner of results.

Negotiation of Lease or Purchase Terms

- Real estate starts preliminary purchase or lease negotiations.
- Real estate and construction departments award AE contract.
- Real estate arranges construction and AE site visit.
- Real estate, RF engineering, AE, and construction visit site.

Approval of Final Lease or Purchase Agreement

- Real estate obtains lease exhibits.
- Real estate obtains 60 percent drawings from AE.
- Company reviews 60 percent drawings.
- RF engineering performs evaluation of site for FAA compliance.
- Company reviews and accepts lease exhibits.

- Real estate receives preliminary lease from landlord or purchase agreement from broker.
- Real estate obtains 90 percent drawings from AE.
- Real estate reviews lease alongside 90 percent drawings with landlord or purchase agreement with broker.
- Real estate submits lease or purchase agreement to wireless systems company management.
- Wireless systems company management approves lease or purchase agreement.
- Request permit application from office of local building department inspector.
- Prepare building permit application.
- Obtain permits from building department/inspector.

Obtaining Building Permits When Zoning Issues Exist

- Real estate awards attorney.
- Real estate obtains the required variance applications.
- Real estate has the zoning variance application signed by the landlord or wireless company management.
- Real estate makes request to be placed on the planning board agenda.
- Real estate assists attorney, AE, and RF engineering with prezoning.
- Real estate, AE, attorney, and RF engineering attend work session.
- If needed, real estate instructs the AE to revise the plans and/or application.
- Real estate, AE, attorney, and RF engineering attend a prep meeting.
- Real estate and RF engineering obtain an EMF report.
- *Wait one week (from prep meeting)**: Real estate, AE, attorney, and RF engineering attend planning board meeting.
- *Wait one month:* Real estate, AE, attorney, and RF engineering attend architect review board meeting.

*Intervals estimated and will differ depending on the board's agenda and the local ordinance requirements.

Also, real estate, AE, attorney, and RF engineering attend zoning board meeting.

- *Wait one month:* Real estate, AE, attorney, and RF engineering attend planning board meeting. Then real estate reports to wireless operator's management on planning board's accept or reject decision. Finally, real estate receives formal planning board resolution.
- *45-day appeal process:* If the planning board resolution was favorable, then real estate reports those results to the wireless operator's management. Real estate then instructs AE to make whatever design changes are necessary to comply with planning board's ruling. Real estate then submits a building permit application to building inspector's office. The application is accompanied by the planning board's ruling and the newly modified cell site design.
- Building department issues building permits.

Q. *What is a tower lease?*

A. A *tower lease* is used when it is necessary to install a wireless communication system on a preexisting tower. The key issues that need to be factored into the lease involve the location of the antennas, the number of permitted antennas, maintenance, and accessibility to not only the tower but also the equipment room or shelter associated with the radio equipment. Two other key issues that need to be addressed are the location of the radio equipment and its power supply and the security measures available at the site. In addition, it must be possible to make the site comply with all government regulations. Finally, a contingency procedure to be followed must exist in the event that another tenant wishes to install equipment on the same premises if his or her equipment is apt to create interference with the first operator's equipment.

Q. *What is a roof-top lease?*

A. A *roof-top lease* is used when it is necessary to install either radio equipment or antennas or both upon the roof of a building. A roof-top lease may be a sublease if the operator is dealing with a roof-top management firm. Some of the issues associated with a roof-top lease that should be addressed up front are, of course, the accessibility to the equipment and antennas on a 24-hour basis and the location of the equipment and its power supply. The antenna placement needs to be secured in advance to ensure that the optimal position that is available is obtained. Another issue to address in the lease is a contingency procedure to be followed in the event that another tenant wishes to install equipment on the same premises if his or her equipment is apt to create interference with the first operator's equipment. The procedures must ensure that the other tenants' antenna system does not block or hinder the proposed or existing antenna structure secured in the first operator's lease. More times than not, in the wireless industry a roof-top lease is part of an interior lease agreement.

Q. *What is an interior lease?*

A. An *interior lease* is written to cover a *tenant improvement*—that is, a situation in which the radio equipment will be occupying space within a building. This type of lease should address the placement of the antennas, the feed-line runs, the power supply, and the HVAC requirements. In addition, interior leases should provide for parking for a technician's vehicle and elevator access if the room is not on the ground floor. The interior lease is used more commonly in urban environments.

Q. *What is an exterior lease?*

A. An *exterior lease* is used in situations in which the equipment is to be located outside in its own cabinet or shelter. In this case no modification to the existing building is needed except perhaps an antenna installation. Some of the more common issues that need to be addressed in an exterior lease pertain to access to the equipment, parking for operations personnel, and the availability of a generator hookup.

Q. *What is a shelter?*

A. In wireless communications systems, a *shelter* is a self-contained structure that houses the wireless operator's telecommunication—usually radio—equipment. Some of the items normally found in a shelter are the radio equipment itself, batteries, an HVAC system, rectifiers, and telecom demarc equipment. Figure 6.2 is a picture

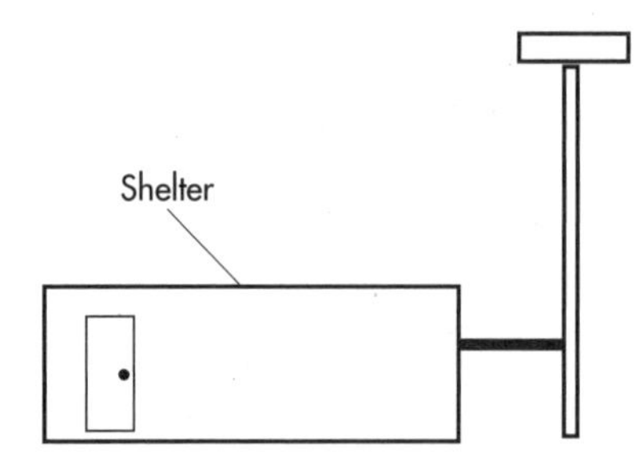

Figure 6.2 Shelter.

of a shelter. The sizes of shelters are quite varied within any given wireless communications market.

Q. *What is a telco closet?*

A. A *telco closet* is a room, which is usually the size of a broom closet, that houses telecommunication equipment for the building. The telco closet is usually the demarc point for the building for all the telephony services, and in a larger building it is usually also the location of 66 blocks.

Q. *In what way do aesthetic considerations affect installations?*

A. Aesthetics pertain to the real estate issues in wireless system setups—specifically with the placement and physical appearance of the wireless system's antennas. Often an antenna's appearance plays a critical role in where it can be mounted in that a placement on a building or structure that proved to be satisfactory during the testing phase may be an aesthetically impossible placement and thus unusable from an engineering and cost perspective. More often than not, the aesthetics of an antenna structure are overlooked until the antenna is installed, whereupon its appearance becomes a major issue to local residents. It is advisable that in the early stages of the design the aesthetic impact of the antenna system be addressed to maximize the building's use while at the same time reducing postimplementation problems.

Q. *What is a stealth installation?*

A. The term *stealth* in the wireless industry pertains to the building of facades that are meant to disguise an antenna system. Several approaches to hiding an antenna system completely have been proposed and tried. However, that extreme solution is not always necessary, and the majority of wireless installations can implement a pseudo-stealth method by cleverly placing antennas so that they meet the design requirements, minimize installation costs, and provide the least visual impact. The use of any type of stealth disguises for an antenna system usually results in a manyfold increase in the overall installation cost for the wireless facility.

Q. *How are clock towers used to disguise antennas?*

A. Some wireless operators have resorted to using clock towers as a way to obtain the local permits necessary to install their wireless facility. The clock tower method is a stealth technique, and it is an unusual and extreme approach to obtaining a cell site for a given area. Before implementing this strategy, operators should be aware that the installation costs associated with this approach are exorbitant and therefore it should be used only after other designs have been carefully considered.

Q. *What is a tree?*

A. A *tree* in most contexts refers to a woody perennial plant in a backyard, but a *tree* in the wireless industry refers to a monopole tower that is designed to look like a southern pine tree so that it blends into a surrounding landscape. Trees are used increasingly in the wireless industry to meet the aesthetic concerns of communities. It is worth noting that the shelters located at the base of the trees could be improved aesthetically.

Q. *What is a site survey form?*

A. A site survey form is a document that is used during the initial investigation of a potential cell site to determine its viability. There are many variants of the form. However, Fig. 6.3 provides a sample form that can be used in most situations to convey the technique information about the site back to the engineering community.

Q. *What is an encumbrance?*

A. An *encumbrance* is either a right or an interest someone other than the owner has in a particular property. An encumbrance can be one of two types: lien encumbrances and usage encumbrances. A *lien encumbrance* exists when someone has a monetary claim against a property. A *usage encumbrance* exists when someone

Figure 6.3 A site survey form.

Site no.: ____________ Date: / /XX

Address: ______________________________________
____
____

1. Obstructions (direction and type):

2. Terrain:

3. Colocation (who and where):

4. Equipment location:

5. Land-use issues:

Comments:

Roof and antenna sketch:

has placed a restriction on the use of the property and the restriction is in the deed. Two other types of encumbrances are easements and encroachments.

It is important to note that many times the encumbrances are found in a simple title search for the property. Encumbrances such as judgments or liens can stymie a real estate transaction, so it is advisable to discover any that exist before the real estate purchase or leasing process is far along.

In a reverse of the preceding situation, an operator may need to conduct a title search to be certain that a particular encroachment needed to use a site for wireless communication equipment actually exists. For example, the operator may need to cross someone else's property to reach the cell site, in which case he or she would need an easement.

Q. *What is a lien?*

A. A lien is a monetary claim against a property.

Q. *What is an agent?*

A. An *agent* is a person or company that is authorized to act for or in place of another person or company. Usually authority is granted to an agent via a letter that stipulates the conditions in which the agent may act. A telecommunications firm may hire a real estate firm to locate a potential cell site and then negotiate a lease. Generally, an operator establishes guidelines for the agent's negotiations with the property owner.

Q. *What are air rights?*

A. *Air rights* are rights to use the open space above another person's property. Air rights are an interesting issue that sometimes arises in lease negotiations and municipal taxation. Some communities, for example, are trying to raise revenue by asserting some type of ownership of the air space over their land area.

Q. *What is a title abstract?*

A. A *title abstract* is an executive summary of the title for a particular parcel of real estate. Its purpose is to provide a condensed version of the title and to identify the particular salient points that need to be considered in the lease or potential purchase of the property.

Q. *What is an assignment?*

A. An *assignment* is the transfer of the interest in a particular property from party A to party B. An assignment is commonly associated with a lease, but it could also be applied to a bond or mortgage. For example, if a landlord were to sell a building in which an operator was renting space, the landlord may try to assign the operator's lease to a new owner. It is advisable to anticipate a possible lease assignment situation to avoid having to renegotiate a lease with a new owner who may not honor the terms of the original lease.

Q. *What is a binder?*

A. A *binder* is an agreement to which money has been attached for the purpose of insuring the good faith or intent of the originating party to complete the purchase or lease of the property. Wireless operators use binders sometimes as a method of securing a facility and location for its antennas prior to the execution of the lease. In other instances, prospective landlords have required binding agreements when they are leasing tower space or have no prior operating experience with the new carrier. Note that if at all possible binders should be avoided because they impact cash flow.

Q. *What is a breach of contract?*

A. A *breach of contract* is simply a violation of the terms and conditions in a contract. An example of a breach of contract is a failure to make timely lease payments for a facility. A breach of contract can result in serious consequences—a rentor, for example, a wireless operator, can be evicted for nonpayment of rent.

Q. *What is a real estate broker?*

A. A *real estate broker* is someone who buys, sells, or leases real estate for another party and who is paid a fee for doing so. In the wireless communications industry, real estate brokers help expedite the build program, and the fees they are paid for their services are paid by the landlord or the seller of the building. In some cases, a landlord may not have had a previous agreement with the broker prior to the lease negotiations. Brokers can be a value to the organization, but sometimes they are used without management's direct knowledge. If a company does not want to use the services of a broker, one way the company can find out if a broker is involved is to check the contact sheet for the site visits.

Q. *What are building codes?*

A. *Building codes* are the regulations that are set forth by the state and local authorities to ensure that the structural and electrical requirements for a given building are met. Each town has their own set of building codes, and these building codes determine what is required when building a site. An example of a building code is the Americans with Disabilities Act that spells out what types of access must exist to and from a site. Another building code concerns whether the site is staffed or unstaffed. If it is staffed, then toilet facilities may be required; if it is unstaffed, they may not be.

Q. *What are bylaws?*

A. *Bylaws* are rules and regulations that govern an association. Some examples of associations are condominiums and coops. One way in which bylaws may affect a wireless facility is that the operator must abide by an association's bylaws if a site is located in a condominium. Any modifications the operator wishes to make to the site must be approved by a vote of the condominium association. Adhering to that requirement can introduce delays in the schedule if the association meets on a monthly basis. In addition, some bylaws prohibit commercial applications within their boundaries.

Q. *What is a certificate of occupancy?*

A. A *certificate of occupancy* (CO) is a piece of paper, or document, that is issued by the local municipality stating that the building or facility complies with the local and state building codes as well as the health and safety codes. The CO permits the facility to be occupied; without it, a person is not allowed to occupy the location, even if he or she is just performing maintenance. This is a critical document, but it is often overlooked by operators since it is obtained after the site is operational.

Q. *What are common elements?*

A. In a condominium or association ownership arrangement, a *common element* is a piece of property or a facility that can be used by all the residents of the association. Two examples of a common element are a tennis court and a roof top. If a wireless operator is working with an association, the operator must evaluate the RF emissions into common elements, or areas. In addition, the wireless operator must be careful to not affect the aesthetic qualities of a common element.

Q. *What is a condominium?*

A. A *condominium,* commonly called a *condo,* is a situation in which the operator owns the unit or apartment outright in addition to owning a share in the common elements. Condominiums are governed by an association.

Q. *What is a contract?*

A. A *contract* is an agreement between two or more parties to do or not to do certain things. In the wireless industry, a contract could be a lease or a purchase agreement.

Q. *What is the cubic-foot method?*

A. The *cubic-foot method* is a means of estimating building cost on a per-cubic-foot basis. This measurement can guide decision making in real estate matters between operators and property owners.

Q. *What are delinquent taxes?*

A. *Delinquent taxes* are unpaid, or past-due, taxes. This is a problem that could affect wireless operators if a building in which their equipment is housed in rented space is owned by a landlord who owes tax payments to the local government. Should the local government foreclose on the property, the wireless operator could be evicted.

Q. *What is a dominant tenement?*

A. A *dominant tenement* is a property that includes in its ownership the right to use an easement through another person's property for a specific purpose. This is an unusual situation and may involve access to and from a property that may be landlocked with no direct egress without the use of the easement. One example could be access to and from a water tank.

Q. *What is an easement?*

A. An *easement* is the right to use someone else's property for a specified purpose. An example of an easement is a right-of-way. Utility companies often use easements through private property. It is important for wireless companies to determine if there are easements in place as needed for utilities, like power and telco, to be delivered to the property. There have been cases in which a separate lease has been signed at a higher rate than the lease for the facility for the sole purpose of allowing power and telco the ability to reach the facility once it has been built.

Q. *What is eminent domain?*

A. *Eminent domain* is the right of a federal, local, or state government, or even a quasi-public body like a utility, to acquire a property for the public's use such as to build railroad facilities. This acquisition is achieved through the condemnation process. Eminent domain has not been asserted in wireless communications situations. However, it does impact wireless operators if their facilities are located on condemned property. Thus, when securing property, it is important to ascertain the government's plans for the surrounding area.

Q. *What is an encroachment?*

A. An *encroachment* is an advance into property owned by someone else. For example, if a fence or hedge extends beyond its property line into another property, that is considered an encroachment. Wireless operators need to be careful that their installations will not encroach on someone else's property.

Q. *What is an encumbrance?*

A. An *encumbrance* is a claim against property that can potentially diminish its value. Some examples of encumbrances are mortgages, liens, easements, or encroachments. The search for encumbrances should be conducted as part of the initial real estate acquisition phase to ensure there are no hidden surprises several months into the process or even after the property is in use for wireless operation.

Q. *What is execution?*

A. An *execution* is the official signing and delivery of a legal document like a lease.

Q. *What is a fully executed lease?*

A. A *fully executed lease* has been signed by both the lessee and lessor with no changes made by either party. It is the legal document that details how and where the property will be used. This document is normally needed prior to obtaining a building permit. There are cases in which a landlord does allow a building permit to be obtained prior to having a fully executed lease; however, every time this has occurred in my experience, the result has been a minidisaster due to the resulting complete lack of room to maneuver during negotiations.

Q. *What is foreclosure?*

A. A *foreclosure* is a procedure that is followed when there is a default on payments and the property that is pledged as security is then sold to satisfy those debts. A foreclosure could affect a wireless operator who had an installation in a rented space in a building that had been pledged as security by the landlord for another business action. It is important to know if the property that is being used is potentially at risk due to other business failures that may occur. It is important to note that a multiyear lease is an asset against which a landlord may borrow money.

Q. *What is a general contractor?*

A. A *general contractor,* or GC, is a construction specialist. For wireless operators, it is the GC who enters into a contract with either the landowner or the lessee for an installation construction project. Sometimes a landlord has specific GCs they prefer to use, and sometimes other constraints exist such as that union (or nonunion) labor must be used on a particular project.

Q. *What is a general lien?*

A. A *general lien* is a situation in which a creditor has the right to have all the debtor's property sold to satisfy the debt owned to him or her. This could be important in wireless operations since a bad business deal made by the landlord could result in the property's being sold.

Q. *What is a gross lease?*

A. A *gross lease* is a lease in which the landlord agrees to pay all the property expenses normally incurred such as taxes and regular maintenance. Wireless operators generally use this type of lease.

Q. *What is a gross rent multiplier?*

A. A *gross rent multiplier* is a method used to determine the value of a property. The value is calculated by multiplying the overall income for the property by the prevailing interest rate.

Q. *What is a ground lease?*

A. A *ground lease,* as its name states, is a lease for the ground only. Thus the tenant owns a building or constructs a building over land owned by someone else. The ground lease is typically used for green-field applications.

Q. *What is an improvement?*

A. In real estate transactions, the term *improvement* means a structure that is erected on a property that improves its value. Some examples of improvements are a fence, driveway, porch, or shelter.

Q. *What is an involuntary lien?*

A. An *involuntary lien* is an attachment that is imposed upon property without the consent of its owner. Some simple examples of involuntary liens are tax liens and a special assessment that can be originated by a condominium association.

Q. *What is a judgment?*

A. A *judgment,* as its name implies, is a formal judgment, or decision, by the courts regarding one person's claims against another person. In the wireless industry, judgments affect whether a building permit is granted or denied.

Q. *What is a junior lien?*

A. A *junior lien* is an attachment that is subordinate to another attachment. The relative subordination of liens is based on their filing dates and types.

Q. *What is a lease?*

A. A *lease,* one of the most-used vehicles for obtaining wireless facilities, is a written, or oral, contract between a landlord and a tenant that transfers the right to use the property for a specific length of time in exchange for rent. The leases that are typically used for wireless installations are written documents to which are attached lease exhibits that specify the details of the site's use and in particular the lease hold improvements that are to take place.

Q. *How many years does a typical lease cover?*

A. A typical lease length is 5 years with three renewable options that are renewable at the discretion of the wireless operator. The initial 5 years plus three 5-year extensions give the lease a potential life of 20 years. Very few leases are signed for less than 5 years due to the extensive investment needed to improve the facility for installation purposes. There are, of course, exceptions to this, but they are very specific in that the facility is not envisioned to be usable in the near future and is due to be replaced with a more permanent location within the specified period of time.

Q. *What is a lessee?*

A. A *lessee* is a tenant. For example, a wireless operator who rents property from someone else is a lessee.

Q. *What is a lessor?*

A. A *lessor* is a landlord. For example, the owner of the property in which a wireless company installs its equipment is a lessor.

Q. *What is a master plan?*

A. A *master plan* is a comprehensive plan that prescribes optimal land-use patterns for a *particular area.* Usually a master plan is developed by the local government so as to guide long-term physical development of the whole community or a given area within it. Wireless operators should consult the master plan for the area in which they plan to install a cell site to make sure that the property they are buying or leasing is not slated for development by the municipality.

Q. *What is a mechanic's lien?*

A. A *mechanic's lien* is an attachment placed on a property by a contractor or material supplier who has provided either labor and/or material for a building on the property. A mechanic's lien can be applied by a contractor to a given property and its landlord if the wireless operator fails to pay their bills to the contractor for legitimate work.

Q. *What is a metes-and-bounds description?*

A. A *metes-and-bounds description* is the legal written definition of the boundaries of a particular property or parcel of land. The description is derived by starting from a defined point and following the boundary of a property, using directions and distances from the deed for the property, finally arriving back at the origin.

Q. *What is a month-to-month tenancy?*

A. A *month-to-month,* or *at-will, tenancy* is an informal rental agreement between a tenant and a landlord. This situation exists sometimes during transition periods. The month-to-month tenancy does not have to be a month-to-month arrangement—it could in fact be a three-month interval.

Q. *What is a net lease?*

A. A *net lease* is a rental contract which stipulates that the tenant pays not only for the rent but also for some, if not all, of the operating costs associated with a property. Additional operating costs for a property could be taxes, utilities, and/or maintenance on repairs. It may not be advisable for wireless operators to sign a net lease because it could be a vehicle with which a property owner may attempt to pass the fully loaded cost of the property onto the operator.

Q. *What is an option?*

A. An *option* is an agreement by a property owner to keep open for a defined period of time the opportunity to either lease or purchase a certain property. Options are commonly used in the wireless industry as a method of securing properties in advance of their actual use. The use of an option is sometimes beneficial during the land-use acquisition or zoning review processes that are prerequisite to obtaining a necessary building permit. Options, however, typically have an expiration or termination date that should be noted when they are written. In several cases a zoning process has been well under way when the wireless operator's option to buy or rent the property expired, causing the facility to be abandoned after much time and effort, not to mention money, had already been expended.

Q. *What is quiet enjoyment?*

A. *Quiet enjoyment,* a term that appears commonly in many leases, is the right of a person, either the owner or lessee, to use the property without interference of any kind. This issue arises for wireless operators as it pertains to the physical asthetics, the antennas, and noise caused by the HVAC system or fans from a wireless cell site. The quiet-enjoyment issue is not normally invoked by the wireless operator but rather by other building tenants.

Q. *What is rent?*

A. *Rent* is the price paid at periodic intervals by a tenant of a property to the landlord for the use of the property. Wireless operators pay rent as agreed upon in the lease with any appropriate cost-of-living adjustments included. The issue of who pays the rent can be daunting for a new operator because it must be handled on a near-automatic basis. In other words, the landlord does not remind the operator to pay the rent.

Q. *What is a right-of-way?*

A. A *right-of-way* is the legal permission for someone to pass over or through someone else's land according to the nature of the easement that has been granted. Wireless operators may need to negotiate rights-of-way with neighboring property owners when they are developing a green-field cell site. A right-of-way may be the only means available to ensure that utilities such as telephone and electricity may be delivered to a particular site.

Q. *What is a setback?*

A. *Setback* is the amount of space a local zoning regulation requires between a lot line and a building line. This regulation is very important to wireless operators who must often place a shelter on a property or improve an existing structure perhaps involving the installation of stairs for entering and exiting the wireless facility.

Q. *What is a specific lien?*

A. A *specific lien* is an attachment that affects only a certain specific parcel and/or piece of property. A specific lien could be filed against a wireless operator by a contractor who is trying to collect for services rendered. This might or might not be an issue with a landlord depending on the situation at hand.

Q. *What is a survey?*

A. A *survey* is the process by which a parcel of land is measured and its area ascertained and shown on a map indicating the boundary measurements. Often a survey is performed before construction begins on a shelter or a cabinet. It is also required that a survey accompany a building permit application, in which case the level of detail required for the survey depends on the local building inspector's requirements.

Q. *What is a tax lien?*

A. A *tax lien* is an attachment filed against a property for its owner's unpaid taxes. A tax lien assessment takes priority over all other liens that might exist on a property. Wireless operators need to look for liens during the initial and later title searches for property they intend to rent or purchase.

Q. *What is a tenant?*

A. A *tenant* is someone who holds or possesses land or tenements by any kind of right or title. A wireless operator paying rent on a lease is a tenant, and he or she is afforded certain rights in that capacity. The rights afforded a tenant are different from state to state and should be checked and noted before leasing a property, with special attention to the rights of corporations as well as how they apply to a corporation as tenants.

Q. *What is a title?*

A. A *title,* with regard to real estate, is evidence that the apparent owner of land is the actual owner. Wireless operators should determine that the person he or she is negotiating with is the actual owner or is an agent of the actual owner. It is possible for an unscrupulous individual to offer property for use and attempt to obtain money when it does not actually belong to him or her. Also, a title will indicate any outstanding liens that have been placed against the property.

Q. *What is a title search?*

A. A *title search* is a process by which the public records are searched with regard to the property in question to determine its ownership and any encumbrances that may exist on it. It is strongly advised that before a wireless operator enters into a five-year lease or a lease with renewable options that he or she conduct a title search. The cost of the title search is trivial in comparison to the financial investment that will be made to use the property.

Q. *What is a tort?*

A. A *tort* is basically a wrongdoing by one person against another for which relief may be obtained in the form of damages or an injunction.

Q. *What is a townhouse?*

A. A *townhouse* is a hybrid form of real estate ownership in which the owner has a fee-simple title to the living unit and the land below it plus a fractional interest in the common elements. Wireless operators should be aware that when a cell site is located on a common element, all the townhouse owners within that community are then involved as compared to a single landowner. Other interesting points about townhouses are that the individual units are normally not allowed to pursue any commercial activities and that any change must go before the whole community for a vote.

Q. *Why does PCS technology require more towers than cellular and SMR technologies?*

A. The reason that PCS technology requires more towers than cellular or SMR is that PCS operates at a higher frequency—about 7 dB more attenuation. This necessitates having two to three times the number of cell sites needed for cellular and SMR systems depending on which engineering department is speaking.

Q. *Can cellular, SMR, and PCS providers colocate on the same tower or roof top?*

A. The short answer is yes they can. There are, of course, technical issues that have to be addressed as well as antenna real estate issues. Some of the technical issues pertain to out-of-band emissions and intermodulation products besides EMF emissions. The primary antenna real estate issues concern the obstruction of the existing or proposed antenna structure by another operator or any necessary expansion of the building's HVAC system.

Q. *Where is information available pertaining to questions that might be asked by citizens regarding radio systems?*

A. The best locations to obtain information with which to answer citizens' questions regarding wireless facilities are trade magazines like *Microwave News* and, of course, the FCC's Web page, which is very good: http://www.fcc.gov.

Q. *How is a lease acquired?*

A. A lease is acquired usually by the real estate department. Some of the steps involved in leasing a property are the following:

- Negotiation terms with landlord.
- Compile lease exhibits.
- Conduct title search.
- Write lease memo if required.
- Define local jurisdiction hierarchy.
- Conduct environmental studies if required.
- Work closely with legal department, and provide them with lease language as needed.
- Obtain signed and executed lease from wireless operator's management and landlord.

7

Radio Frequency

This chapter addresses the most commonly asked questions pertaining to radio equipment and radio frequency. The questions that are posed and answered are those most likely to be asked of or inside the RF engineering department of a wireless communications company. While the number of radio questions that could possibly be asked is vast, the issues covered in this chapter have been chosen because they are the questions most common within the wireless industry.

Q. *What is compression?*

A. The term *compression* is used to describe the 1-dB compression point, which is also called *receiver blocking,* or just *blocking.*

Q. *What is a transceiver?*

A. A *transceiver* is a radio that can both transmit and receive radio signals. A transceiver is found in cell sites as well as in subscriber units.

Q. *What is a radio?*

A. A *radio* is a device that can receive electromagnetic signals. Some radios, *transceivers,* can also transmit electromagnetic signals.

Q. *What is demodulation?*

A. *Demodulation* is the reverse of modulation, and it is performed by a receiver or the receiver portion of a radio system. The method used for demodulation depends on the technology platform utilized in that the demodulation process and circuitry is different for analog and digital systems. Also there are numerous variations in how the demodulation process takes place depending on the type of analog or digital system used. It is important to note that the demodulation process that is used must be the complement of the modulation process used in order to extract the information. For example, if the signal is an analog FM signal, then an analog FM demodulator is needed.

Q. *What is a receiver?*

A. There are many types of *receivers* that are utilized for wireless communication. The type of receiver should be selected according to the way in which it is desired for the information content to be received, which should be accomplished by the most efficient method. The receiver and the receiver system utilized by a wireless system are crucial elements of the network. Specifically, the receiver's job is to extract the desired signal from the plethora of other signals and noise that exist in its environment. A basic receiver block diagram is shown in Fig. 7.1 for reference.

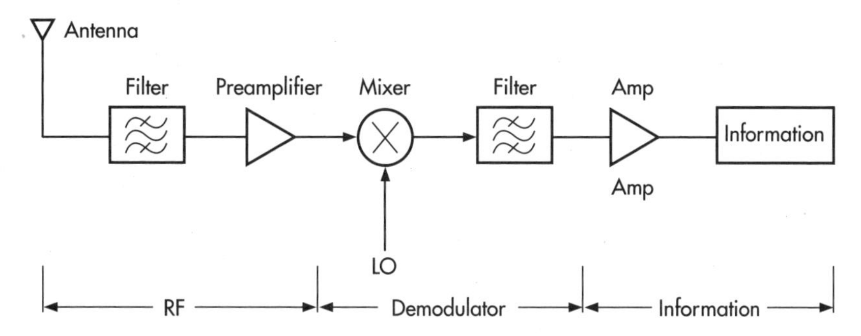

Figure 7.1 Basic radio receiver.

Q. *What are the components of a receiver?*

A. The components of a receiver are shown in the block diagram in Fig. 7.1. The components that make up each of the block diagrams is different depending on the modulation format that is desired to be extracted as well as the frequency and bandwidth of operation.

Q. *What are some receiver design issues?*

A. There are numerous receiver design issues that need to be taken into account when specifying receiver requirements. Some of the top electrical performance and cost drivers for a receiver are the following:

1. Frequency range
2. Dynamic range
3. Phase noise
4. Tuning resolution
5. Tuning speed
6. Sensitivity
7. Distortion (gain and phase)
8. Noise
9. Form
10. Size
11. Cost

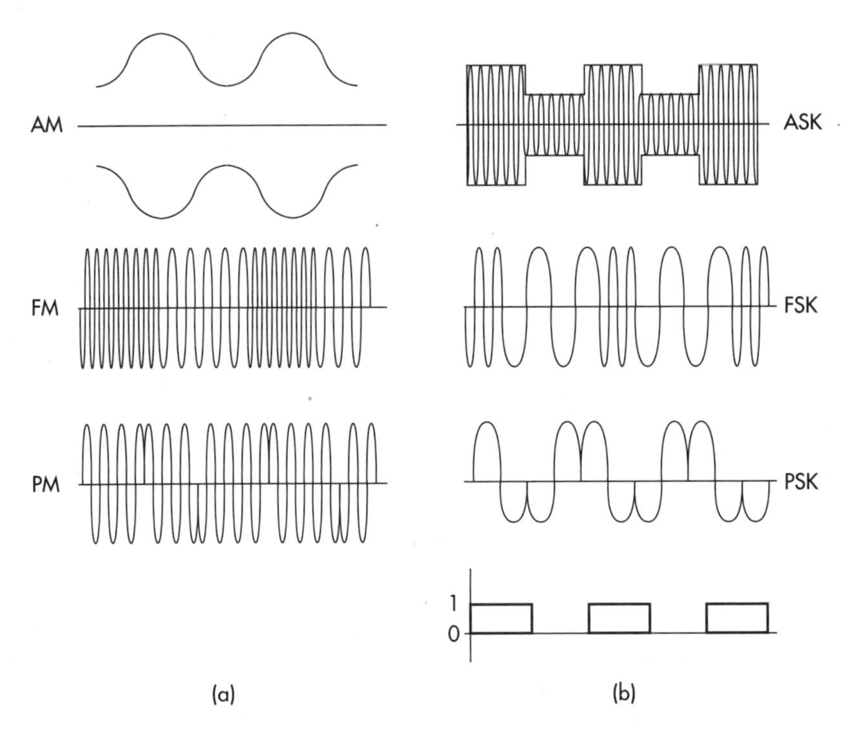

Figure 7.2 Modulation formats. (*a*) Analog. (*b*) Digital.

Q. *What types of modulation are commonly in use?*

A. There are numerous types of modulation formats that are used commonly. However, all modulation formats are derived from one or a combination of the three most basic forms of modulation. The modulation types are shown in Fig. 7.2 for reference.

Q. *What types of receivers are commonly in use?*

A. There are three general types of receivers: amplitude modulation (AM), frequency modulation (FM), and phase modulation (PM). Many receivers are a combination of the three types of basic receiver elements, and they are assembled based on the technology platform utilized for the wireless system.

Q. *What is an amplitude modulation receiver?*

A. The most basic type of radio receiver is the *amplitude modulation* (AM) *receiver,* which utilizes, needless to say, amplitude

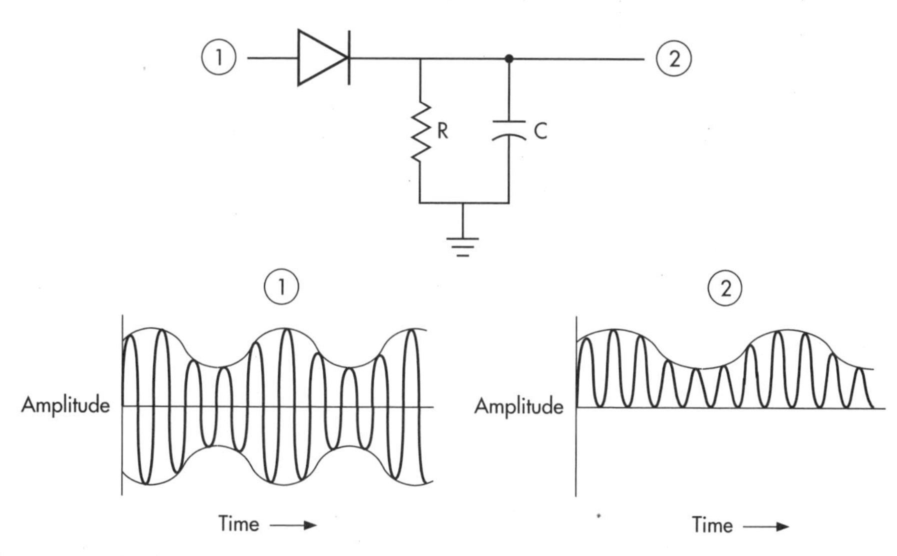

Figure 7.3 Envelope detector. $f_c = 1/(2\pi RC)$.

demodulation. There are many variants in the way an AM signal can be demodulated; however, the most basic method relies on an envelope detector. The envelope detector, as shown in Fig. 7.3, is extremely easy to construct.

Q. *What is frequency modulation?*

A. *Frequency modulation* (FM or *angular modulation*) is one of the more common forms of radio system design. FM is utilized extensively for cellular communication, and it is often referred to as "analog," although that term is only partially correct in this context. Utilization of FM for a radio transmitter and receiver has the advantages of being more noise immune, but this comes at the expense of equipment complexity.

There are many types of radio systems that use FM for transporting the information content. Some of the communication systems utilizing FM are AMPS, TACS, NAMPS, SMR, and various forms of two-way communication systems besides commercial FM radio services. The usual bandwidths for an FM system can range from 200 kHz for a commercial FM broadcast system to 30 kHz for AMPS all the way to 10 kHz for NAMPS. Again the bandwidth that is utilized for the system depends on numerous factors

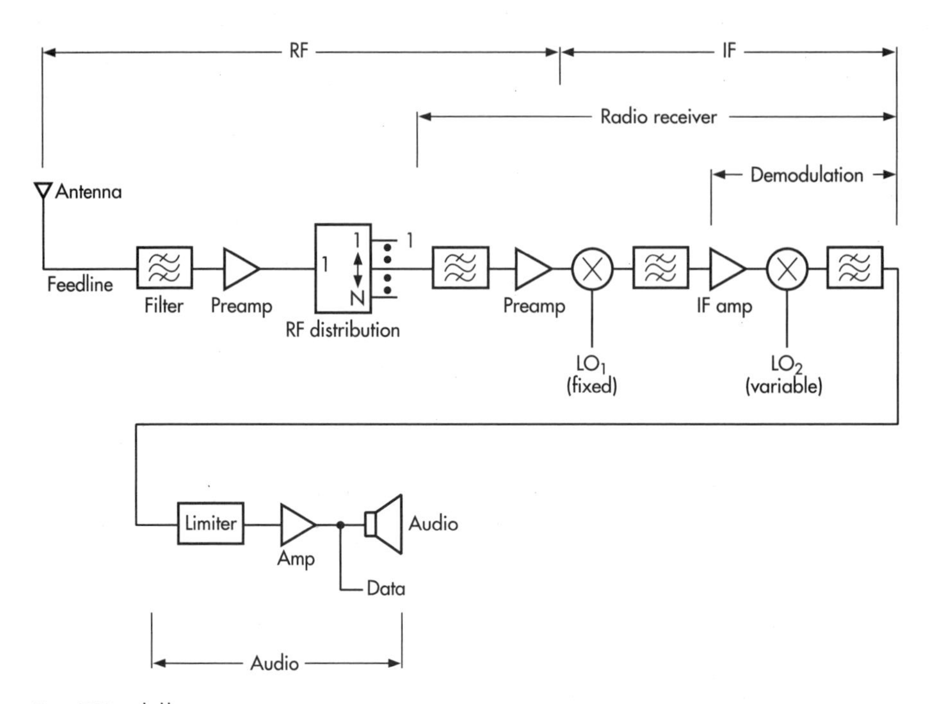

Figure 7.4 Basic double conversion receiver.

including available spectrum and information content desired to be sent and received. The available spectrum and information content have a direct impact on the cost of the system.

An FM receiver configuration is shown in Fig. 7.4 for reference.

Q. *What is phase modulation?*

A. *Phase modulation* (PM) *receivers* offer many advantages over pure AM or FM receivers. Specifically, the PM receiver is able to be more spectral efficient due to its modulation method. PM lends itself to digital communication and is one of the primary communication formats for IS-54 and IS-136. Other variants to PM modulation are used for CDMA systems. A generalized block diagram of a PM receiver is shown in Fig. 7.5.

Q. *What is a preamplifier?*

A. A *preamplifier* is usually the first active component in a communication systems receive path. The basic function of any RF pre-

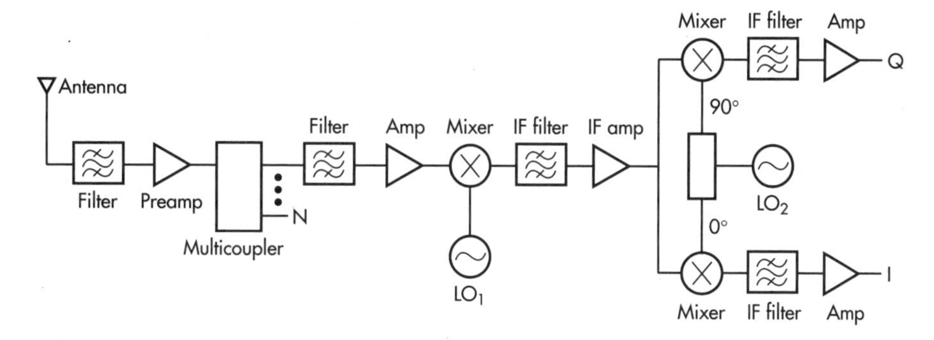

Figure 7.5 PM receiver.

amplifier is to increase the signal-to-noise ratio of the received signal. The preamplifier receives the desired signal at the lowest level in any of the receive stages for the communication site. Since the RF preamp receives the desired signal at the lowest level of any receive stage in the cell site's receive path, any noise or other disturbances introduced in this stage has a proportionally greater effect. (See Fig. 7.6.)

Q. *What is a low-noise amplifier?*

A. A *low-noise amplifier* (LNA) is any amplifier that has a low noise figure, usually less than 3 dB, although this figure is not defined by industry. The low-noise amplifier is also called an LNA or a *preamplifier.*

Q. *What is a multicoupler?*

A. A *multicoupler* is a device which ensures that received signals are routed to the appropriate receivers. The multicoupler enables one receive antenna to be connected to many radios. The multicoupler itself normally has the preamplifier included as part of the configuration. (See Fig. 7.7.)

Q. *Where is a multicoupler used?*

A. A multicoupler is used in the receive path for a cell site between the antenna system and the radio system. In a Lucent cell site the antenna interface frame (AIF) serves as the multicoupler.

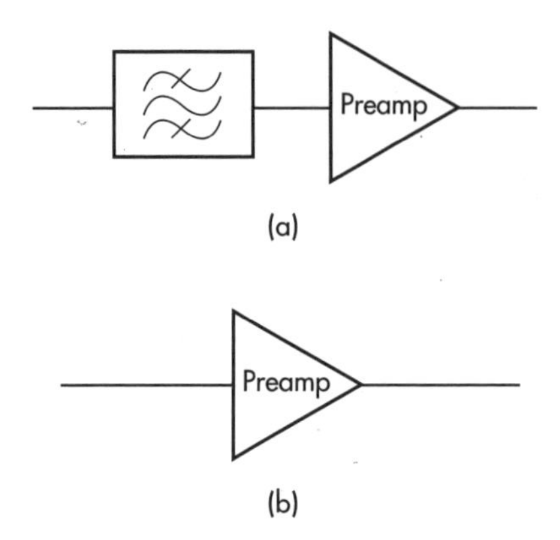

Figure 7.6 Preamplifier. (*a*) Filtered. (*b*) Nonfiltered.

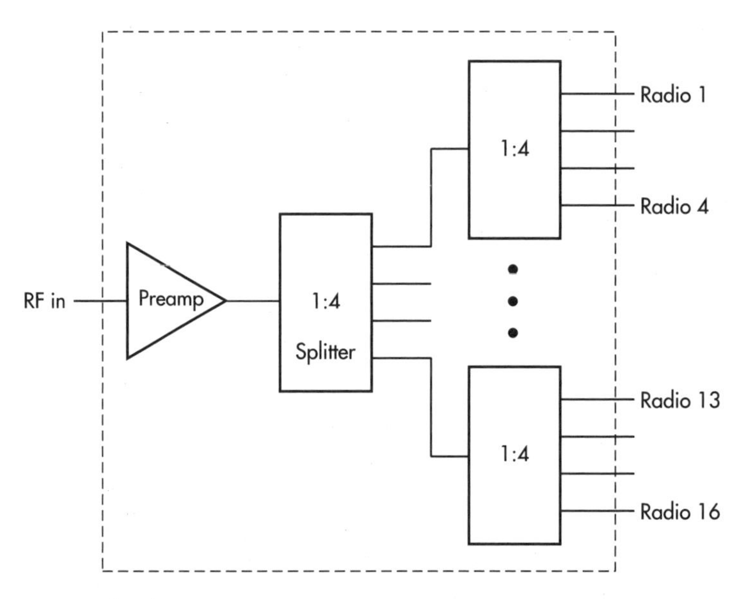

Figure 7.7 Multicoupler.

Q. *What is a radio receiver?*

A. The *radio receiver* is the actual physical device that converts the RF energy into a usable form. The radio receiver can have one or multiple receive paths connected to it. Usually there are two paths connected to the radio receiver in a cell site and only one path for a mobile or portable unit. (See Fig. 7.8.)

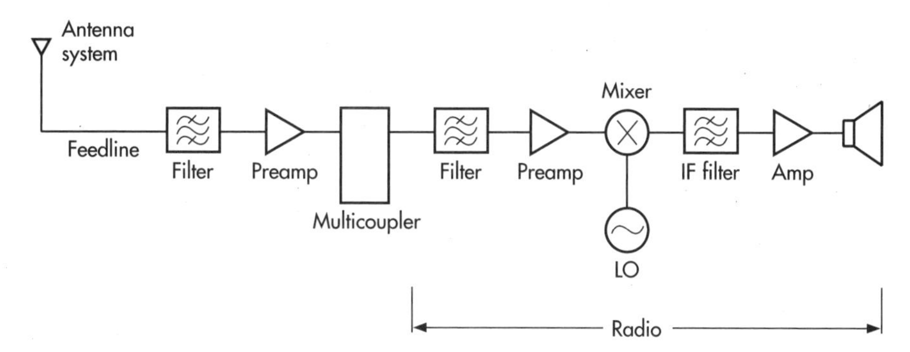

Figure 7.8 Radio receiver system.

Q. *What is a preamp?*

A. A *preamplifier* (or *preamp*) for the radio is the first amplification stage in the actual radio itself. The purpose of the amplifier in the receive path of the radio receiver is to help set the noise figure and sensitivity of the radio itself. The preamp is also called an LNA.

Q. *What is a down converter?*

A. A *down converter* located in a receiver is a mixer. The purpose of the mixer is to down convert the incoming RF spectrum containing the information content into an IF output ideally without adding noise or intermodulation products to the initial signal.

Q. *What is a mixer?*

A. *Mixers* are sometimes described in textbooks as "multipliers." A mixer normally has three ports, and it is a vital component in either up or down converting the information into another frequency. For a receiver the mixer can either up or down convert the frequency; however, the standard method that is utilized is to down convert the frequency so that it can be processed at an intermediate frequency band instead of the RF band. For a transmitter the mixer will normally up convert the signal in frequency.

Q. *What is a multiplier?*

A. A *multiplier,* also called a *mixer,* is found in a radio system used for either up or down conversions.

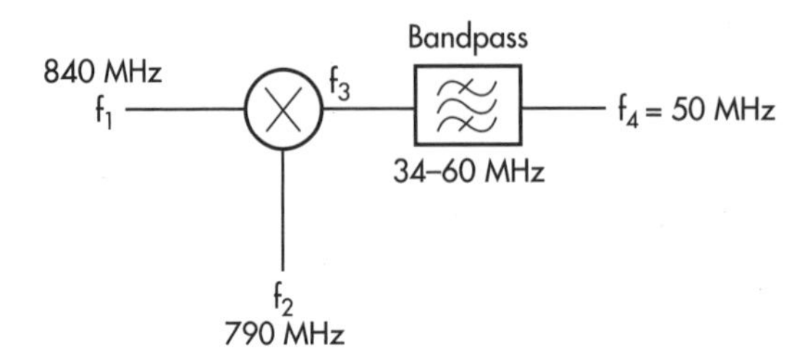

Figure 7.9 Receiver mixer. f_1 = source frequency, f_2 = local oscillator, f_3 = resultant frequency, f_4 = intermediate frequency.

Q. *How does down converting take place?*

A. An example of the down-converting process is shown in Fig. 7.9. In the diagram a source signal is coming into the mixer, which is f_1 and is represented by the signal entering the mixer on the left. The source signal is combined with the second frequency f_2, which is the local oscillator frequency. The combination of the two frequencies is the output f_3. It is interesting to note that the mixer has many products as a result of the two signals' combining. To remove the undesired products, the output of the mixer f_3 is then passed through a bandpass filter. A low-pass filter could also be used for this postprocess effort, however, the bandpass filter will help reduce any unwanted noise energy that could be present. The output frequency is the resultant frequency and is some lower frequency usually referred to as an *intermediate frequency* (IF) and is represented here as f_4. The IF has only one component output from the down-converting process in this example.

Q. *What is the intermediate frequency?*

A. The *intermediate frequency* (IF) is part of the radio system's path either for the receiver or transmitter. Normally the term an "IF for a radio" means the receiver portion of the radio.

An intermediate frequency is best described as the sign that is the result of the down conversion or part of the up-conversion process. For example, if the radio were to transmit at, say, 893 MHz, then a possible IF could be 23 MHz.

Q. *What is the IF stage for a radio?*

A. The *IF stage* is an important part of the receiver chain because this is where the postprocessing of the information takes place. Most of the amplification in the receiver takes place at the IF level. The IF stage can take on several variants depending on the technology used. However, the basic premise is the same in that the signal is at a lower frequency range where it can be post-processed more easily.

Q. *What is IF selectivity?*

A. *IF selectivity* is probably the most, or one of the most, important specifications for a receiver. To prevent interference between channels at the receiver, IF selectivity is used to obtain the necessary interference protection. The selectivity of an IF section is a measure of the total response of all the IF stages, if there are several involved with the process. The selectivity of the receiver must be sufficient to allow the desired modulated signal to be amplified uniformly across the desired band while rejecting all the unwanted energy. The selectivity of a receiver is usually defined in terms of its Q.

Q. *Why is diversity used in a radio?*

A. Diversity, when it is applied at the radio receiver, can minimize the negative effects of signal fading that often occur in a mobile environment. It is important to note that the smaller the bandwidth for the information content to utilize, the more susceptible to fading a communication system will be. To eliminate some of the negative effects of signal fading, two diversity techniques can be used for a radio receiver: max ratio combining and select.

Q. *What is the difference in the application of the principle of diversity for antennas versus radios?*

A. To achieve diversity in an antenna system, the antennas are positioned such that they will receive signals from the source by two distinct and somewhat correlated paths. In contrast, to achieve diversity in a radio receiver, the radio compares the signals brought by the antennas and then either selects the best signal (select diversity) or combines the signals (max ratio combining).

Q. *What is diversity combining?*

A. *Diversity combining* is the use of several methods to improve the quality of the received signal at the receiver independent of the antenna diversity scheme. (Note, however, that the diversity scheme utilized for the antenna has a direct impact on how effective the diversity-combining technique will be.) There are two diversity-combining techniques discussed here: max ratio combining and select have advantages and disadvantages.

Q. *What is max ratio combining?*

A. *Max ratio combining* (or *max ratio*) is a technique by which the received signal from one branch is combined with the signal from another branch to increase the overall signal level. The number of branches used for max ratio combining is normally two, but it can at times be three. A diagram of IF–level max ratio combining is shown in Fig. 7.10.

Q. *What is diversity gain?*

A. *Diversity gain* is the actual gain in decibels that results from using a diversity scheme in the receiver. The typical two-branch diversity receive system has a maximum gain of 3 dB while a three-branch system has a maximum gain of 4.7 dB. The gain experienced is achieved only when using max ratio combining.

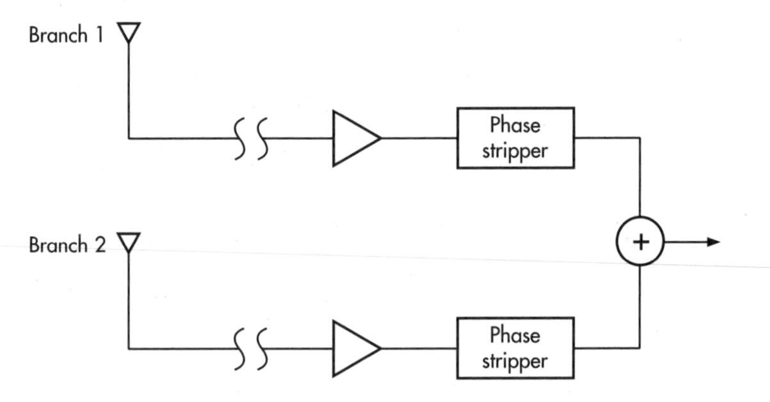

Figure 7.10 Max ratio combining.

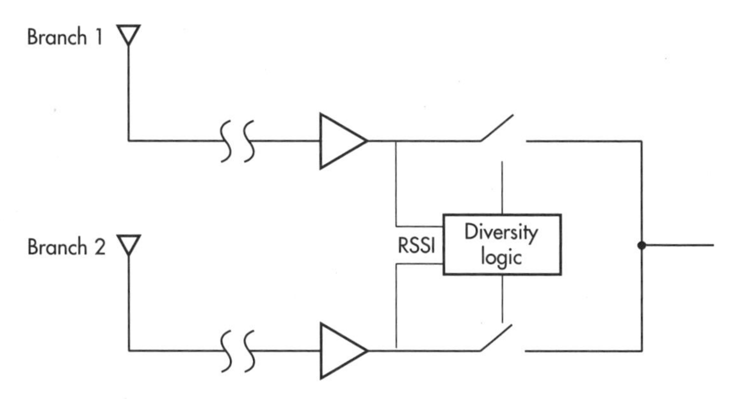

Figure 7.11 Select diversity.

Q. *What is select diversity?*

A. *Select diversity* is a technique by which the branch with the strongest received signal strength indication (RSSI) in the receive system is chosen as the path for eventual demodulation and use by the receiver itself. The selection process is continuous and is meant to overcome the fades that take place in the environment. The number of branches used for select diversity is usually two, but it can at times be three. The number of branches used is driven by cost since most of the overall improvement that can be achieved is accomplished with two branches. A diagram of a select diversity scheme is shown in Fig. 7.11.

Q. *By what criteria is radio receiver performance evaluated?*

A. Radio receivers are evaluated to determine if they can successfully extract information content in a usable form within certain environmental conditions. The criteria that are used to judge a radio receiver's performance are listed below for reference:

- Sensitivity
- Selectivity
- Dynamic range

Q. *What is a minimal discernible signal?*

A. A *minimal discernible signal* (MDS) is a measure of the weakest signal level, in decibels, milliwatts, or decibels, microvolts, that a receiver can detect from background noise. The MDS is a measure of sensitivity, which incorporates the bandwidth of the system.

Q. *What is sensitivity?*

A. *Sensitivity* is the responsiveness of a receiver to a weak radio signal. A receiver's sensitivity is important in the overall performance of the receiver itself and the entire network. Sensitivity is also a critical element in the link budget for any system. The relationship for receiver sensitivity is defined below:

$$\text{Sensitivity} = 10 \log_{10} (kTB) + 10 \log_{10} (\text{bandwidth, Hz}) + NF$$

$$f_c = 840 \text{ MHz} \qquad T = 25°\text{C} \qquad k = 1.38 \times 10^{-23} \text{ J/K}$$

$$10 \log_{10} (kTB) = -174 \text{ dBm/Hz}$$

$$\text{Sensitivity} = -174 \text{ dBm/Hz} + 10 \log_{10} (\text{bandwidth}) + NF$$

where k = Boltzmann's constant; T = temperature, K; and B = bandwidth, Hz.

Q. *How is receiver sensitivity calculated?*

A. A calculation for receiver sensitivity is shown below for both an AMPS voice channel and a CDMA channel. The AMPS channel is 30 kHz wide, and the CDMA channel is 1.23 MHz wide.

$$\text{Sensitivity}_{\text{analog (30 kHz)}} = -174 \text{ dBm} + 10 \log_{10} (30 \text{ kHz}) + 4$$
$$= -125.22 \text{ dBm}$$

$$\text{Sensitivity}_{\text{CDMA (1.23 MHz)}} = -174 \text{ dBm} + 10 \log_{10} (1.23 \text{ MHz}) + 4$$
$$= -109.10 \text{ dBm}$$

Q. *What is selectivity?*

A. *Selectivity* is the receiver's ability to reject unwanted signals. The degree of selectivity is largely driven by the filtering system within the receiver, which protects the radio from off-channel interference. The IF portion of the receiver affords the most selectivity. The greater the selectivity, the better the receiver is able to reject unwanted signals from entering into it. Note, however, if the receiver is too selective, it may not pass all the desired energy.

Q. *What is the dynamic range?*

A. *Dynamic range* is the range over which an accurate output will be produced in a receiver. The lower level is called the *sensitivity level,* and the upper level is called the *degradation level.* The sensitivity level is determined by the noise figure, the IF bandwidth, and the method of processing. However, the degradation level is determined by whichever of the components in the receiver reaches its own degradation level first. Therefore, it is important to understand all the components in the receiver path because the first component to degrade will define the dynamic range of the system. Figure 7.12 is a chart that can be used to determine the dynamic range of a radio system.

Q. *What is the spurious free dynamic range?*

A. The *spurious free dynamic range* (SFDR) is a measurement of the performance of the radio as the desired signal approaches the noise floor of the receiver, providing an overall receiver SNR or

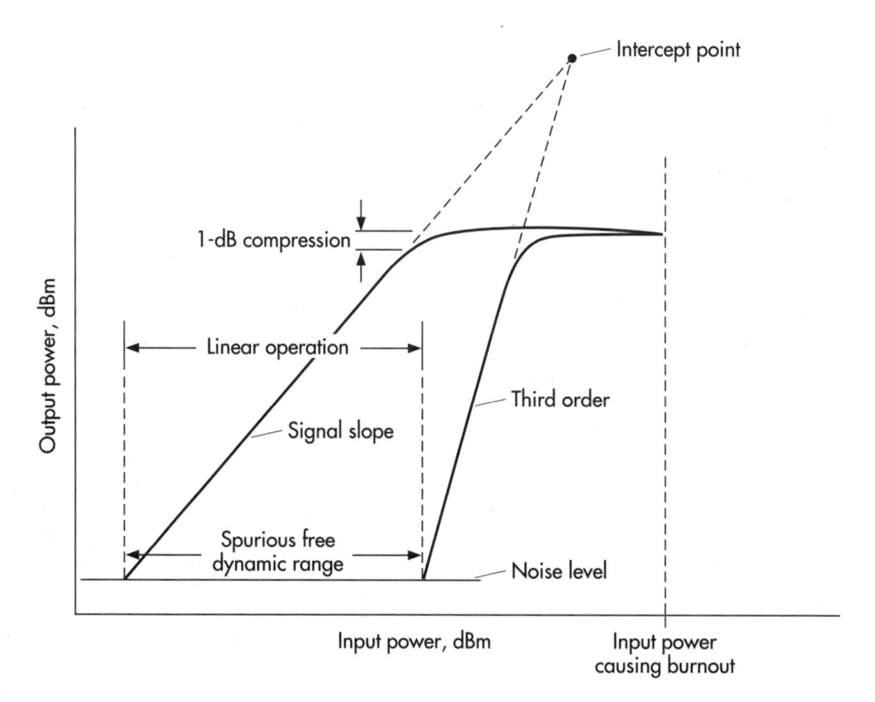

Figure 7.12 Dynamic range.

BER. For example, if a radio can accurately digitize signals from −13 to −104 dBm in the presence of multiple signals, the DR = −91, and it implies an SFDR of 95 to 100 dB.

SFDR is a very important specification because, when the site is near other radio transmitters, the SFDR is a direct indication of how much the signal interferes with adjacent channels.

Q. *What is the signal-to-noise ratio?*

A. The *signal-to-noise (S/N) ratio* (SNR), expressed in decibels, describes a receiver's performance in the presence of noise. Note, however, that the signal-to-noise ratio does not include distortion products. To determine the signal-to-noise ratio, the example below may be followed. If $S = -100$ dBm and $N = -117$ dBm, then:

$$\frac{S}{N} = 17 \text{ dB}$$

Q. *What is the signal-to-noise and distortion?*

A. The *signal-to-noise and distortion* (SINAD) is a ratio, expressed in decibels, that describes the receiver's performance in the presence of noise. The SINAD is usually a reference value used in determining a receiver's sensitivity—that is, 12 dB SINAD. The SINAD is a measure of the voltage ratio expressed in decibels at the audio output of the receiver. The equation that is used to determine the SINAD is shown below:

$$\text{SINAD} = 20 \log \left(\frac{\text{signal} + \text{noise} + \text{distortion}}{\text{noise} + \text{distortion}} \right)$$

Q. *What is C message weighting?*

A. *C message weighting* is one of several methods used to measure the SINAD. C message filtering has a frequency range of between 300 and 3000 Hz. Therefore, when measuring for SINAD without C message weighting, any signaling tones that are used in radio communication could distort the readings.

Q. *What causes signal distortion?*

A. Signal distortion is caused by unwanted signals that appear at the output of any device in the RF path. A common place for distortion to occur is at the output of an amplifier. Note that for purposes of this discussion, the distortion is that which occurs at the preamp for the communication site.

Q. *What is intermodulation distortion?*

A. *Intermodulation distortion* (IM or IMD), the most common form of distortion, is the product of several signals mixing together. The amount and levels of intermodulation distortion are a direct result of the number of signals that are available to be mixed at any location. Intermodulation products for a second- and third-order mix for two signals are shown below in Example 7.1.

Example 7.1

$$\left.\begin{array}{l} A + B \\ A - B \\ B - A \end{array}\right\} \text{Second order}$$

$$\left.\begin{array}{l} 2A + B \\ 2A - B \\ 2B + A \\ 2B - A \end{array}\right\} \text{Third order}$$

$\therefore$ if $A = 925$ MHz, $B = 870$ MHz

Order		MHz
$A + B$	$925 + 870$	1795
$A - B$	$925 - 870$	55
$B - A$	$870 - 925$	NA
$2A + B$	$2(925) + 870$	2720
$2A - B$	$2(925) - 870$	980
$2B + A$	$2(870) + 925$	2665
$2B - A$	$2(870) - 925$	815

Q. *What is capture?*

A. *Capture,* a process used in FM receivers, is the complete suppression of the weaker signal by the strongest of several cochannel signals. For example, FM capture is experienced every day in car radios when the radio is receiving signals from two different radio stations on the same channel. When one of the radio channels is dominant, or captured, the other is not heard, but during that transition period, both can be heard because neither signal has yet been captured by the receiver.

Q. *What is noise?*

A. In a radio system, *noise* is a form of signal degradation, and it is unwanted. Noise occurs in the forms of *short* and *thermal noise.* Noise should not be confused with interference since they are different, although they both degrade the receiver's ability to perform at its optimal level.

Q. *How does noise affect a radio system?*

A. Noise directly affects a communication system's overall performance. All receivers need to have a certain *C/I* or E_b/N_o value to perform properly. If the overall noise that the receiver experiences or has to deal with increases, the desired signal needs to be increased in strength, without increasing the noise content, to ensure the proper ratio is maintained.

Q. *What is a noise figure?*

A. A *noise figure* is a fundamental measure of a receiver's performance. This measurement should be made at a predetermined location for the receiver itself. A noise figure for a receiver degrades—that is, increases—with each successive stage in the receive path. A common point at which to measure the noise figure is the audio output for a receiver, but in digital radios, there is no audio output and the measurement point is then the IF output.

Q. *What is the equation for noise figure?*

A. The equation used to calculate the noise figure is shown below. Please note that the noise factor (F_N) values are expressed as linear, not log:

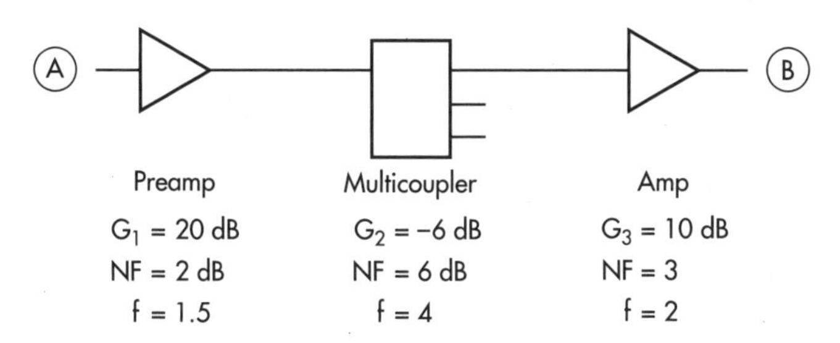

Figure 7.13 Noise representation.

$$\text{Noise figure, dB} = 10 \log (F_N)$$

$$\text{Noise factor} = F_N = \frac{S_{in}/N_{in}}{S_{out}/N_{out}}$$

Q. *How is the noise figure calculated?*

A. The noise figure of the receive system is directly related to the overall receiver sensitivity. The noise figure for the system is calculated as shown in Fig. 7.13 and is normally set by the first amplification stage in the receive path.

Example 7.2

$$f_B = f_1 + \frac{f_2 - 1}{G_1} + \frac{f_3 - 1}{G_1 \cdot G_2}$$

$$= 1.5 + \frac{4 - 1}{100} + \frac{2 - 1}{(100)\,(.25)}$$

$$= 1.5 + \frac{3}{100} + \frac{1}{25}$$

$$= 1.57$$

$$NF_{A \to B} = 10 \log_{10} (f_B) = 1.958 \text{ dB}$$

Q. *How do you calculate noise temperature?*

A. Thermal noise is a direct result of kinetic energy and is directly related to the temperature. Noise temperature is represented with respect to thermal noise, which is expressed in terms of decibels per hertz. The thermal noise is therefore often expressed as noise power per hertz:

$$kTB = (1.38 \times 10^{-23}\ \text{J/K})\ (290\ \text{K})\ (115)$$
$$= 4.002 \times 10^{-21}\ \text{J/s} = 4.002 \times 10^{-21}\ \text{W}$$

$$\text{Noise power per hertz} = 10 \log_{10}\left(\frac{4.002 \times 10^{-21}\ \text{W}}{10^{-3}\ \text{W}}\right)$$
$$= -173.97\ \text{dB/Hz}$$

where k = Boltzmann's constant, 1.38×10^{-23} J/K; T = temperature, K (290°); and B = bandwidth, Hz.

Q. *What is the noise figure equation?*

A. The noise figure equation is shown below. Please note that the values are linear, not log:

$$F = F_1 + \frac{F_2 - 1}{G_1} + \frac{F_3 - 1}{G_1 \times G_2} + \cdots + \frac{F_x - 1}{G_1 \times G_2 \cdots G_x}$$

Q. *What is the relationship between the receiver sensitivity and the noise figure?*

A. The relationship between the receiver sensitivity and the noise figure is best shown in the following equation:

$$\text{Sensitivity, optimal} = -174\ \text{dBm/Hz} + 10 \log_{10}(\text{receiver bandwidth, Hz}) + \text{noise figure}$$

However, a rule of thumb is that for every decibel of improvement in the noise figure for the first stage of the receiver system, the equivalent decibel improvement is ported to the receiver's sensitivity.

Q. *What is the relationship between receiver sensitivity, receiver bandwidth, and receiver noise figure?*

A. The relationship between receiver sensitivity, receiver bandwidth, and receiver noise figure is best shown using the following example of an AMPS or TDMA system as compared to a GSM and CDMA system.

EXAMPLE 7.3

$$N = 10 \log_{10} kTB$$

where k = Boltzmann's constant, 1.38×10^{-23} J/K; T = temperature, K (290°); and B = bandwidth, Hz.

$$\text{Thermal noise} = -174 \text{ dB/Hz } (290 \text{ K})$$

$$\text{Sensitivity (optimal)} = -174 \text{ dB} + 10 \log_{10}(B) + NF$$

BW	30 kHz, dB	200 kHz, dB	1.25 MHz, dB
Bandwidth (dB) noise	44.77	53.01	60.96
$-174 + 10 \log_{10}(B)$	−129	121	113
NF	1.95	1.95	1.95
	127.05	119.05	111.05

The biggest contributor to the sensitivity of a receiver is the bandwidth of the signal. The narrower the bandwidth, the more sensitive the receiver can become. However, the reduced bandwidth also means that there is a higher potential for fades.

Q. *What is the 1-dB compression point?*

A. The *1-dB compression point* is where the power gain for the receiver is down 1 dB from the ideal gain. That is, if the input signal goes up by 2 dB and the output goes up by only 1 dB, then this is the point at which the 1-dB compression occurs.

The *1-dB compression point* is commonly used to define the performance of a particular receiver, or, rather, amplifier, in the receiver itself. When the "1-dB compression point" is a reference to the entire receiver, however, the 1-dB compression point can mean an individual amplifier, like the preamp.

Note that the 1-dB compression point is also referred to as the *blocking point* for the receiver. The blocking occurs in that the weaker signals are not amplified properly, possibly leading to their being blocked from being detected. The 1-dB compression point is

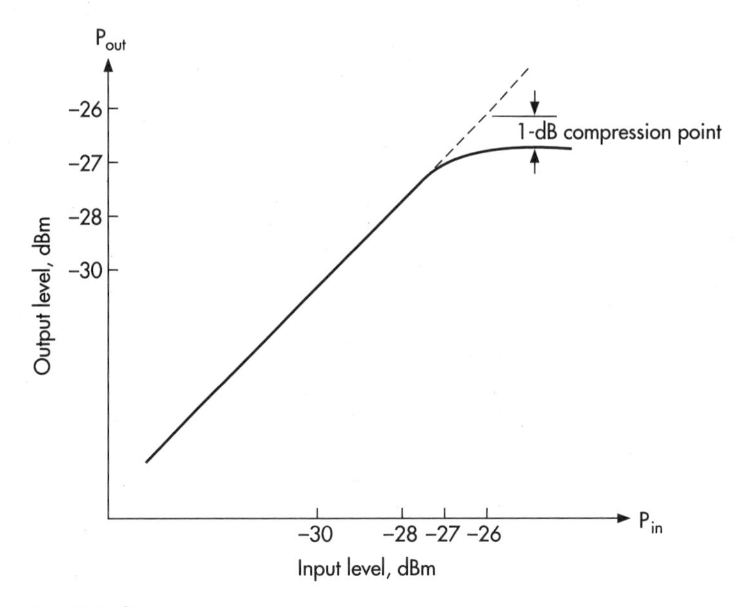

Figure 7.14 1-dB compression point.

part of the component for determining the receiver's overall dynamic range. The 1-dB compression point is shown in Fig. 7.14.

Q. *What is blocking?*

A. The term *blocking* is another means of describing the 1-dB compression point for a radio amplifier. The term *blocking* is also used to describe the denying of subscriber access to a system because it is already in use.

Q. *What is the third-order intercept point?*

A. The *third-order intercept point* (IP3) is a figure of merit for a receiver. Specifically, the IP3 value directly determines the receiver's dynamic range. Furthermore, the third-order intercept value combined with the second-order intercept (IP2) determine the receiver's linearity.

It should be noted that the IP3 value is a theoretical value, and it is achieved by extrapolating the third-order curve. In addition, the IP3 value is frequency dependent, and based on the selection of the test frequencies, a different IP3 value may be achieved.

However, most IP3 tests are set up to produce a product that falls within the first IF of the receiver.

Basically the IP3 value increases by 3 dB for every 1-dB increase for the desired signal. The slope of the IP3 line is therefore 3:1, as shown in Fig. 7.15. The IP3 value is important for determining the receiver's performance because the presence of two larger signals can generate spurs caused by nonlinearities, or rather distortions, that can override the weaker desired signals.

Q. *What is phase noise?*

A. *Phase noise,* measured in decibels, is a type of signal degradation that is caused by the receiver's phase and frequency perturbations

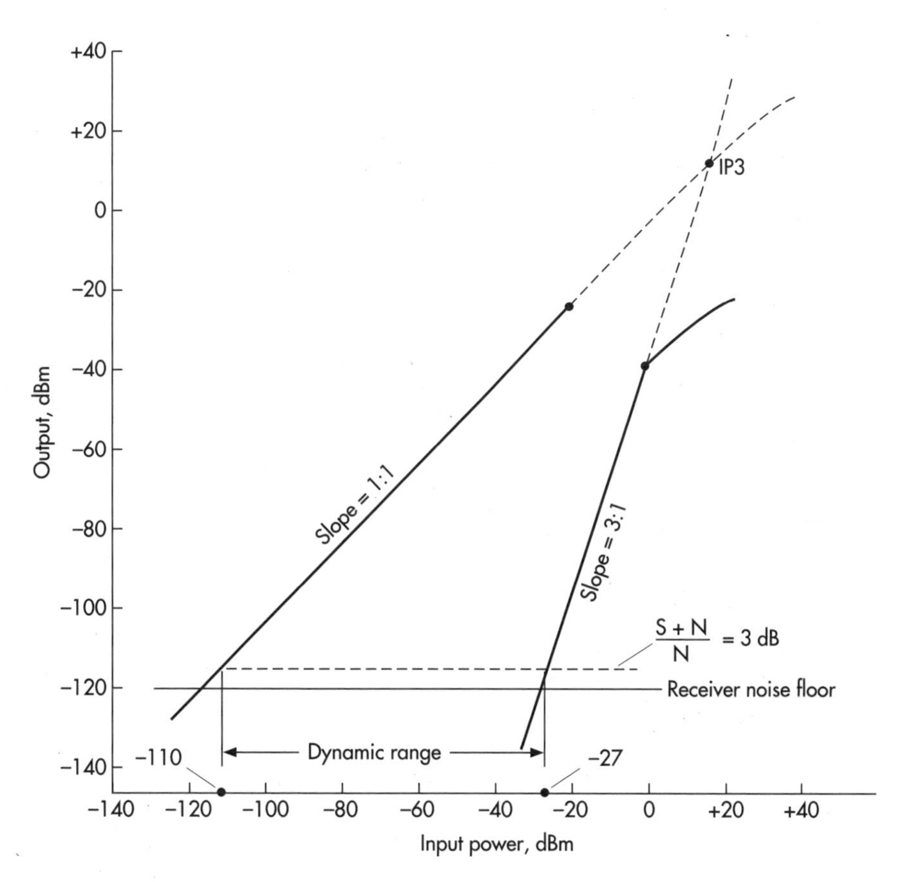

Figure 7.15 Dynamic range = −110 − (−27) = 83 dB.

that are added to the initial input signal. The effect of phase noise is to distort the initial signal so that it either degrades the initial signal or degrades the receiver's sensitivity.

Q. *What is desense?*

A. *Desense* is technically a reduction in the receiver's overall sensitivity, which is caused by artificial or natural *radio frequency interference* (RFI). Desence occurs when a very strong signal begins to overload the front end of the receiver and makes the detection of weaker signals more difficult. Any form of RF energy, whether it is artificial or natural in origin, has the potential to adversely affect the receiver.

Q. *What is radio frequency interference?*

A. *Radio frequency interference* (RFI) is any RF signal that adversely affects the receiver's ability to perform properly.

Q. *What is modulation?*

A. *Modulation* is the process of incorporating information onto an RF carrier for the purpose of conveying voice and data information from one location to another. Without being able to physically connect them together, it is necessary to send the information by another method. There are three primary methods for modulating a signal: AM, FM, and PM. Most communications systems utilize a combination of these three to achieve their modulation format. It is important to note that the modulation takes place at the sending portion, the transmitter, and the associated demodulation takes place at the receiver.

Q. *What are the basic elements of a radio system?*

A. The general components of a radio system are shown in Fig. 7.16.

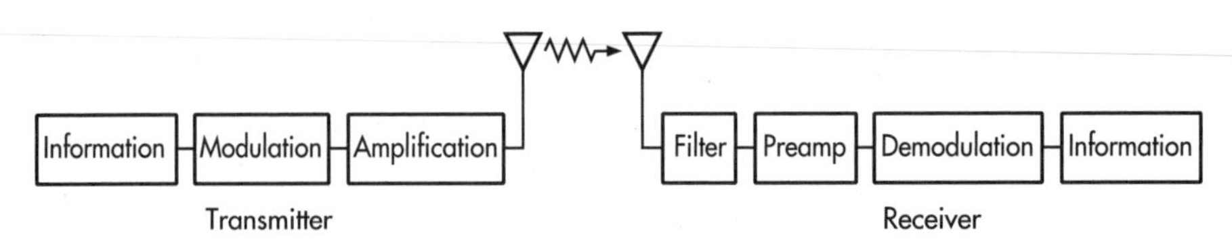

Figure 7.16 Basic radio system.

Q. *How is the modulation method chosen?*

A. The choice of modulation and demodulation methods to be utilized for a radio communication system is directly dependent upon the information content desired to be sent, the available spectrum to convey the information, and the cost. The fundamental goal of modulating any signal is to obtain the maximum spectrum efficiency, or rather information density per hertz.

Q. *What is the difference between AM, FM, and PM modulation?*

A. Figure 7.17 highlights the differences between *amplitude modulation* (AM), *frequency modulation* (FM), and *phase modulation* (PM) in terms of their impact to the electromagnetic wave itself.

Q. *Where is amplitude modulation used?*

A. Amplitude modulation has many good qualities; however, this form of communication is not utilized in cellular communication

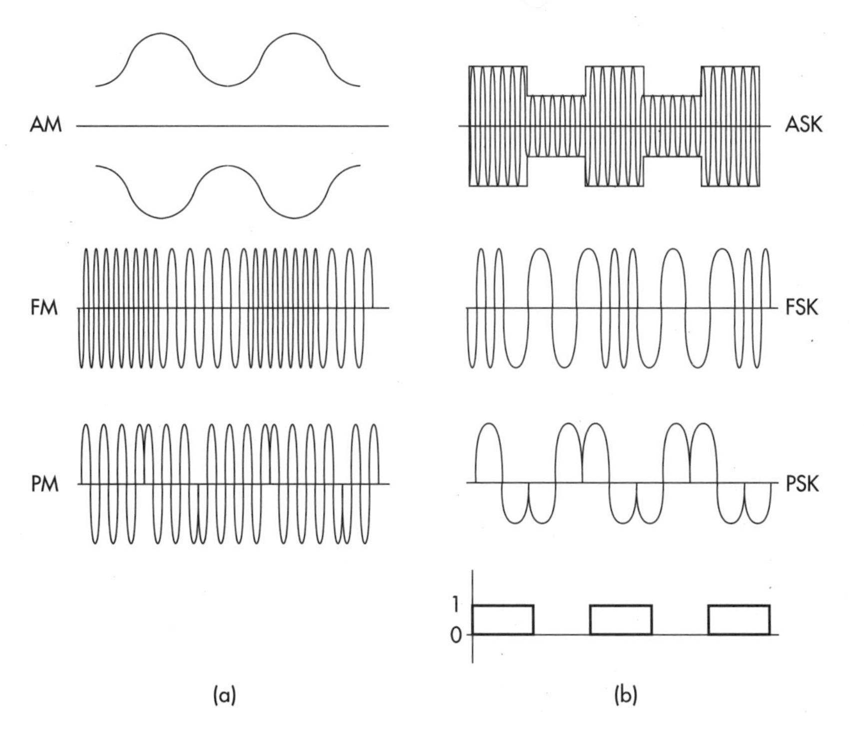

Figure 7.17 Modulation formats. (*a*) Analog. (*b*) Digital.

primarily because it is more susceptible to noise. AM is used for commercial broadcast purposes. However, a form of amplitude modulation, *quadrature amplitude modulation* (QAM), is used for iDEN systems.

Q. *Where is frequency modulation used?*

A. Frequency modulation is utilized for AMPS and TACS analog communication. FM is utilized for analog cellular communication because it is more resilient to interference. Cellular systems utilize a channel bandwidth of 30 kHz for FM. GSM also utilizes a form of FM, referred to as GMSK.

Q. *Where is phase modulation used?*

A. Phase modulation is used in TDMA and CDMA communication systems. There are many variations of phase modulation methods. Specifically, many digital modulation techniques rely on modifying the RF carrier's phase and amplitude.

Q. *What are some types of AM?*

A. There are many forms of amplitude modulation that are used in wireless communications ranging from simple AM to ASK, SSB, DSB, VSB, and QAM.

Q. *What are some types of FM?*

A. There are many forms of frequency modulation that are used in wireless communications ranging from simple FM to frequency shift keying (FSK), narrow-band FM, wideband FM, minimum shift keying (MSK), and gaussian minimum shift keying (GMSK).

Q. *What is preemphasis?*

A. *Preemphasis* is a technique utilized in FM to improve the *S/N* of a signal. Preemphasis is accomplished prior to actual modulation, and it is done at the audio level. In radio communication higher frequencies in the audio band are less likely to experience noise and other distortions. One way to take advantage of this is to amplify the higher-frequency signals that are in the audio pass band. The *preemphasis circuit,* also referred to as a *compander,* is shown in Fig. 7.18.

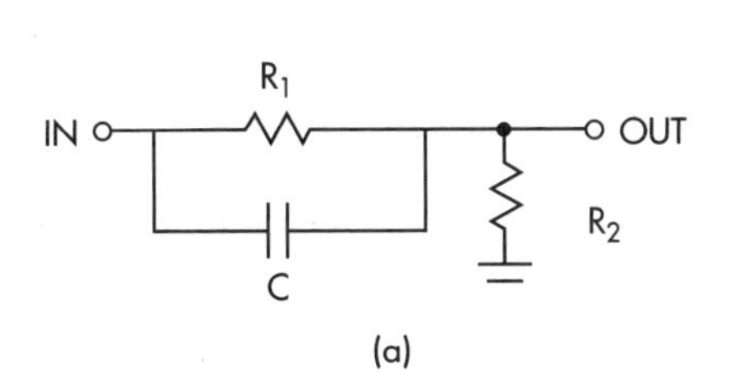

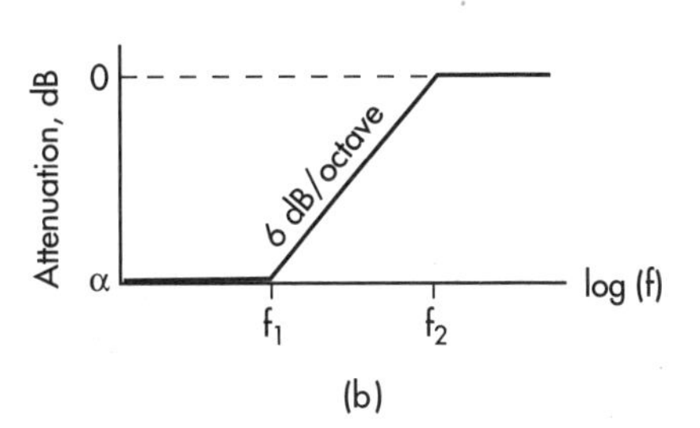

Figure 7.18 (*a*) Compander circuit. (*b*) Preemphasis.

Q. *What is a compander?*

A. A *compander* is a circuit that is used in an FM system to improve the *S/N* ratio of the signal, and it is the physical implementation of the preemphasis process. A compander circuit is shown in Fig. 7.18.

Q. *What is deemphasis?*

A. *Deemphasis* is a technique utilized in FM to improve the *S/N* of the signal. Deemphasis is used in the demodulation process, and it has the exact opposite effect as the preemphasis process that occurs at the beginning of the modulation process.

Deemphasis is achieved at the receiver, and it is exactly opposite from the preemphasis process. Specifically, the deemphasis is where the lower-frequency signals are amplified more than the higher-frequency audio signals. The deemphasis takes place after the demodulation takes place since the deemphasis process is at the audio level. The *deemphasis circuit,* also referred to as an *expander,* is shown in Fig. 7.19.

Q. *What is an expander?*

A. An *expander* is a circuit that is used in a frequency demodulation system to improve the *S/N* ratio of the signal, and it is the physical implementation of the deemphasis process. An expander circuit is shown in Fig. 7.19.

Q. *What is frequency shift keying?*

A. *Frequency shift keying* (FSK) is another form of frequency modulation. FSK is used for sending data from one point to another by

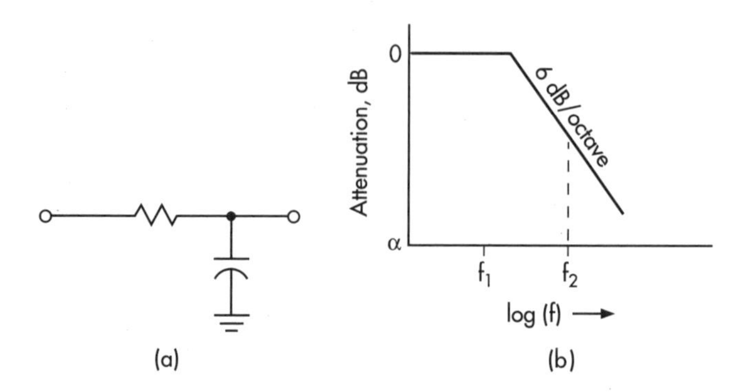

Figure 7.19 (*a*) Expander. (*b*) Deemphasis.

changing the frequency at which the carrier or information content operates. FSK can be used to send simple binary information by sending information as one of two different frequencies. FSK also can be sent by more than two different frequencies, say, four discrete frequencies, representing a four-level FSK system.

Q. *What is minimum shift keying?*

A. *Minimum shift keying* (MSK) is an FSK signal that has a modulation index of 0.5, which means, therefore, that the frequency shift is twice the data rate. MSK has the unique distinction of being referred to as many different forms of modulation. For example, MSK is also referred to as *fast FSK* (FFSK) and *continuous phase frequency shift keying* (CPFSK), and a variant of *offset quadrature phase shift keying* (OQPSK).

Q. *What is gaussian minimum shift keying?*

A. *Gaussian minimum shift keying* (GMSK) is another modulation technique that is a derivative of the MSK method for modulation. GMSK is the same as MSK except that it uses a gaussian low-pass filter prior to modulating the signal. The objective of using the filter is to improve the bandwidth efficiency for the signal.

Q. *What are some types of phase modulation?*

A. Some of the phase modulation formats are *phase shift keying* (PSK), bipolar phase shift keying (BPSK), quadrature phase shift keying (QPSK), differential quadrature phase shift keying (DQPSK), and offset quadrature phase shift keying (OQPSK).

Q. *What is quadrature phase shift keying?*

A. *Quadrature phase shift keying* (QPSK), one form of digital modulation, has four different phase states to represent data. The four phase states are arrived at through different *I* and *Q* values. Utilizing four phase states—that is, quadrature—allows each phase state to represent 2 data bits.

Q. *What is differential quadrature phase shift keying?*

A. *Differential quadrature phase shift keying* (DQPSK) is similar to QPSK, but DQPSK does not require a reference from which to judge the transition. Instead, DQPSK's data pattern is referenced to the previous DQPSK's phase state.

DQPSK has four potential phase states with the data symbols, which are defined relative to the previous phase state, as shown in the table below:

Symbol	DQPSK phase transition, degrees
00	0
01	90
10	−90
11	180

Q. *What is Pi/4 differential quadrature phase shift keying?*

A. *Pi/4 differential quadrature phase shift keying* (Pi/4 DQPSK) modulation is very similar to that of DQPSK, but the Pi/4 DQPSK phase transitions are rotated 45 degrees from that of DQPSK. Like DQPSK, Pi/4 DQPSK has four transition states, which are defined relative to the previous phase state shown in the table below:

Symbol	Pi/4 DQPSK phase transition, degrees
00	45
01	135
10	−45
11	−135

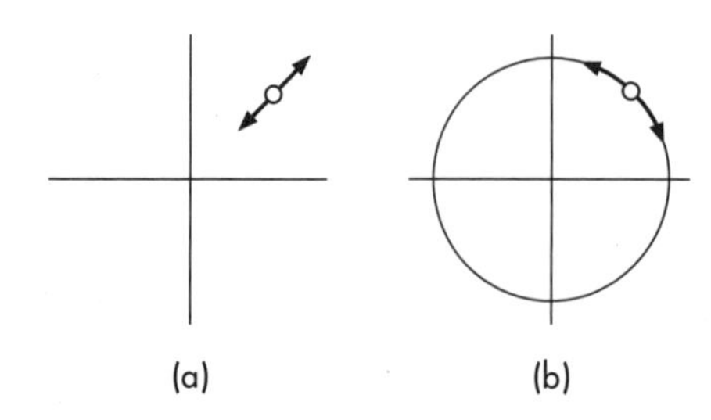

Figure 7.20 *I&Q* diagrams. (*a*) AM. (*b*) FM.

Q. *What is an I&Q diagram?*

A. An *I&Q diagram* is used to represent phase and amplitude modulation for a digitally modulated signal. The *I&Q* diagram shown in Fig. 7.20 utilizes vector notation for representing the actual *I* and *Q* values.

Q. *What is a vector diagram?*

A. A *vector diagram* plots the *I* components as a function of *Q*. The primary purpose of utilizing a vector diagram for analysis of a digitally modulated signal is to view the transitions among the various states in a quadrature-modulated signal. The transition status can be used to determine the overall modulation quality of the signal being viewed. If there is little error with the signal, the locations on the vector diagram that represent symbol points are easily definable, and the variation transitions' trajectories intersect closely at each of these points. (See Fig. 7.21.)

Q. *What is a constellation diagram?*

A. A *constellation diagram* is utilized to show the relationship among the different amplitude and phase states of the modulated signal by displaying the error vector at the symbol sample time (Fig. 7.22). The *error vector* is the difference between the theoretical symbol location and the actual symbol location on the constellation diagram.

Q. *What is an eye diagram?*

A. An *eye diagram* is one method used to analyze a digital communication system. For quadrature modulation, the eye diagram

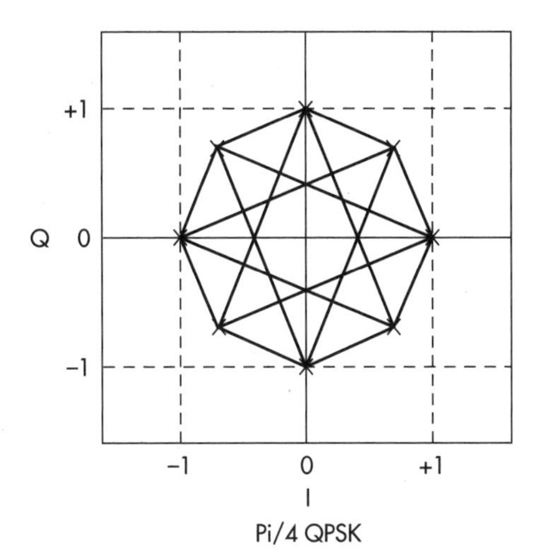

Figure 7.21 Vector diagram.

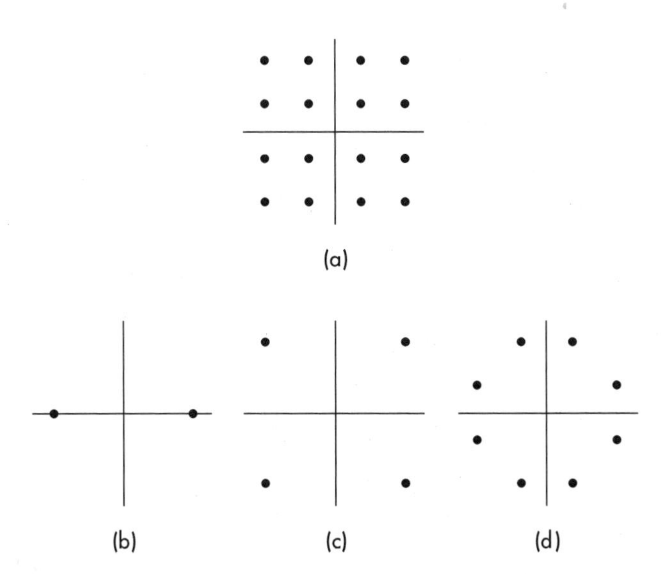

Figure 7.22 Constellation diagrams. (*a*) 16 QAM; $m = 16$. (*b*) BPSK; $m = 2$. (*c*) QPSK; $m = 4$. (*d*) 8 PSK; $m = 8$.

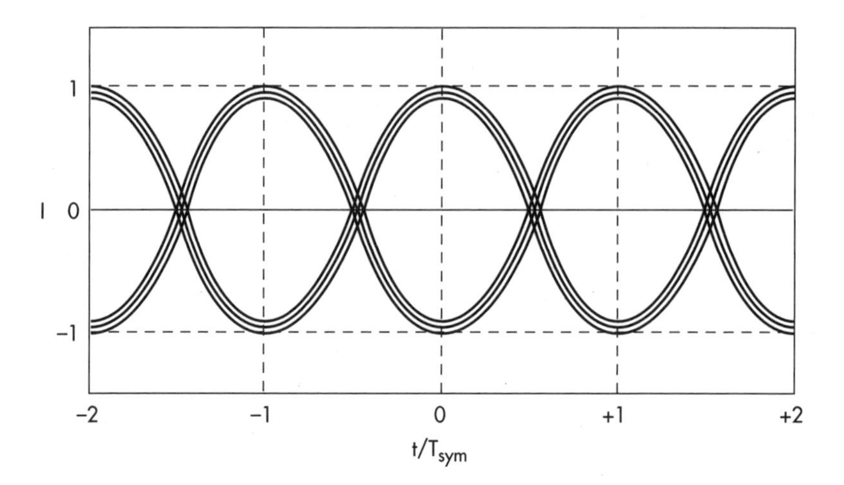

Figure 7.23 Eye diagram.

plots both the *I* and *Q* components as a function of time. Utilizing the eye diagram, the phase and amplitude errors of the system can be determined (Fig. 7.23). As the number of errors in phase and amplitude increase, the eye pattern closes.

Q. *What is an eye diagram used for?*

A. An eye diagram is used to determine a digital system's performance, and it is used as shown in Fig. 7.24. Specifically, the height of the eye is a direct indication of how well the signal is behaving in the presence of noise in the network. The width of the eye is an indication of the signal's immunity to clock phase errors. A decrease in the width is an indication that the sampling

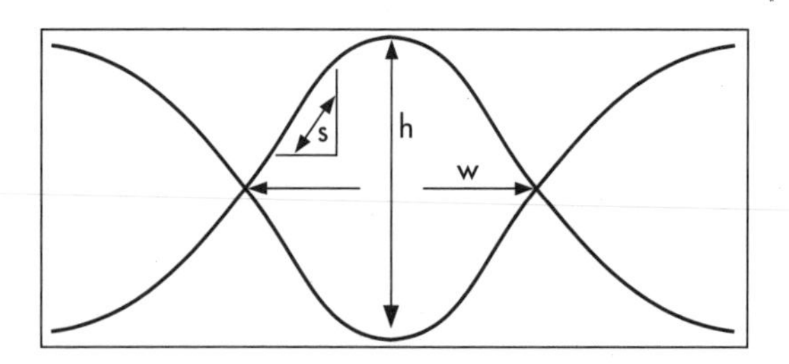

Figure 7.24 Eye diagram. h = vertical opening, w = width of eye, s = slope of rising edge.

that the system is using is not sufficient. And the slope *s* is an indication of the timing jitter. The steeper the slope, the less immune the signal will be to changes in the clock sampling time.

Q. *What is Rx?*

A. Rx is an abbreviation that is used to represent a receiver, the receive path, or a component that is involved in the receiver or its link budget.

Q. *What is Tx?*

A. Tx is an abbreviation that is used to represent a transmitter, the transmit path, or a component that is involved in the transmitter or its link budget.

Q. *What is a transmitter?*

A. A *transmitter* is an amplifier that is used in a communication system. There are many types of transmitters, or amplifiers, in a communication system, and most of them are located in the receiver for the communication cell site. However, the fundamental difference between the amplifiers in the transmit portion and those in the receiver lies in the total power they are able to amplify and deliver to the desired load.

Q. *What is an AM transmitter?*

A. An *AM transmitter* changes the amplitude of the carrier as a function of the information content. An AM transmitter block diagram is shown in Fig. 7.25.

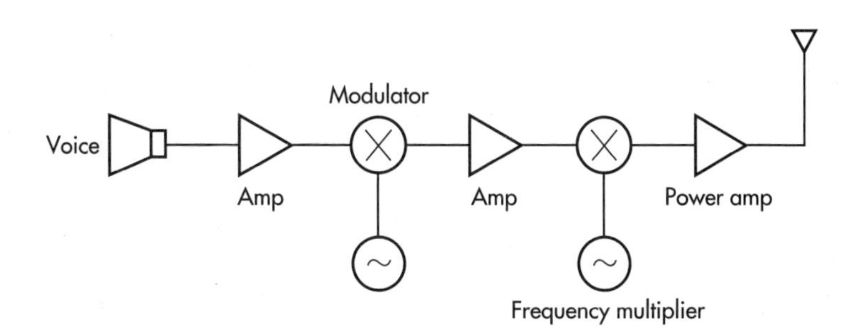

Figure 7.25 Amplitude modulation.

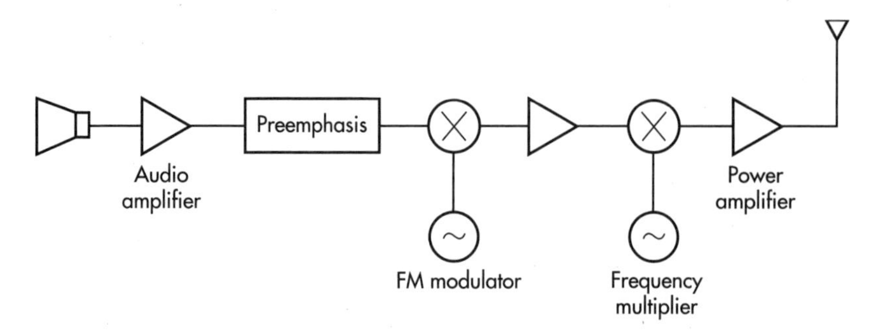

Figure 7.26 Frequency modulation.

Q. *What is a FM transmitter?*

A. An *FM transmitter* modulates the information content by changing the frequency of the carrier as a function of the information content. An FM transmitter block diagram is shown in Fig. 7.26.

Q. *What is a PM transmitter?*

A. A *PM transmitter* places the information onto the carrier much as it is done with FM; however, the modulation is achieved through the adjustment of the phase of the information that rides on the carrier. A PM transmitter block diagram is shown in Fig. 7.27.

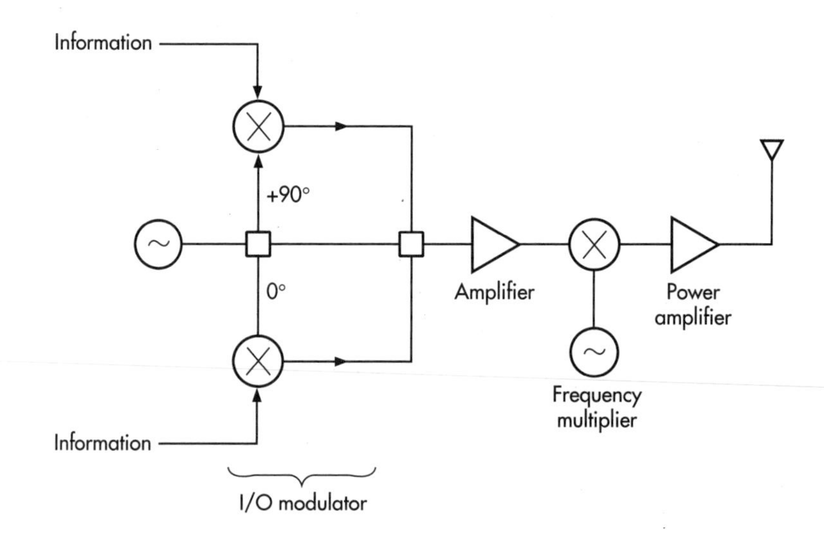

Figure 7.27 Phase modulation.

Q. *What are the types of amplifier classes?*

A. There are numerous types and classifications of amplifiers. Each of the classes defined has advantages and disadvantages. The type of amplifier chosen needs to meet the requirements set forth in the design objective, maximize output while consuming as little power as possible, and at the same time pass the desired modulation format with no distortion over its frequency and power operating range. The primary amplifier classes are the following:

- Class A
- Class B
- Class AB
- Class C

Q. *What is a class A amplifier?*

A. A *class A amplifier* conducts throughout the entire 360 degrees of the sine-wave input into the amplifier. Class A amplifiers offer high linearity with respect to the transfer function of the amplifier. The class A amplifier will remain linear as long as the amplifier does not saturate, properly dissipates its power, and operates within the defined linear range of the transfer function for the amplifier.

Q. *What is a class B amplifier?*

A. A *class B amplifier* conducts for only half, or rather 180 degrees, of the sine-wave input. The class B amplifier is typically not used because it relies on a push-pull amplifier design. The push-pull amplifier design does not afford linearity across its defined operating range. The class B amplifier has the advantages of offering greater output power than the class A amplifier, and it is more energy efficient. However, the class B amplifier is known to exhibit crossover distortion, which is why it is not a desired amplifier for RF.

Q. *What is a class AB amplifier?*

A. *Class AB amplifiers* are used in many cases where a single transmitter is needed. The class AB amplifier does not exhibit the amount of crossover distortion experienced in a class B amplifier, and it is not as linear as a class A amplifier. The class AB amplifier is finding more and more use in wireless applications.

Q. *What is a class C amplifier?*

A. The *class C amplifier* does not conduct throughout the entire 360 degrees of the sine wave but only a little less than 180 degrees. Class C amplifiers are utilized throughout the wireless industry, and they are more power efficient than the A, B, or AB amplifiers. The class C amplifier will distort the signal, but frequency and phase modulation techniques do not need linearity for the amplification.

Q. *What is a feed-forward amplifier?*

A. A *feed-forward amplifier* (also called an LAC or *linear amplifier*) is a type of amplifier that exhibits linearity for amplifying many RF channels at the same time. The feed-forward amplifier allows channel combining at a low power level and relies on one massive amplifier at the final stage to amplify all the signals together. The complexity of the feed-forward amplifier is great since it must cancel out unwanted signals that are generated as a result of the combining and amplification process.

Q. *When does overmodulation occur?*

A. *Overmodulation* happens in the amplification of a signal when, during the amplification, the signal is overmodulated causing the initial signal to distort.

Q. *What is an isolator?*

A. An *isolator* is a device that conducts RF energy in one direction and rejects, or rather highly attenuates, energy in the opposite direction. The isolator is normally placed between the transmitter and the antenna system. (See Fig. 7.28.)

The purpose of the isolator is to isolate the transmitter from the antenna and minimize the potential for intermodulation products to be generated. The isolator also acts as a balanced load for a transmitter, ensuring maximum power transfer to the antenna system.

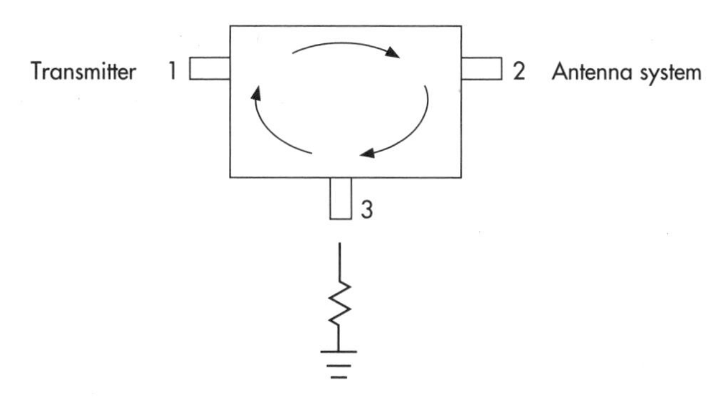

Figure 7.28 Isolator.

Q. *What is a circulator, or gyrator?*

A. A *circulator,* or *gyrator,* is the part of the isolator without the load.

Q. *What are the components of an isolator?*

A. An isolator consists of a circulator and load, and it is a three-port device. One port is for the input of the signal, another is for the load, and the third is for the system being connected to.

Q. *Is it possible to better isolate a receiver by using an isolator?*

A. No, the isolation cannot be increased by the use of an isolator if it is used in the receive path. The isolator will pass energy in one direction and impede it in another. Hence the isolator will not provide isolation for the receiver except to block all energy from possibly entering the receiver, including the desired signal, depending on how it is installed.

Q. *What are some factors to consider when selecting an isolator?*

A. Several factors that need to be considered when selecting an isolator are the following:

1. Power rating
2. Isolation requirements, decibels
3. Ambient temperature and heat for the application
4. Worst-case reflected power at output of isolator port

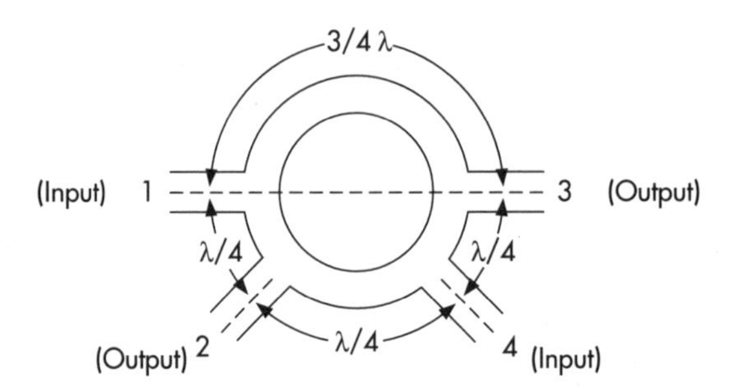

Figure 7.29 Hybrid ring combiner. No coupling between 1-4 and 2-3.

Q. *What is a hybrid combiner, or hybrid?*

A. A *hybrid combiner,* or *hybrid,* combines discrete RF signals from two or more paths onto one path. (See Fig. 7.29.)

Q. *When is a hybrid combiner used?*

A. Often a hybrid combiner is used to combine signals when the isolation between frequency channels is not sufficient to ensure proper operation of the transmitter. The hybrid combiner also enables multiple signals to be combined at a low energy level for later amplification by a linear amplifier. Also the hybrid can be used for receive as well as transmit paths.

Q. *What is a cavity?*

A. A *cavity* is a selective filter that is used to combine different RF channels together or to make up a filter system. A cavity is shown in Fig. 7.30.

Q. *Where are cavities used?*

A. Cavities are utilized in a communication site to provide isolation for the transmitter, ensure spectral purity, and to facilitate combining multiple carriers onto a single antenna system. The arrangement of cavities—that is, their physical location and frequency of operation—determine what their function is.

In cellular communication systems the cavity filter is associated with the transmit portion of the cell site. In two-way operations

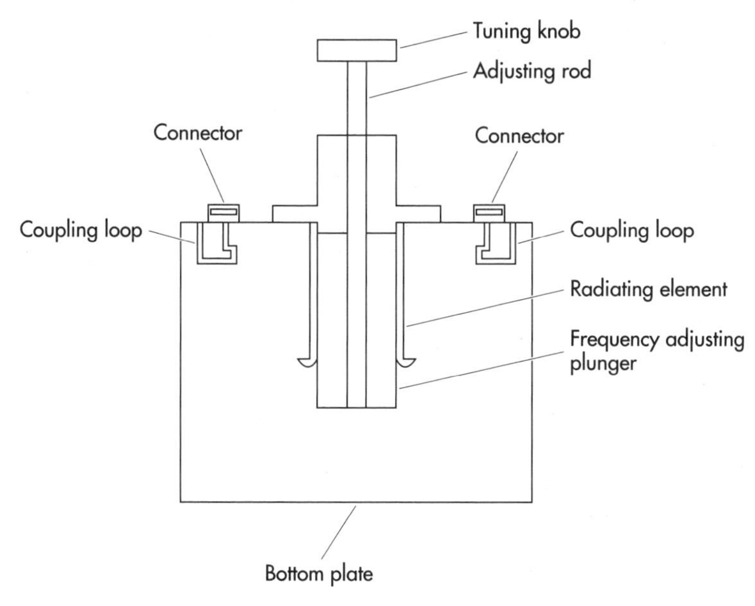

Figure 7.30 Cavity.

the cavity filter can be used on the receive portion of the system as a method of obtaining high selectivity for a single channel.

Q. *What is an autotune combiner, or autotune?*

A. An *autotune combiner,* also called an *autotune,* is a device that allows multiple RF channels to be combined onto one antenna system. Autotune combiners are used extensively in cellular systems that do not have a linear amplifier. The autotune combiner also facilitates frequency retunes.

Q. *What is a duplexer, or diplexer?*

A. A *duplexer,* or *diplexer,* is a valuable piece of hardware at a communication site. The duplexer effectively allows a single antenna to be used by both the transmit and receive portions of a communication system. Most of the subscriber units that transmit and receive at the same time utilize a duplexer. Communication base stations often deploy a duplexer when the number of antennas needs to be reduced for any one of a variety of reasons. The duplexer is a combination of a transmit filter and receive filter, which are joined together into one package (Fig. 7.31).

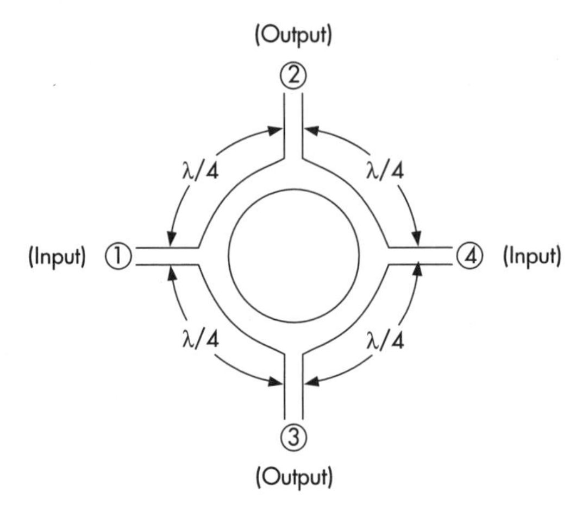

Figure 7.31 Hybrid ring combiner.

Q. *What is a triplexer?*

A. A *triplexer* is a special type of duplexer that can have two different transmit ports and one receive port. A triplexer is often used for a dual-band system.

Q. *What is a cross-band coupler?*

A. A *cross-band coupler* is a device that allows different communication systems to share a coaxial cable. The use of a cross-band coupler allows multiple bands to be combined with minimal insertion loss. In addition, cross-band couplers enable VHF and UHF systems to be combined into one. (See Fig. 7.32.)

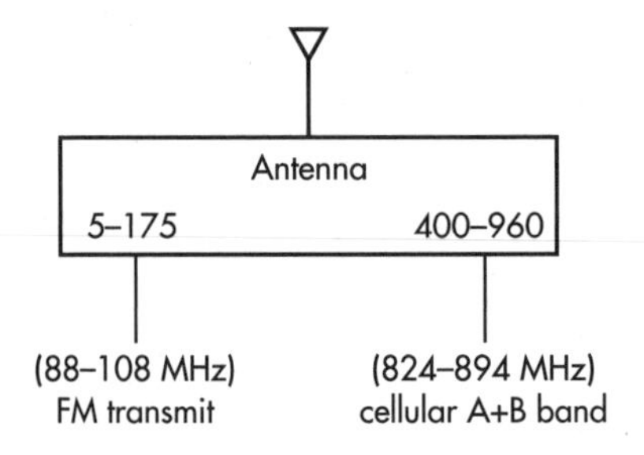

Figure 7.32 Cross-band application.

Q. *Where are cross-band couplers used?*

A. One location where cross-band couplers are used is in the installation of tunnel coverage systems. They are also used to reduce the number of cables needed for a tower installation.

8

Filters

This chapter addresses many of the commonly asked questions that pertain to the use of the filters within a wireless communication system. Filters play an integral role in the design and operation of a radio and communication system. The purpose of a filter is to allow the desired energy or information to pass undistorted either in phase, amplitude, or time while simultaneously and completely suppressing all other energy. There are many types of filters that can be deployed successfully throughout a system.

With the proliferation of wireless communication, especially in the 800-MHz and 1.9-GHz bands, the use of filters has been receiving more attention than it has in the past. The following questions and answers will provide some immediate assistance in this ever-demanding arena.

Q. *What is a filter?*

A. When applied in radio communication, a *filter* is a device that is used to allow certain frequencies to pass through unimpeded and to reject all the others. There are numerous types of filters used in radio communication systems. Filters are located in both the transmission and reception ports of the system's equipment. They are also found in the landline portion of the communication system.

Q. *What types of filters are used in wireless communication systems?*

A. There are four general classifications of filters that are used throughout all of radio communications. However, it should be noted that there are many combinations of the general filter classes listed.

1. Low-pass filter
2. High-pass filter
3. Band-pass filter
4. Band-reject filter (notch filter)

Q. *What is a harmonic filter?*

A. A *harmonic filter* is either a low-pass or a band-pass filter depending on the filter's application. The objective of the harmonic filter is to remove from transmission all the harmonics except the carrier and its sidebands. Figure 8.1 is a representation of a harmonic filter utilizing a low-pass filter. The harmonic filter is usually located between the output of the transmitter and the antenna system.

Q. *What is a band-pass filter?*

A. A *band-pass filter* is used to allow only those signals that are desired to pass through it. Specifically, a band-pass filter passes frequencies in the pass-band region and attenuates frequencies above and below that region. The band-pass filter is designed to separate the low- and high-pass filter characteristics, and then cascade them as shown in Fig. 8.2. The bandpass filter is usually located on the receive path for a cell site.

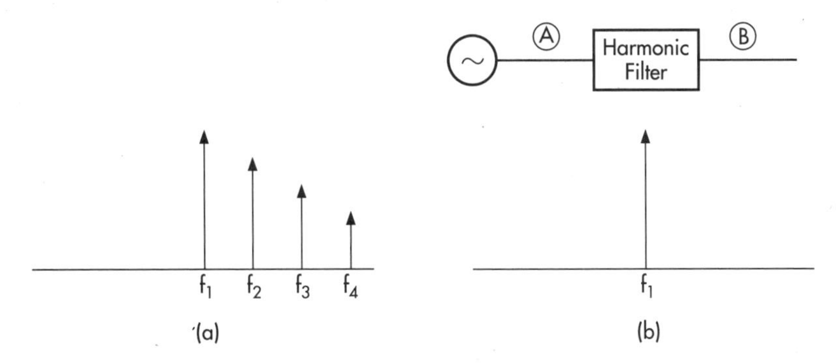

Figure 8.1 Harmonic filter. (*a*) f_1 = 880 MHz, $2f_1$ = 1760 MHz = f_2, $3f_1$ = 2640 MHz = f_3, $4f_1$ = 3520 MHz = f_4. (*b*) Filter output.

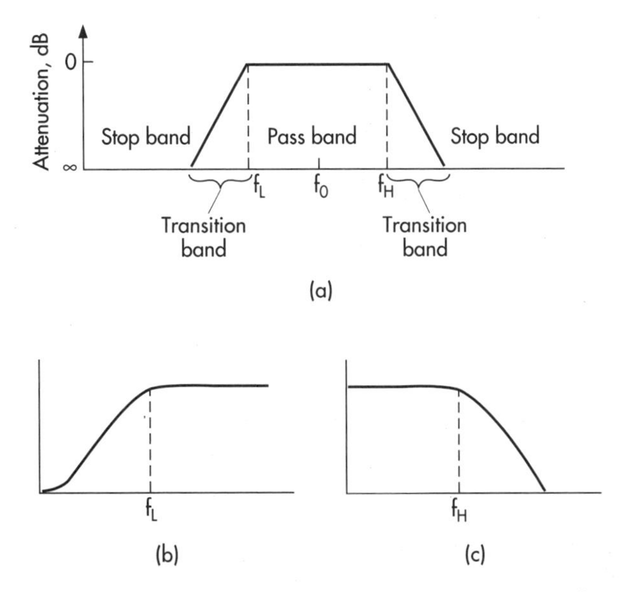

Figure 8.2 Band-pass filter. (*a*) f_0 = center frequency, f_L = lower cutoff, f_H = higher cutoff. (*b*) High-pass filter. (*c*) Low-pass filter.

Q. *What is a low-pass filter?*

A. A *low-pass filter* passes signals from zero frequency, dc, to a certain cutoff frequency and rejects all signals whose frequencies are beyond the cutoff frequency. Hence, the name "low-pass filter" because it passes all the signals below a certain frequency without any impairment. A simple low-pass filter is the RC circuit shown in Fig. 8.3.

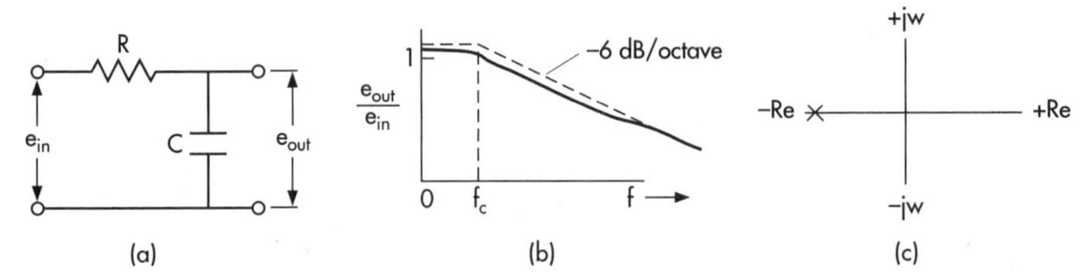

Figure 8.3 (*a*) RC low-pass filter. (*b*) Bode plot. (*c*) Low-pass–filter polar plot.

Q. *What is a high-pass filter?*

A. A *high-pass filter* will not impede any signals that are above a certain frequency and will impede signals that are below a certain frequency. A representation of a high-pass filter is shown in Fig. 8.4. The high-pass filter will attenuate—that is, reject—all the signals that are below the cutoff frequency and pass all the frequencies above the cutoff frequency. The high-pass filter is the RC circuit shown in Fig. 8.4.

Q. *What is a notch, or band-reject, filter?*

A. A *notch*, or *band-reject, filter* is used to selectively remove unwanted signals from the communication system. Specifically, the band-reject filter attenuates the signal frequencies within the center, or notch, region and passes frequencies above and below the notch. Figure 8.5 represents a band-reject filter. Notch filters are used in many cases to get rid of unwanted signals when one system is colocating with other wireless systems. The notch filter is often used in conjunction with a band-pass filter.

Q. *What are the characteristics of a low-pass filter?*

A. The characteristics of a low-pass filter are high-frequency-interference rejection, band limiting, and harmonic suppression. Ideally the low-pass filter will allow all the signals that are below a certain frequency to pass through the filter and attenuate the rest. The low-pass filter will pass signals from dc to the cutoff frequency. The low-pass filter is found in Tx filters and at the output of mixers, to mention but two possible locations. Therefore, if the low-pass filter has a pass band between 0 MHz (dc) and 894 MHz, the band-pass filter will allow signals at 890 MHz to pass unaffected while attenuating, say, a signal at 900 MHz.

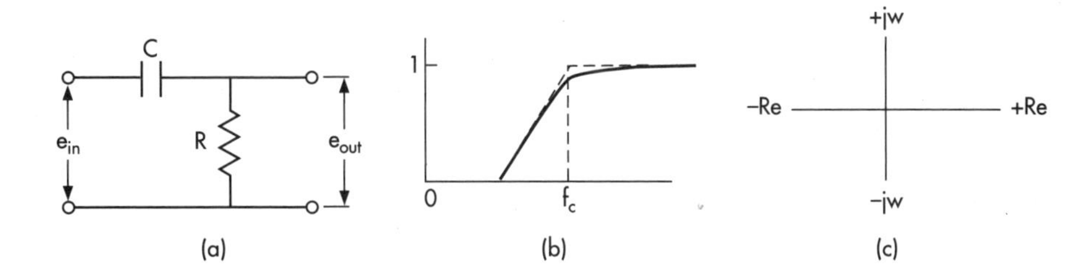

Figure 8.4 High-pass filter. (*a*) RC filter; $e_{out}/e_{in} = S/[(1 + S/(2\pi RC)]$. (*b*) Bode plot; corner frequency $(f_c) = 1/(2\pi RC)$. (*c*) Polar plot.

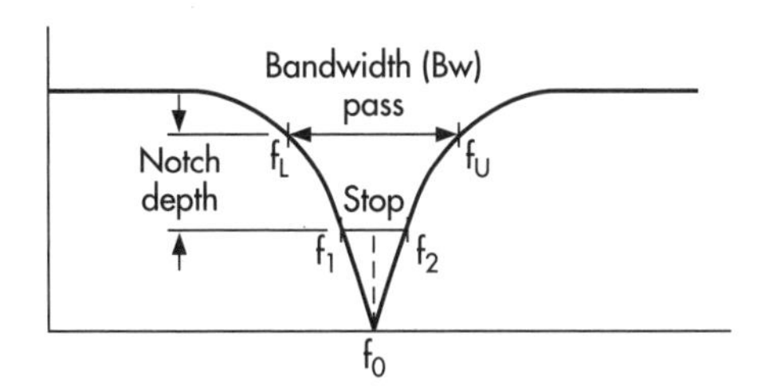

Figure 8.5 Band-reject filter (notch). $f_0 = \sqrt{f_L f_U}$, $Q = f_0/Bw$, AS = pass band/stop band.

Q. *What are the characteristics of a high-pass filter?*

A. The characteristics of a high-pass filter are band limiting, noise reduction, interference elimination, and broadcast signal conditioning. Ideally, the high-pass filter will pass signals that are above a specified frequency and attenuate all the ones that are below that cutoff frequency. High-pass filters can be found in communication systems that are broadband but require the elimination of signals that are lower in operation. Therefore, if the high-pass filter has a pass band starting at 806 MHz, the high-pass filter will allow signals at 815 MHz to pass unaffected while attenuating, say, a signal at 800 MHz.

Q. *What are the characteristics of a band-pass filter?*

A. The characteristics of a band-pass filter are band limiting, comb filter, and interference elimination. Wireless systems utilize band-pass filters extensively in their radio systems. Often there are numerous band-pass filters within a cell site or even a radio. An ideal band-pass filter will pass only those signals that are within the pass band of the filter itself. Therefore, if the band-pass filter has a pass band between 824 MHz and 846.5 MHz, the band-pass filter will allow signals at 840 MHz to pass unaffected while attenuating, say, a signal at 850 MHz.

Q. *What are the characteristics of a notch, or band-reject, filter?*

A. The characteristics of a notch, or band-reject, filter are selective frequency rejection, noise reduction, and interference elimination. Notch filters are used in communication sites to remove unwanted energy when the use of a band-pass, high-pass, or low-pass filter by itself will not provide the required attenuation. An ideal notch filter will attenuate all the signals within its pass band and not affect the rest of the signals. Therefore, a notch filter has a *notch band* between 845 and 846.5 MHz, and it will allow a signal at 847 MHz to pass unaffected while attenuating, say, a signal at 846 MHz.

Q. *What is Q?*

A. Q represents the selectivity of a filter. The higher the Q, the more the filter will reject unwanted signals from being passed through it. An ideal filter has an infinite Q. The Q of a filter is defined as the ratio between the center frequency and the bandwidth of the filter.

Q. *What is selectivity?*

A. The *selectivity* of a filter, represented as its Q, is a key attribute of the filter. The higher the selectivity of the filter, the better it rejects unwanted signals from being passed through it. Ideally, a filter should be extremely selective and allow only the desired signals through, and it should not distort the desired signal.

Q. *What are the key criteria that apply in evaluating a filter?*

A. The key criteria for evaluating a filter are listed below:

Frequency response
Insertion loss
Pass-band ripple

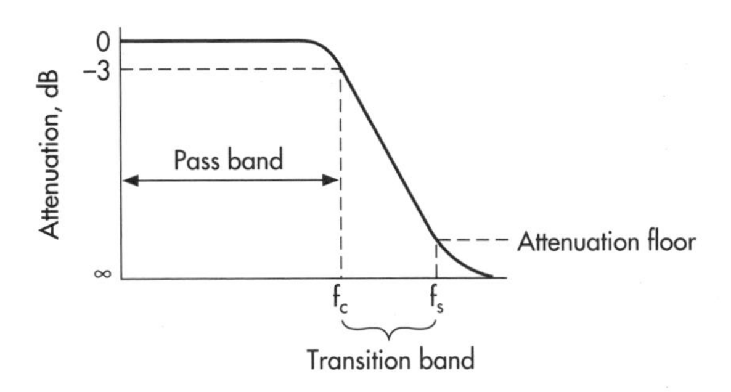

Figure 8.6 Depiction of filter's ability to reject out-of-band energy. Shape factor (SF) = f_s/f_c, where f_c = cutoff frequency, f_s = stop frequency.

Attenuation floor
Shape factor
Phase error
Group delay
Selectivity, Q
Temperature stability

Q. *What is the frequency response characteristic of a filter?*

A. The *frequency response characteristic* of a filter defines which frequencies will be passed and which ones will be attenuated. The components that make up the frequency response characteristic of a filter involve the pass band, the cutoff band, the transition band, and the stop band. (See Fig. 8.6.)

Q. *What is the pass band?*

A. The *pass band* is the frequency zone in which the filter allows some signals to pass while discriminating against, or attenuating, other signals. The pass band is normally defined as the area of the filter that experiences the lowest level of attenuation—ideally 0 dB—and that has a characteristic low-level ripple. Figure 8.7 shows an example of the pass band for a filter.

The pass band is one of the most important filter criteria, and it is usually the characteristic on which the RF engineer focuses. The pass band is a key performance indicator for the filter; however, it must be stressed that it is only one of the components that

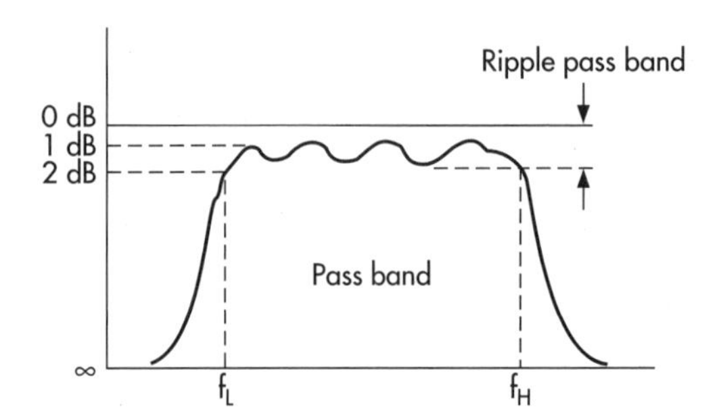

Figure 8.7 Pass-band ripple.

define the filter's response, and decisions made solely on the pass-band capabilities could be made in error.

Q. *What is the cutoff frequency?*

A. The *cutoff frequency* for a filter is simply the location in the filter's frequency response band where 3 dB of attenuation occurs as compared to the pass band. The cutoff frequency is used to define where the filter begins to remove the unwanted energy. Ideally, the filter will have 0 dB of attenuation in the pass band and infinite attenuation elsewhere. If the cutoff frequency should not be within the pass band, and equally important, the cutoff frequency should not be too far from the end of the pass band. Figure 8.8 helps show where the cutoff frequency is and its relationship with other key filter functions.

Q. *What is pass-band ripple, or ripple?*

A. *Pass-band ripple,* or *ripple,* is a variation in gain, or rather the insertion loss that occurs in the pass band of a filter. A plot of the gain versus frequency response for a filter will show its ripples across the pass band. It is important to note that all filters have some ripple and that only the ideal filter has no ripple. Typically, a filter will have 1 to 2 percent ripple within its pass band. If the ripple is too great, then the variation between channels or parts of the channel used for communication may cause degradations either in that channel or with the rest of the system. Figure 8.9 illustrates filter ripple.

Q. *What is the stop-band edge for a filter?*

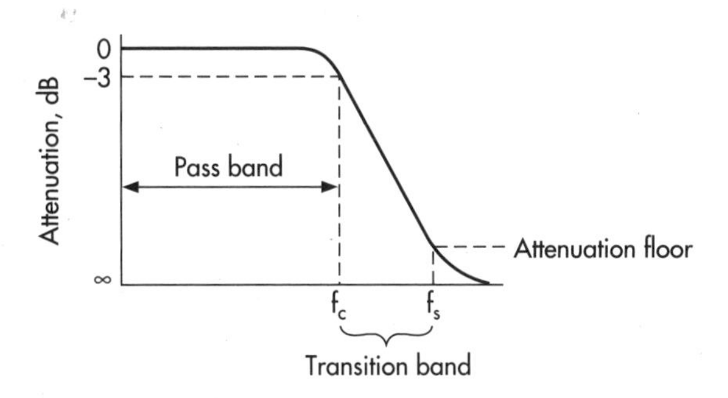

Figure 8.8 Depiction of filter's ability to reject out-of-band energy. Shape factor (SF) = f_s/f_c, where f_c = cutoff frequency, f_s = stop frequency.

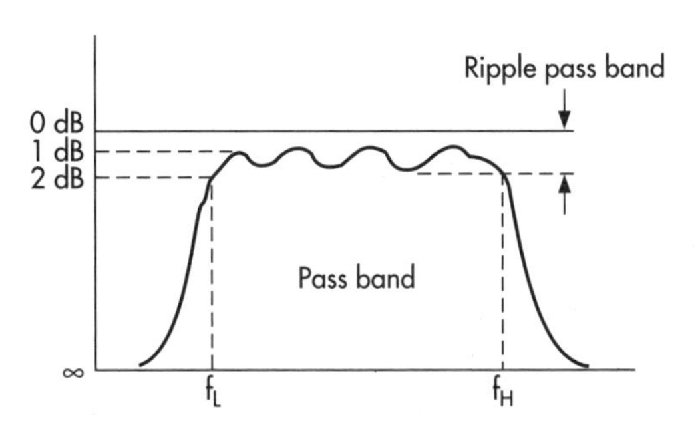

Figure 8.9 Pass-band ripple.

A. The *stop-band edge* for a filter is the highest frequency F_r at which the pass-band ripple occurs. The stop-band edge is also the transition point where a small increase in frequency gives a large increase in attenuation. Often receiver front ends usually specify the stop-band attenuation at the upper and low frequencies in which it operates. For a transmitter, the stop-band frequency is normally specified with respect to the receive band for the receiver.

Q. *What is attenuation, or insertion loss?*

A. *Attenuation,* also called *insertion loss,* is the reduction, that is, attenuation, that a signal goes through when passing through a device. For example, if a filter has an attenuation of 3 dB, it means that at that frequency, the filter will reduce the signal by half its original power.

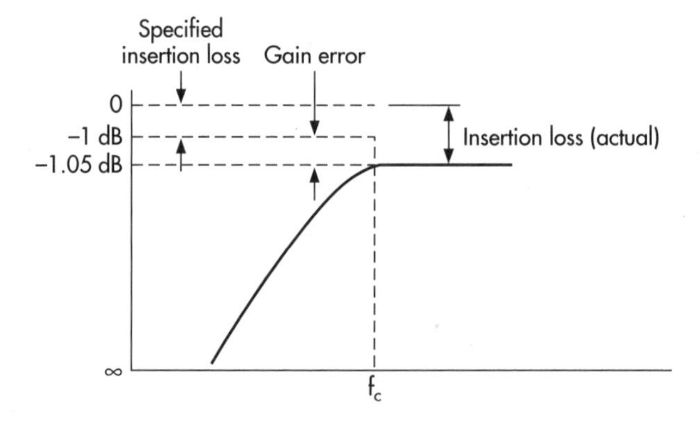

Figure 8.10 Pass-band ripple.

Insertion loss, or attenuation, for a filter is a measure of how much the signal—whether desired or undesired—is reduced in power as it passes through the filter. Ideally the insertion loss for a filter in the pass band should be 0 dB, and there should be infinite attenuation everywhere else. The insertion loss, however, is typically referenced to the pass band of the filter itself, and it is an important metric since this defines how much loss the filter will impose upon the signal as it traverses through the filter itself. It should be noted that any passive filter will have some insertion loss, and this insertion loss also equates to a noise figure for the filter itself. Figure 8.10 depicts the insertion loss and shows how it is represented in a filter plot.

Q. *What is phase error?*

A. *Phase error,* or *phase linearity,* is the linearity of the phase shift versus frequency. If there were no phase errors, then the phase linearity line would be a straight line. It should be noted that if there is no phase error, then there is no group delay since the derivative of a constant is 0 and the group delay is the derivative of the phase error.

Q. *What is group delay?*

A. *Group delay* is the time it takes a signal to pass through a filter, and it is measured in terms of a finite length of time for a burst (pulse) to pass. Ideally the group delay for a filter should be constant across the entire pass band of the filter. A group delay that is not constant across the pass band can cause overshoot or ringing in the pass band itself.

Since the group delay for a filter is a derivative of the filter's phase shift, it is mathematically represented as phase versus frequency response. Note that the actual magnitude of the group delay is not important. However, its flatness across the pass band is the key to the filter's performance and is a measurement of distortion.

If the group delay is not constant over the bandwidth of the pass band for the filter, where the desired modulated signal resides in frequency, some form of waveform distortion will take place. The narrower the desired signal in bandwidth, the less likely it will undergo any noticeable group delay. However, the wider the bandwidth of the signal, the more distortion is likely to occur due to group delay. Specifically, the 30-kHz channel used for DAMPS in the United States is not really susceptible to group delay problems. However, the CDMA signal in the IS-95 system, which is 1.288 MHz wide, is susceptible to group delay.

Q. *What are adaptive filters?*

A. An *adaptive filter* is a practical implementation of an ideal filter except that it adjusts to a changing signal condition. The ability of the adaptive filter to adjust its response makes it an excellent candidate for interference rejection, equalization, echo and noise cancellation, and many other similar applications.

Adaptive filters are being experimented with in the wireless community at this time for the purpose of rejecting interference. The technology works in the military environment and shows reasonable promise for commercial use, although its physical size and complexity must be contended with.

Q. *What are some types of adaptive filters?*

A. There are three basic types of adaptive filters, as shown in Fig. 8.11, that can be deployed in a wireless network.

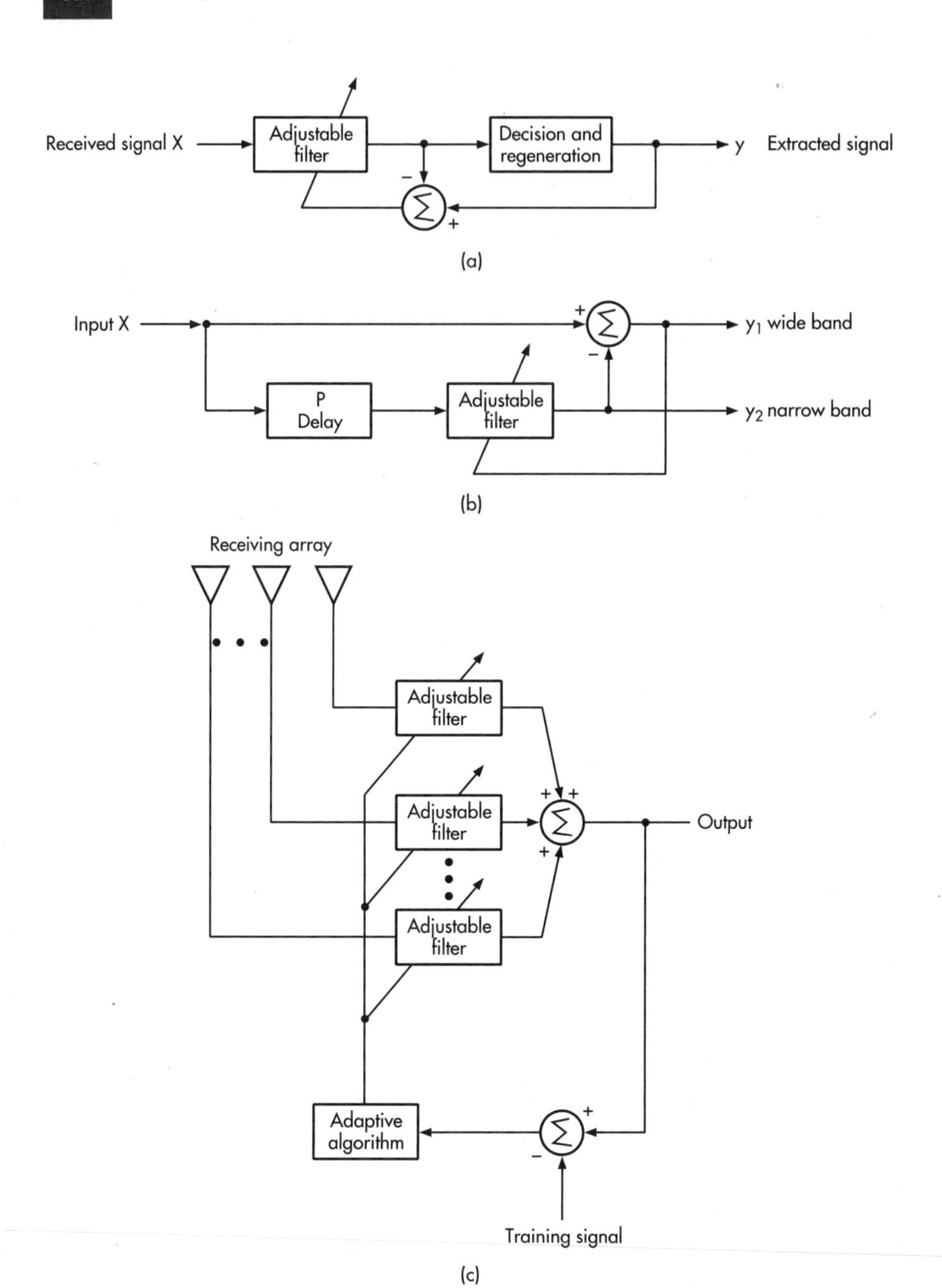

Figure 8.11 (*a*) Decision-directed signal extraction. (*b*) Adaptive whitening filter. (*c*) Adaptive null steering antenna.

Q. *What are superconducting filters?*

A. *Superconducting filters* are a group of special filters that utilize high-temperature superconducting materials to approximate the ideal filter for the receive path of a communication system. The filters can utilize existing cooling systems because the ideal temperature for high-temperature superconductors is around 77 K, and they utilize either thick-film or thin-film superconducting material.

The discovery and use of high-temperature superconductors indicate that the promise of achieving the ideal filter for a communication system may become a reality. The ideal filter has no insertion loss in the pass band but has infinite attenuation everywhere else. The use of high-temperature superconductors means that filters may obtain a near-infinite amount of poles with virtually no loss.

Currently the superconducting efforts are centered in two camps, thick film and thin film. Each has its advantages and disadvantages. The thick-film superconducting filters are able to handle more power and have better intermodulation specifications. However, the thick-film filters are larger in size and, at present, tend to limit themselves to a macro cellular environment. Efforts are presently under way to minimize the thick-film filter's size so that it can be deployed in a micro base station. The thin-film superconductors, while not being able to handle as much power, do enjoy the advantage of physical size. The thin-film technology lends itself immediately to mobile applications and micro base stations.

Q. *What is the pass band for A-band cellular (non-wire line)?*

A. The pass band for A-band cellular in the United States is the following:

Base station Tx: 869 to 880 MHz and 890 to 891.5 MHz

Base station Rx: 824 to 835 MHz and 845 to 846.5 MHz

Q. *What is the pass band for B-band cellular (wire line)?*

A. The pass band for B-band cellular in the United States is the following:

Base station Tx: 880 to 890 MHz and 891.5 to 894 MHz

Base station Rx: 835 to 845 MHz and 846.5 to 849 MHz

Q. *What is the SMR pass band?*

A. The pass band for SMR in the United States is the following:

Base station Tx: 851 to 869 MHz

Base station Rx: 806 to 824 MHz

Q. *What is the Tx band for an air-to-ground transmit station?*

A. The pass band for the AGT transmit system is extremely important to the B-band cellular operators as well as to the A-band operators. The B-band operators have a much more defined problem in that there is not a guardband between the AGT transmitter frequency and that of the B-band receiver frequency, which is often the first CDMA channel used. The AGT Tx band is the following:

Base station Tx: 849 to 851 MHz

Q. *What are the Tx and Rx bands for cellular, SMR, and PCS operators in the United States?*

A. Figure 8.12 is a chart that indicates the Tx and Rx frequency bands for cellular, SMR, and PCS U.S. operators. The chart is constructed with respect to the cell site or rather the base station.

Q. *What are the Tx and Rx bands for cellular, SMR, and PCS mobile and portable units in the United States?*

A. Figure 8.13 is a chart that indicates the Tx and Rx frequency bands for cellular, SMR, and PCS U.S. operators. The chart is constructed with respect to mobile and portable units.

Q. *What is the frequency separation between Tx and Rx frequency bands for a cellular radio channel?*

A. The frequency separation between cellular Tx and Rx channels, since it is a full-duplex system, is 45 MHz. The separation applies to both the A- and B-band operators.

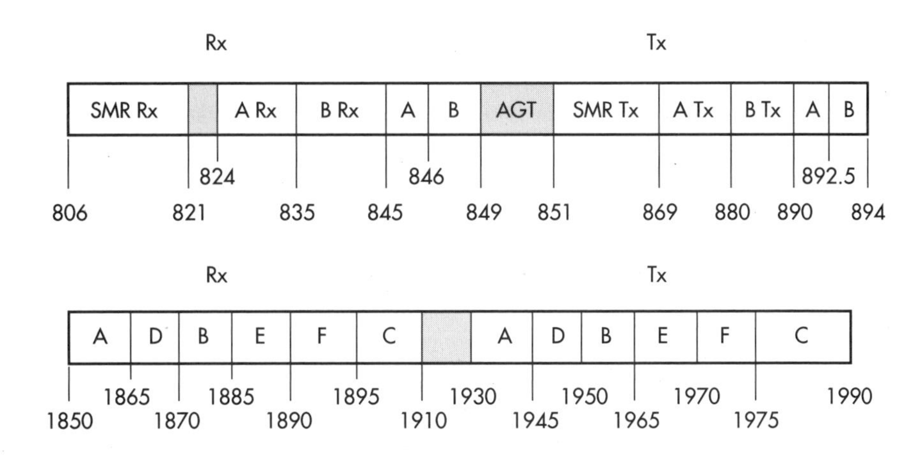

Figure 8.12 Base station Tx and Rx bands for cellular, SMR, and PCS.

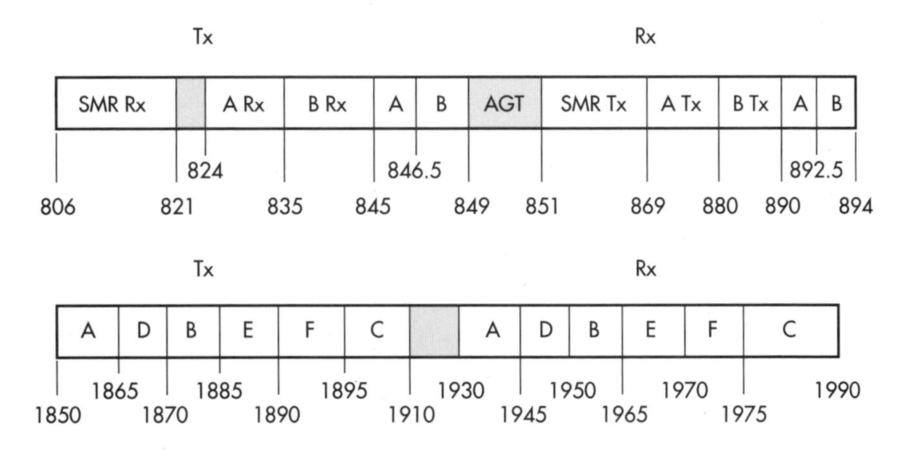

Figure 8.13 Subscriber unit Tx and Rx bands for cellular, SMR, and PCS.

Q. *What is the frequency separation between Tx and Rx frequency bands for SMR radio channels?*

A. The frequency separation between SMR Tx and Rx channels, since it is a full-duplex system, is 45 MHz.

Q. *What purpose does the cell site Tx filter have?*

A. The purpose of the cell site's transmit filter is to suppress the emissions that come from the modulator. The unwanted signals that need to be suppressed could be transmit emissions that fall into the receive band of either one operator's system or another operator's system. The transmit filter for a cell site should ensure that the transmitter emissions remain within the FCC–defined emissions mask. In addition, it is not uncommon for a communication system to install a more selective transmit filter to further reduce the out-of-band energy emissions into other systems even though they are meeting the FCC emissions mask requirements.

Q. *What is the purpose of the cell site Rx filter?*

A. The receive filters at the cell site are meant to prevent receiver overload, suppress the third-order *intermodulation products* (IMPs) below the receiver noise floor, and improve the cell site's selectivity.

Q. *What does an A-band cellular Rx filter system look like?*

A. The A-band receive filter system is best described in terms of a block diagram, as shown in Fig. 8.14. There are two basic types of filters that are used for the A-band cellular receive system. The first system uses a simple band-pass filter while the second one utilizes two band-pass filters that are cascaded. The advantage the cascaded one has over the single band-pass filter is that signals are not suppressed in the desired pass band. However, the additional band-pass filter increases the complexity, size, and, of course, cost.

Q. *What does a B-band cellular Rx filter system look like?*

A. The B-band receive filter system is best described in terms of a block diagram, as shown in Fig. 8.15. There are two basic types of filter configurations that are used for B-band cellular receive systems. The first system utilizes a simple band-pass filter while the second one uses a band-pass filter with a notch filter. The advantage of including the notch filter is that the signals are not suppressed in the desired pass band. However, the additional notch filter increases the complexity, size, and, of course, cost.

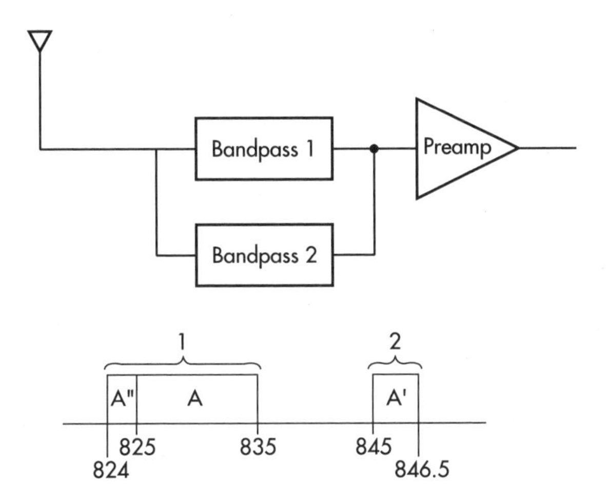

Figure 8.14 Cellular A band.

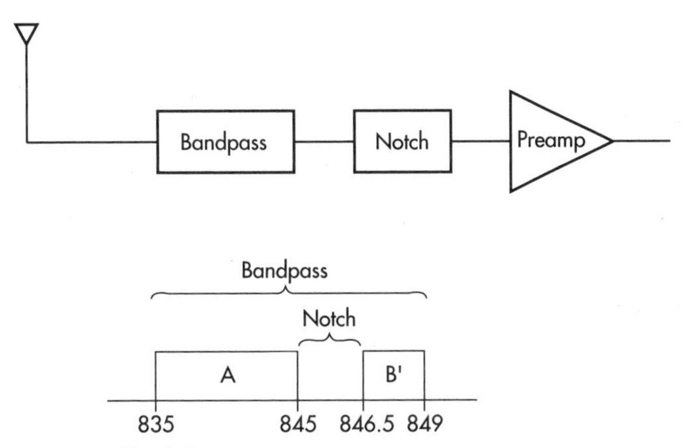

Figure 8.15 Cellular B band.

Q. *How are the filter requirements determined for colocation?*

A. Example 8.1 provides the filter requirements for a communication system involving receiver overload. The example involves a colocation of a B-band cellular operator with an ESMR operator. The example does not consider the issue of composite energy and phase noise to a receiver.

Example 8.1

Tx = 852 MHz	
Rx = 849 MHz	
Tx power	+50 dBm
Rx 1-dB compression	−27 dBm
	77 dB isolation needed

Filter attenuation of Tx in Rx band 20 dB

Isolation	77 dB
Filter attenuation	−30 dB
	47 dB isolation needed

47 dB isolation ≈ 10-ft vertical separation

∴ Filter attenuation	30 dB
Vertical isolation	+50 dB at 10 ft
	80 dB isolation

80 dB > 77 dB

Q. *Can a more selective Rx filter reduce out-of-band emissions from another carrier into the receiver?*

A. Unfortunately, a more selective Rx filter will not reduce out-of-band emissions from another carrier into the receiver. Presumably, this question would arise in a colocation situation or when the offending cell site is near another cell site. The reason a more selective Rx filter will not remove another carrier's out-of-band emissions is that the out-of-band emissions of concern are those components that are in the receive band, and a more selective filter, while helping suppress energy not in the first operator's pass band, will not remove the in-band portion of the signal from the other system. The reason is that the emissions from the other system also fall within the first operator's receive pass band and the receive filter is designed to pass those frequencies.

Q. *What type of filter should be used for reducing out-of-band emissions from one carrier's transmitter into another carrier's receive?*

A. The type of filter that should be used for reducing out-of-band emissions from one carrier's transmitter into another carrier's receive is actually a combination of Rx and Tx filters. The Rx filters for the victim cell site should be as selective as economically possible. The offending carrier should utilize a more selective Tx filter, which is needed to protect their competitor's receiver. One EMSR carrier has made the inclusion of a more selective Tx filter standard installation; however, the out-of-band emission problem at this time still exists for PCS operators in the United States—in particular, the PCS A-band Tx into the C-band Rx band.

9

Antennas

The antenna system for any wireless cell site or subscriber unit is one of the most critical and least understood parts of the system. Many people can point to an antenna and talk about its "gain." However, the particular performance criteria that should be factored into the selection and use of an antenna to maximize its performance are rarely, if ever, used.

This chapter addresses many of the questions commonly asked pertaining to antennas. It is hoped that this chapter contains all the possible questions about antennas that would normally arise in a wireless communication system. As a matter of course, detailed knowledge of how an antenna actually couples the energy between the radio system and the atmosphere, that is, E&M theory, is not as important to understand as what a particular antenna type and its associated performance criteria actually mean in system design and performance.

Q. *What is an electromagnetic wave?*

A. An *electromagnetic wave,* for example, a radio wave, is made up of the electric and magnetic *lines of force,* that in turn make up an electric and magnetic field. These two equal forces, which are oriented at right angles to each other, are commonly called the *E field* (electric) and the *H field* (magnetic). The *E* and *H* fields are of equal importance in supporting the propagation of radio waves. A diagram of an electromagnetic wave is shown in Fig. 9.1.

In general, it is the *E* field that is used in defining an antenna's *polarity.* An antenna that has a *horizontal polarity* is oriented such that the antenna's *E* field is parallel to the earth. An antenna that has a *vertical polarity* is oriented such that its *E* field is vertical, or perpendicular, to the ground. All cellular and PCS antennas, and virtually all mobile radio system antennas, are vertically polarized, meaning that their *E* fields are vertical to the earth. In contrast, television broadcasting antennas are horizontally polarized.

Q. *What is circular polarization?*

A. *Circular polarization* (CP) is produced when equal electrical fields in the vertical and horizontal planes of radiation are out of phase by 90 degrees and rotated a full 360 degrees within one wavelength of the operating frequency. Circulation polarization is used for FM broadcasting and many wireless systems, but it is

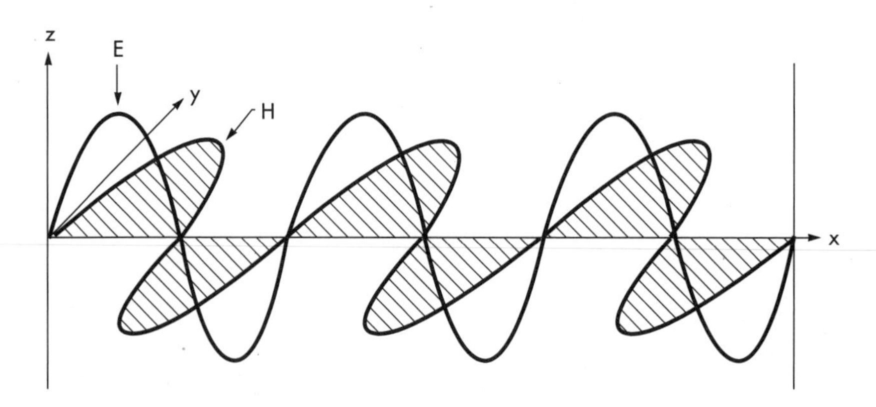

Figure 9.1 Electromagnetic wave.

not used in cellular, PCS, or ESMR systems. Circular polarization is found primarily in microwave point-to-point systems.

Q. *What is an E field?*

A. An *E* field is the electric field of an electromagnetic wave. The *E* field defines the polarization of a wireless system and is therefore often referred to as the *vertical pattern for an antenna.*

Q. *What is an H field?*

A. An *H* field is the magnetic field of an electromagnetic wave. Because its lines of force run horizontally, the *H* field is often referred to as the *horizontal pattern for an antenna.*

Q. *What is antenna polarization?*

A. *Antenna polarization* is the electric charge of the electromagnetic field maintained by the antenna. An antenna's polarization is defined by its *E*-field lines of force, or vectors.

Q. *What is vertical polarization?*

A. *Vertical polarization* is a characteristic of an electric field in which the lines of force are vertical to the ground at 90 degrees. Vertical polarization is the predominant method of polarization used for mobile communication systems. Cellular and PCS systems use vertical polarization.

Q. *What is horizontal polarization?*

A. *Horizontal polarization* is a characteristic of an electric field in which the lines of force are horizontal, or parallel, to the ground. Horizontal polarization is not used for mobile communication; however, it is used to increase isolation and to improve *C/I* performance as needed by control stations in an ESMR/SMR system.

Q. *How are the different types of antennas classified?*

A. There are two classes into which all the various types of antennas fall: The first is the *omni,* or, rather, *omnidirectional, class,* and the other is the *directional,* or, rather, *sector, class.*

Q. *Where are omni antennas used?*

A. Omni antennas are used wherever it is desirable to obtain a 360-degree radiation pattern. Thus omni antennas are found in rural or micro cell sites or in distributed-antenna systems used inside buildings.

Q. *Where are directional antennas used?*

A. Directional antennas are used wherever it is desirable to obtain a refined radiation pattern. A directional radiation pattern is usually needed to facilitate wireless system growth through frequency reuse or to shape a system's contour.

Q. *Where within the radio system are antennas inserted?*

A. Antennas are located at the interface between the radio equipment and the external environment. An antenna can either transmit, Tx, or receive, Rx, electromagnetic signals. In a wireless communication system, the antenna system can consist of a single antenna at a base station that transports electromagnetic signals to a single antenna located at a receiving station, which can be a mobile unit. This simple antenna system is sufficient for establishing and maintaining a communication link.

Q. *How are antennas used at cell sites?*

A. Cell site antennas, which can be either directional or omnidirectional antennas, are used to transmit or receive radio signals generated in wireless communication systems. The number of cell site antennas is directly dependent upon the cell site's configuration, the type of equipment the cell site vendor offers, the technology platform used, the frequency band of operation, and the various land-use constraints in effect at the cell site.

A typical antenna array for a directional, or sector, cell site is shown in Fig. 9.2. The diagram depicts an array of three antennas for each of the three sectors. Note that a GPS antenna is not shown in the figure.

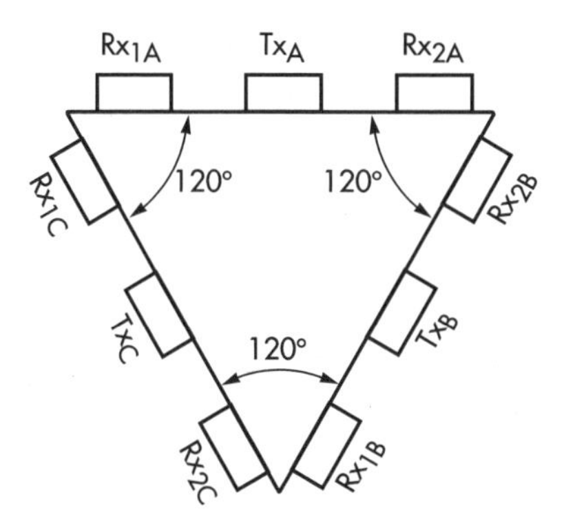

Figure 9.2 Three-sector antenna array.

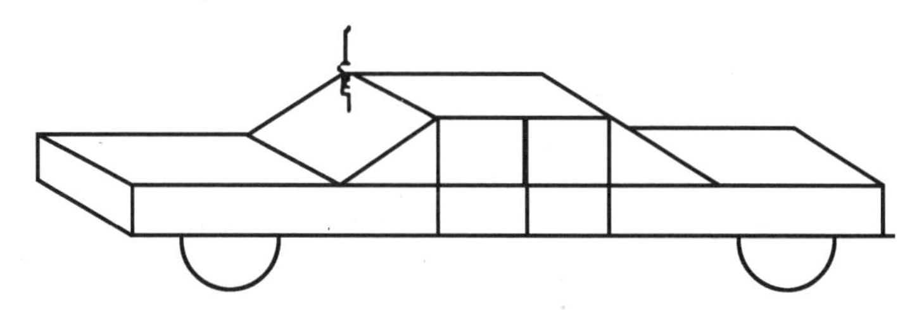

Figure 9.3 Vehicle antenna location.

Q. *What is a vehicle antenna?*

A. A *vehicle antenna* is a duplexed antenna that is used for transmission and reception of radio signals in cellular, ESMR, or PCS wireless networks. Usually there is only one antenna in this type of installation, and it is mounted on the rear window of the vehicle. Figure 9.3 shows the most common location of a vehicle antenna.

Q. *What are some of the problems encountered with vehicle antennas?*

A. Vehicle antennas typically perform very well. However, some problems tend to occur more often than others, some of which are listed below:

- The antenna is missing.
- The cable going from the transceiver to the antenna is crimped.

- The antenna is mounted on a tinted section of the window (rather than on a clear section).
- The connectors have been poorly constructed.
- The antenna is not plumb.

Q. *What are portable antennas?*

A. In cellular applications, the term *portable antennas* usually refers to antennas that are attached to portable phones. This type of antenna in many portable models can be extracted or retracted. The extraction of the antenna generally improves the mobile unit's performance, which results in a high-quality received signal.

Q. *What factors determine the optimal size (height) of an antenna for a particular application?*

A. The physical size of an antenna is calculated in direct relation to the frequency range in which it is designed to operate. Another important factor to be considered is the desired gain. The larger the antenna, the higher its gain will be.

Some unscrupulous antenna manufacturers or distributors will provide an expected gain value but not specify whether the value is being stated in terms of an isotropic antenna or a dipole antenna. It is important to note that a 5-dB gain produced by an isotropic antenna equals a 3-dB gain produced by a dipole antenna of the same height.

A simple rule of thumb to follow is that for every 3 dB of gain desired, the size of the antenna will have to double. Alternatively, reducing an antenna size by half will result in a 3-dB reduction in gain. The above relationship is based of course on proper antenna design.

Q. *What is an isotropic antenna?*

A. An *isotropic antenna* radiates energy uniformly in all directions, and it is therefore used as a reference for all antennas. Needless to say, a truly isotropic antenna exists only in theory, not in fact. In contrast, a dipole antenna commonly has a power gain of 0 dBd, or 2.14 dBi, above that of an isotropic antenna. A diagram of a theoretical isotropic antenna is shown in Fig. 9.4.

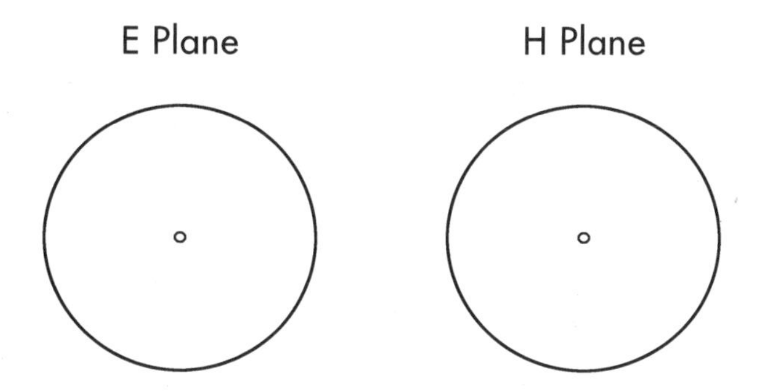

Figure 9.4 Isotropic antenna.

Q. *How is the near field of an antenna calculated?*

A. The near field for an antenna is directly related to the antenna's operating frequency range and the antenna's physical size. The equation that is used to determine where the near-field area begins is the following:

$$\text{Near field} = 2D^2/\lambda$$

The following table shows the relationship between an antenna's physical size and its operating frequency range:

	Near field, ft	
Antenna aperture size, ft	**880 MHz**	**1900 MHz**
1	1.8	3.89
2	7.23	15.39
4	28.95	62.37
8	115.83	249.5

Q. *What is a dipole antenna?*

A. A *dipole antenna* consists of two horizontal rods in line with each other such that their ends (or poles) are slightly separated

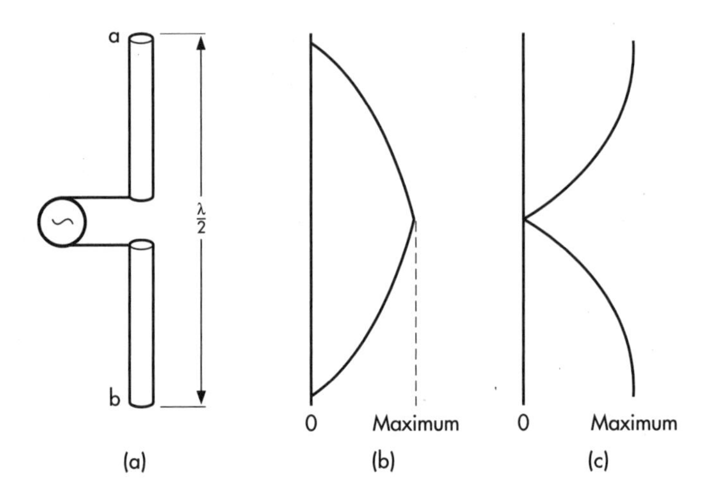

Figure 9.5 Dipole antenna. (*a*) $G = 2.14$ dBi. (*b*) Current (I). (*c*) Electric field (E).

and oppositely charged. It is the smallest self-resonating antenna, and it employs a conductor that is half the wavelength at which the system is supposed to operate. The dipole has a gain of 2.14 dB over that of an isotropic antenna. The dipole antenna is a common reference for any communication system. A diagram of a dipole is shown in Fig. 9.5.

Q. *What is a collinear antenna?*

A. A *collinear antenna* consists of a series of dipole elements that are laid end to end, in a straight line. The elements operate in phase, and the antenna is referred to as a *broadside radiator.* The maximum radiation for the collinear antenna takes place along the axis of the dipole array, and the array consists of a number of parallel elements in one plane.

Both omni and directional antennas may be collinear in design, and many cell site omni antennas are collinear. Figure 9.6 shows a four-element collinear array.

Q. *What is a log periodic antenna?*

A. A *log periodic antenna,* or a *log periodic dipole array* (LPDA) *antenna*, is a directional antenna whose gains, standing-wave ratios (SWRs), and other key figures of merit remain constant

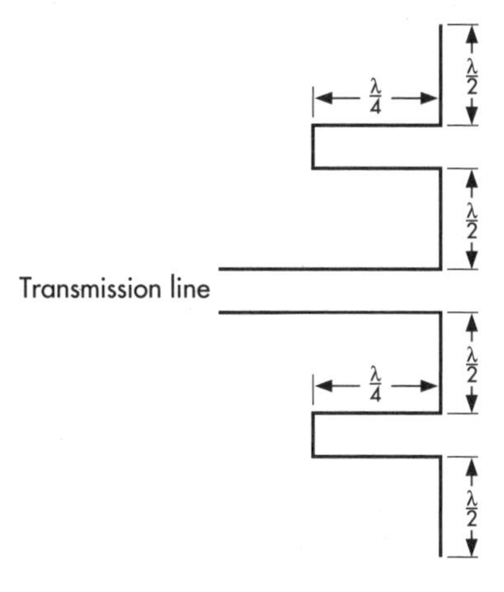

N elements	Gain (dB)
2	1.9
3	3.2
4	4.3

Figure 9.6 Four-element collinear array.

over the operating band. The LPDA is used where a large bandwidth is needed, and its typical gain is 10 dBi. The LPDA is used extensively in wireless operations—such as cellular, PCS, and ESMR systems—due to its excellent performance. Figure 9.7 shows a picture of an LPDA antenna.

Q. *In what types of applications are folded dipole antennas used?*

A. A *folded dipole antenna* is commonly used in two-way communication systems; it is not used in cellular and PCS systems. A diagram of a folded dipole antenna is shown in Fig. 9.8.

Q. *What is a yagi antenna?*

A. A *yagi antenna* is a highly directional and selective antenna that consists of a horizontal conductor of one or two dipole antenna rods connected to a receiver or transmitter and of a set of nearly equal insulated dipole rods parallel to and on a plane with the horizontal conductor. Yagi antennas are used in a variety of applications such as rerads and bidirectional amplifiers. A diagram of a yagi antenna is shown in Fig. 9.9.

Q. *What is a dish antenna?*

A. A *dish antenna* is a directional antenna for which the antenna elements are arrayed in a concave-shaped platform. A broad range of antenna types fit the name "dish antenna." A familiar applica-

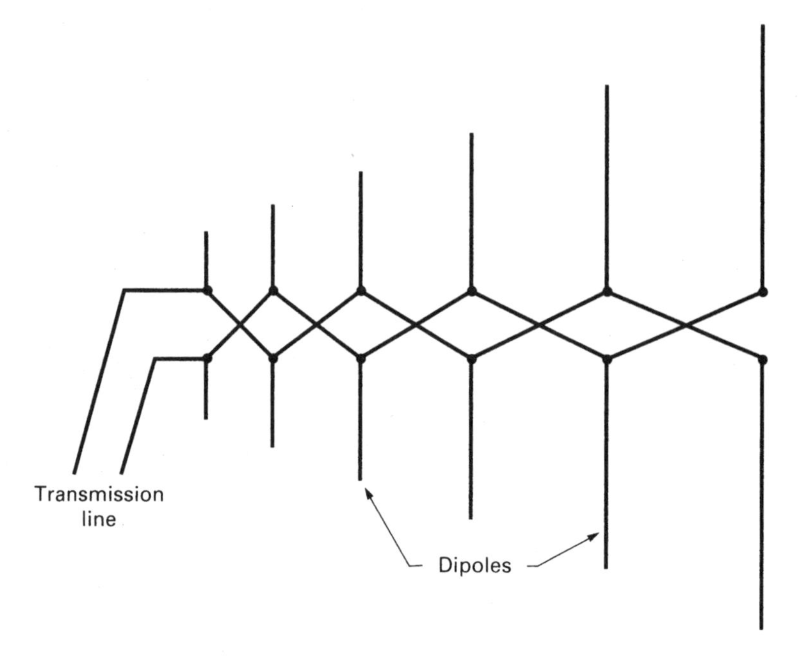

Figure 9.7 LPDA antenna.

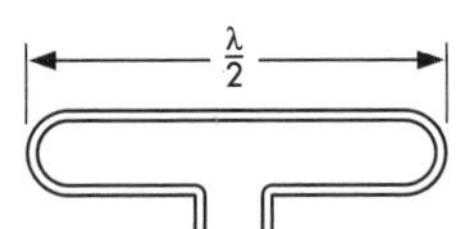

Figure 9.8 Folded dipole.

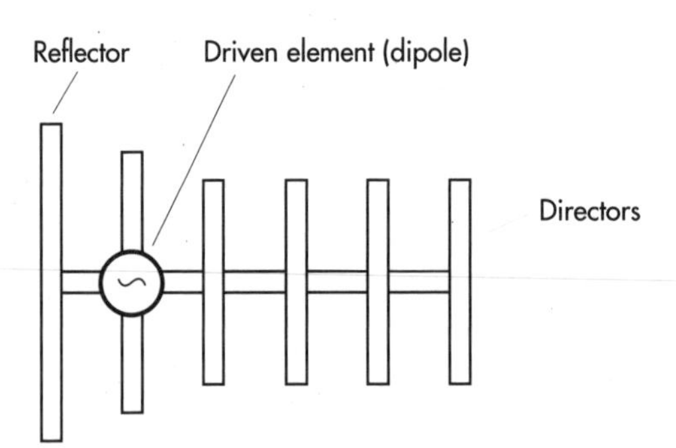

Figure 9.9 Yagi antenna.

Figure 9.10 Panel antenna.

tion of dish technology is found in satellite communications. Dish antennas are also used predominantly in the cellular, PCS, and ESMR systems as reradiators and microwave systems, and in those instances they are used as alternative interconnect tools.

Q. *What is a panel antenna?*

A. A *panel antenna* is a directional antenna for which the antenna elements are arrayed on a flat, rectangular-shaped platform. A broad range of antenna types fit the label "panel antenna." This type of antenna has found widespread use in the cellular and PCS industries. A diagram of a panel antenna is shown in Fig. 9.10.

Q. *What sizes of coaxial cable are commonly used in wireless systems?*

A. The coaxial cable sizes commonly available for use in wireless systems are $^1/_2$ in, $^7/_8$ in, $1^1/_4$ in, and $1^5/_8$ in.

Q. *What is a reflector?*

A. A *reflector* is a physical device that is placed behind an antenna for the sole purpose of increasing the antenna's directionality and gain. Figure 9.11 shows two applications for antenna reflectors.

Q. *What is a corner reflector?*

A. A *corner reflector* has electrical characteristics that resemble those of yagi antennas. Corner reflectors are, in fact, often used in place of yagi antennas because they are so similar. A corner reflector actually has better directivity but at the expense of windload and size. Figure 9.12 shows a corner reflector antenna.

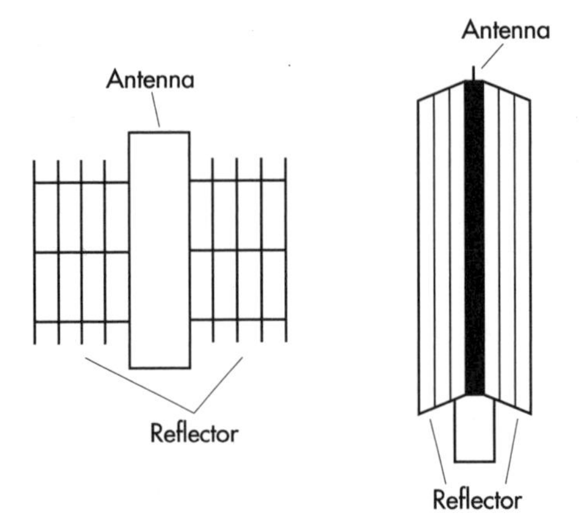

Figure 9.11 Antenna reflectors.

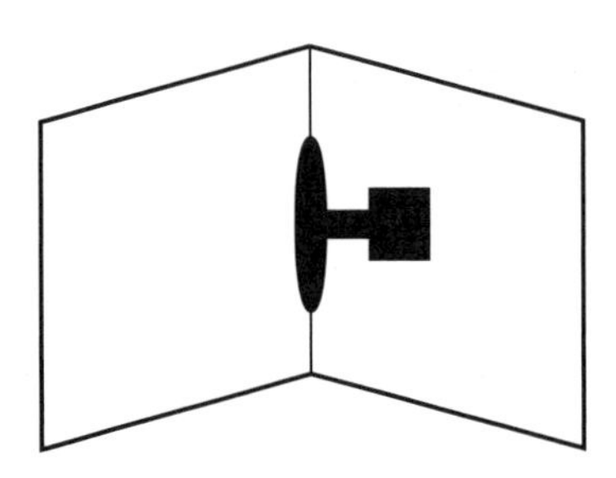

Figure 9.12 Corner reflector.

Q. *What are jumpers?*

A. *Jumpers* are the coaxial cables that connect the feed line for the antenna system to either the antenna elements or the receiver multicoupler for the radio equipment. When they are used to connect the feed line to the antenna, the jumpers are usually the primary maintenance problem in an antenna system. A jumper is usually about 5 ft long, but it could be longer if required.

Q. *What are mounting brackets?*

A. *Mounting brackets* are the physical equipment needed to mount antennas to a structure.

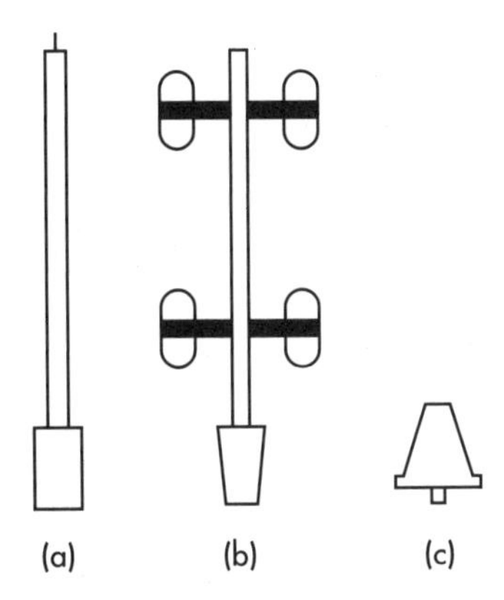

Figure 9.13 Omni antenna. (*a*) Collinear array. (*b*) Folded dipole array. (*c*) Omni (one indoor application).

Q. *What do omni antennas look like?*

A. Figure 9.13 shows some simple omni antennas.

Q. *What do directional antennas look like?*

A. Figure 9.14 shows some of the various directional antennas that are used in wireless communication systems.

Q. *What are the criteria against which antenna performance is compared for evaluation purposes?*

A. The performance criteria for antennas are not restricted to their gain characteristics, physical attributes, and maintenance requirements. Rather, there are many parameters that must be taken into account when evaluating antenna performance. These are collectively referred to as an antenna's *figures of merit* (FOM). The FOM characteristics are listed below:

- Antenna pattern
- Main lobe
- Side-lobe suppression
- Input impedance
- Radiation efficiency
- Horizontal beamwidth

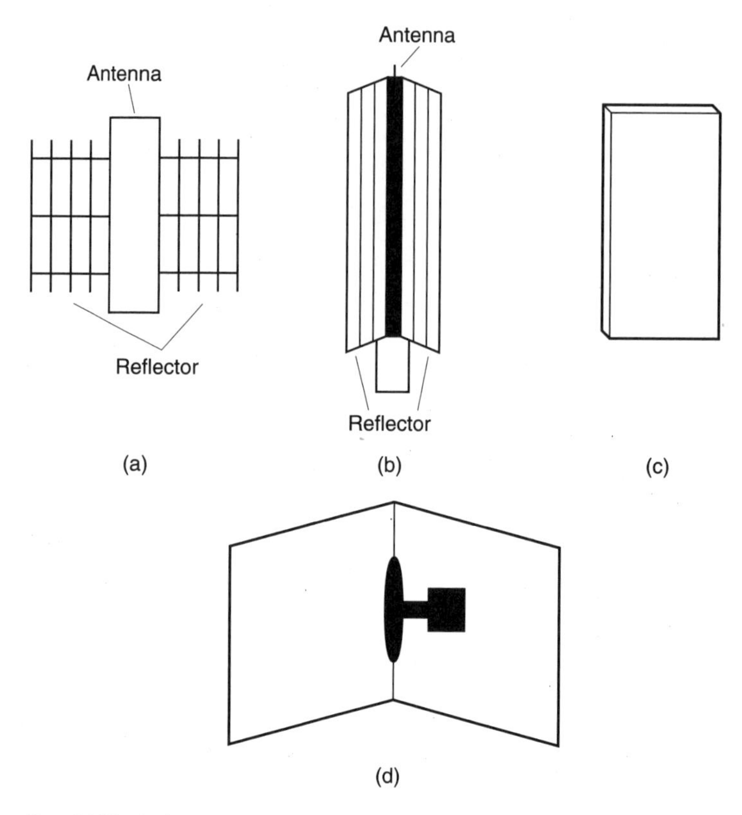

Figure 9.14 Directional antennas.

- Vertical beamwidth
- Directivity
- Gain
- Antenna polarization
- Antenna bandwidth
- Front-to-back ratio
- Power dissipation
- Intermodulation suppression (PIM)
- Construction
- Cost

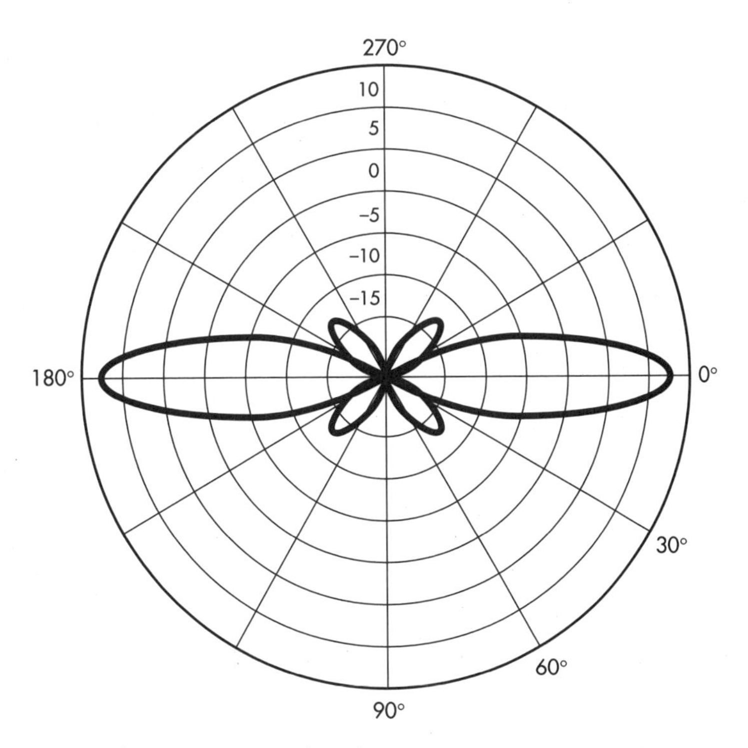

Figure 9.15 Elevation antenna pattern (omnidirectional).

Q. *What is an omnidirectional antenna pattern?*

A. An *omnidirectional antenna pattern,* also called an *omni antenna pattern,* is the path followed by radio waves dispersed by an omni antenna. This path can be represented graphically, as shown in Figs. 9.15 and 9.16, so that the elevation and azimuth antenna characteristics may be plotted.

It is very important to note the reference scale that is used to construct the graph because different scales can lead to different conclusions about the antenna's actual pattern.

Q. *What is an omnidirectional, or sector, antenna pattern?*

A. A *directional,* or *sector,* antenna is the path followed by radio waves dispersed by a directional, or sector, antenna. This path can be represented graphically, as shown in Figs. 9.17 and 9.18,

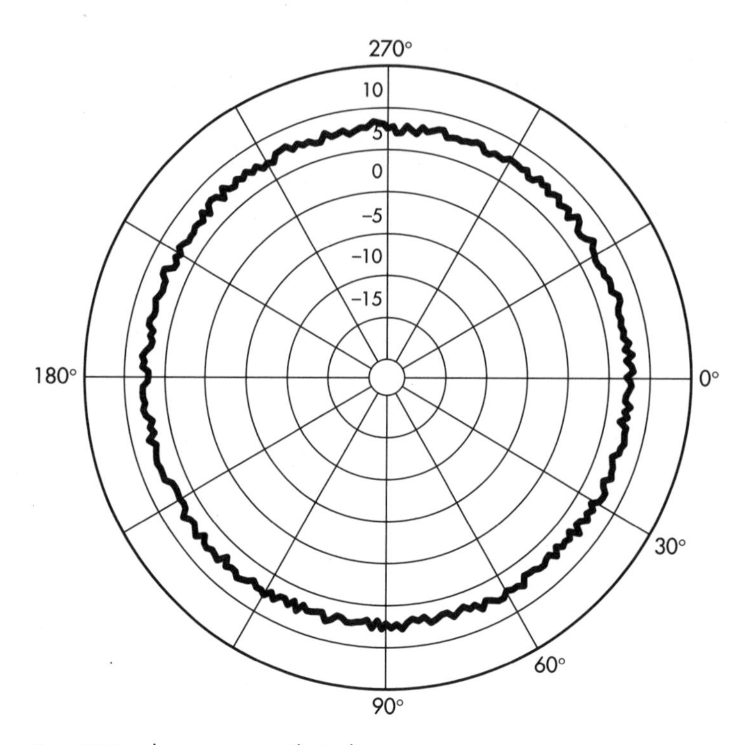

Figure 9.16 Azimuth antenna pattern (omnidirectional).

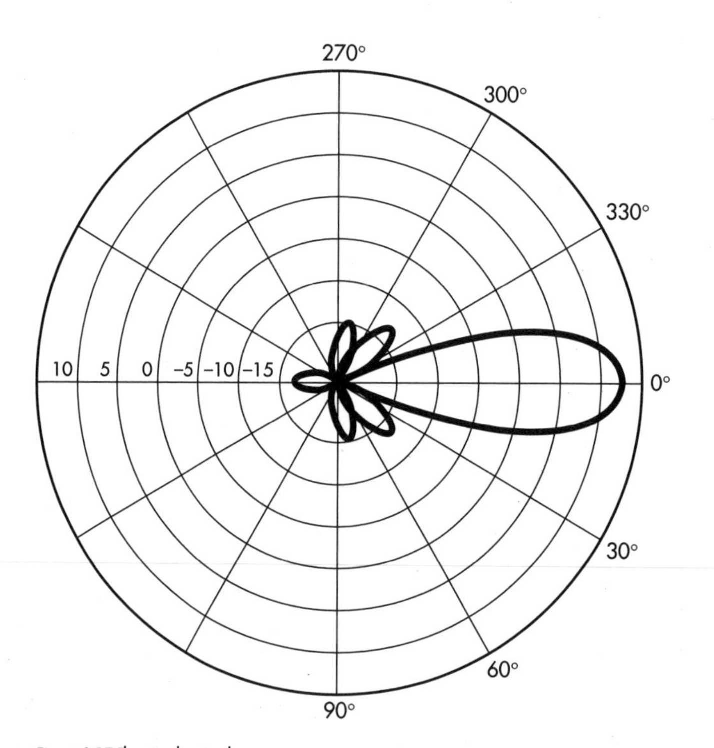

Figure 9.17 Elevation directional antenna.

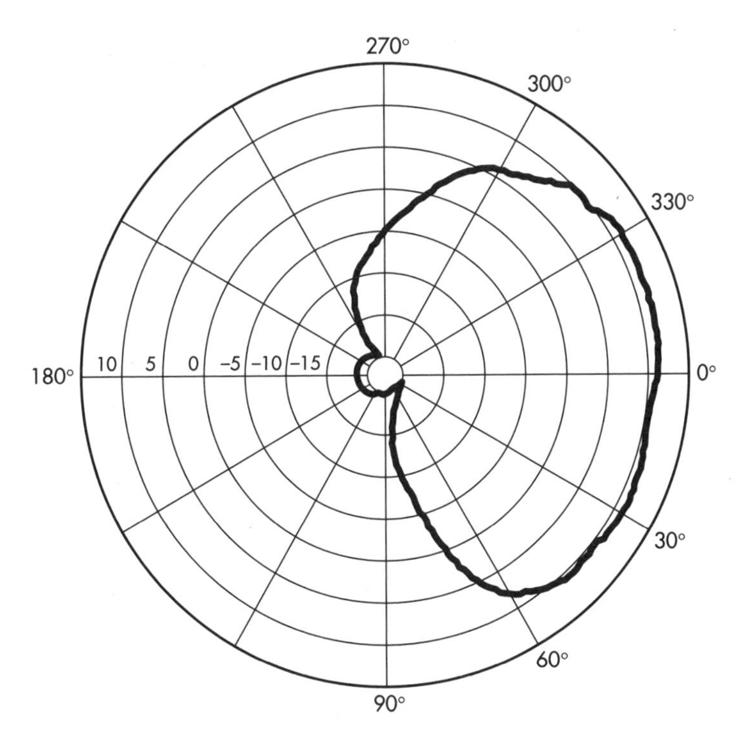

Figure 9.18 Azimuth directional antenna.

so that the elevation and azimuth antenna characteristics may be plotted.

It is very important to note the reference scale that is used to construct the graph because different scales can lead to different conclusions about the antenna's actual pattern.

Q. *What is the main lobe of an antenna?*

A. The *main lobe* of an antenna is its radiation lobe, which has the maximum radiated power in a given direction. The main-lobe direction and polarization reflect the directivity of the antenna itself. If the polarization is vertical, then the main-lobe graphical representation is the antenna's elevation pattern. (See Fig. 9.19.)

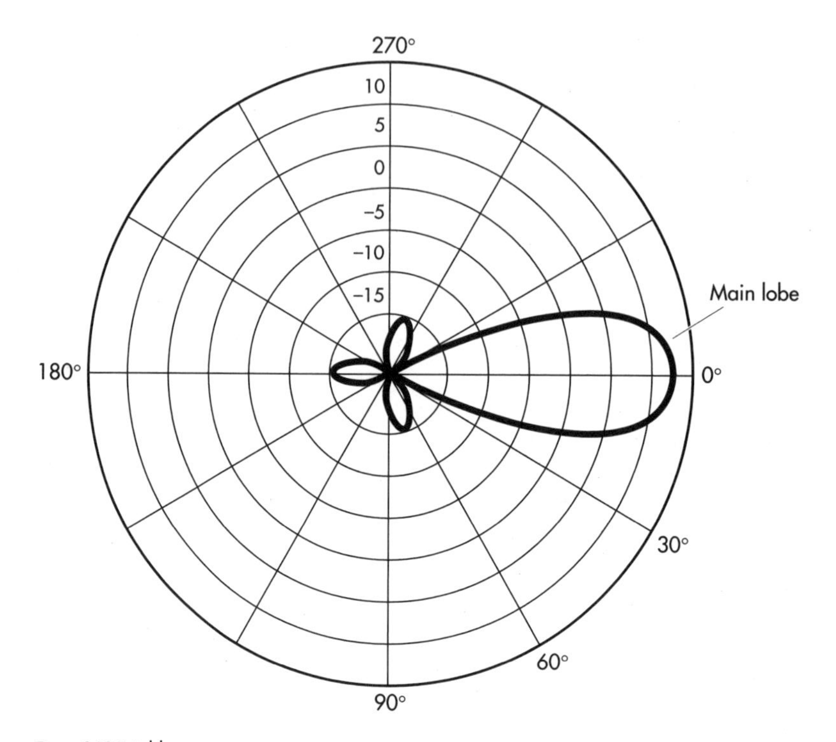

Figure 9.19 Main-lobe antenna pattern.

Q. *What is an antenna side lobe?*

A. An *antenna side lobe* is any radiation lobe that is oriented in a direction other than the main-lobe orientation. Depending on the needs for antenna performance, there may be multiple side lobes to bolster the main lobe. (See Fig. 9.20.) Ideally, however, there should be no side lobes for an antenna.

Q. *What is the antenna beamwidth?*

A. The *antenna beamwidth* is the angular separation between two radio waves that are going in the same direction and that have identical levels of radiation intensity (Fig. 9.21). The half-power point for the beamwidth is usually the angular separation where there is a 3-dB reduction off the main lobe. The beamwidth is an important FOM.

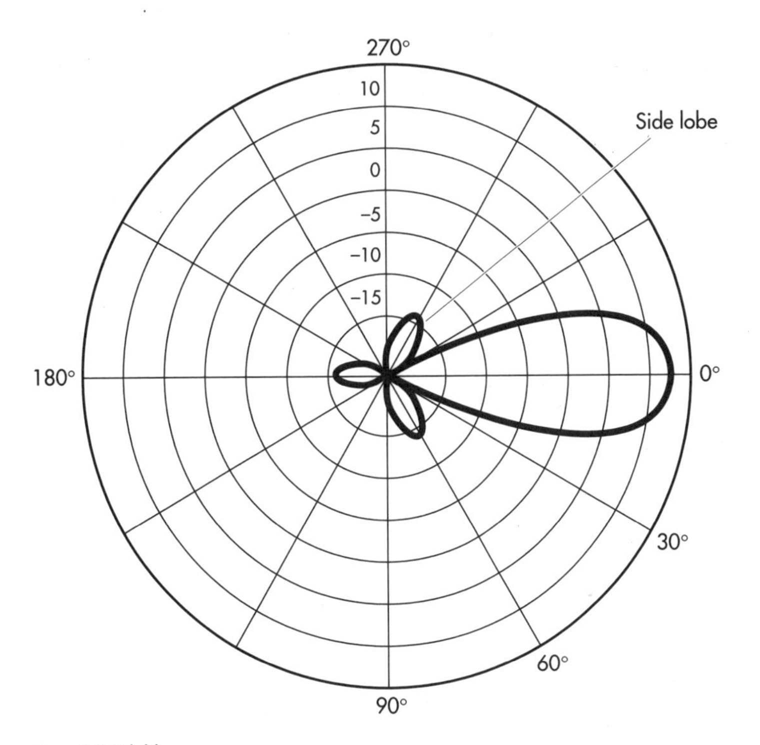

Figure 9.20 Side-lobe antenna pattern.

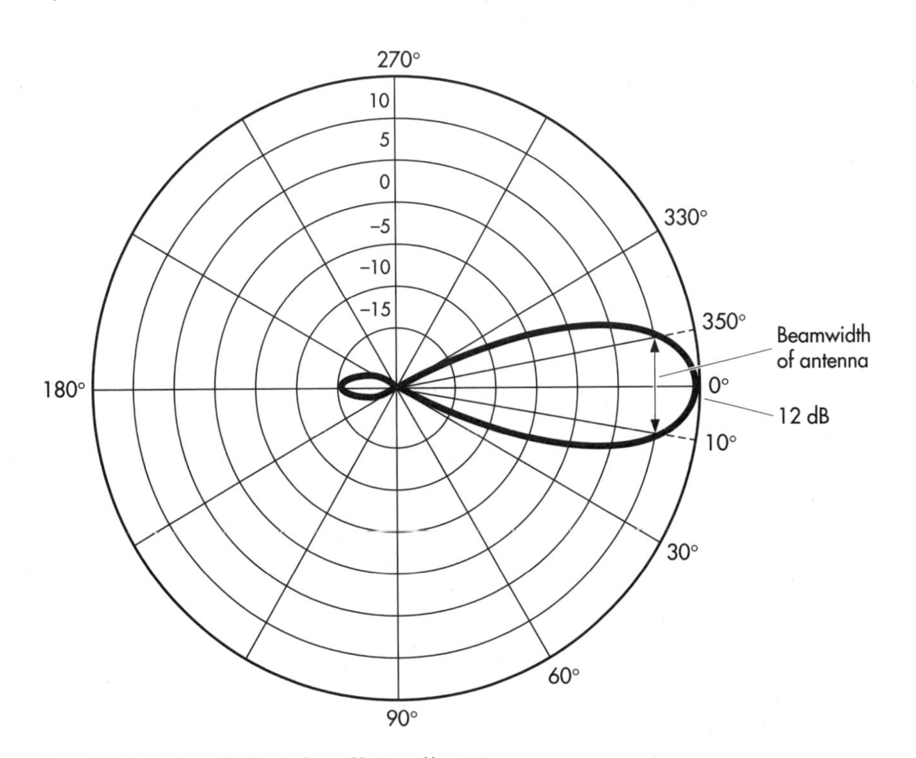

Figure 9.21 Antenna beamwidth/elevation beamwidth. Beamwidth = 20°.

Q. *What is the elevation, or vertical, beamwidth of an antenna?*

A. The *elevation beamwidth* of an antenna, also referred to as the *vertical beamwidth,* is the angular separation between two radio waves that are going in the vertical direction and that have identical levels of radiation intensity (Fig. 9.21). The half-power point for the elevation beamwidth is usually the angular separation where there is a 3-dB reduction off the main lobe. The elevation beamwidth is an important FOM.

Q. *What is the azimuth, or horizontal, beamwidth of an antenna?*

A. The *azimuth,* or *horizontal, beamwidth* of an antenna, also referred to as the *horizontal beamwidth,* is the angular separation between two radio waves that are going in the horizontal direction and that have identical levels of radiation intensity (Fig. 9.22). The half-power point for the azimuth beamwidth is usually the angular separation where there is a 3-dB reduction off the main lobe. The azimuth beamwidth is an important FOM.

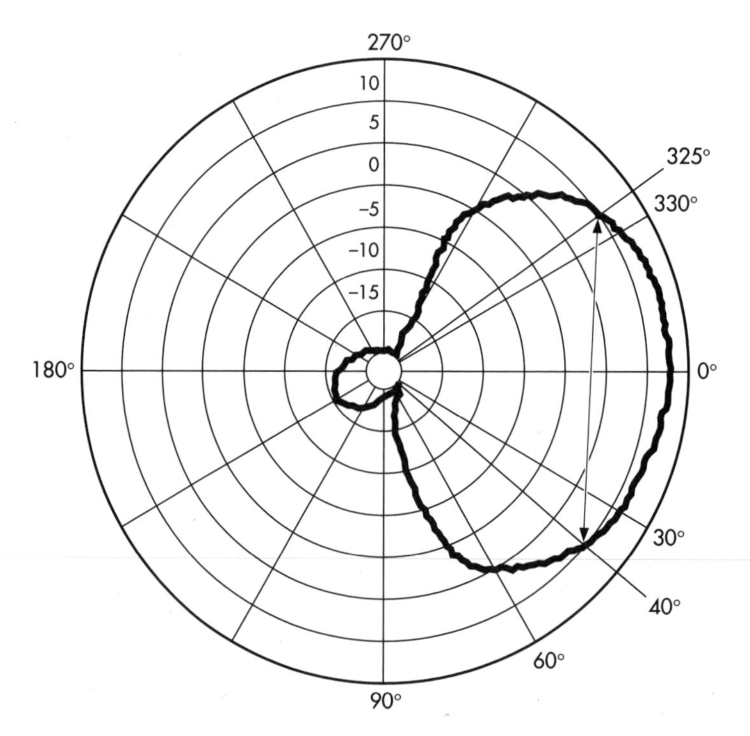

Figure 9.22 Azimuth beamwidth. Beamwidth = 75°.

Q. *What is directivity?*

A. *Directivity* is the ratio of radiation intensity of a radio wave going in a certain direction to that of the radiation intensity averaged over the intensity of the radio waves going in all the other directions. The directivity of an antenna can be improved by using reflectors or adjusting the physical location of the antenna to a leg of a tower.

Q. *What is the gain of an antenna?*

A. The *gain* of an antenna is the ratio of the radiation intensity of the radio waves going in a given direction to radiation intensity of an isotropically radiated signal. The gain of an antenna, in this case 12 dB, is shown in the elevation pattern in Fig. 9.23. Note that from the pattern it is unclear whether this gain is referenced to a dipole or isotropic antenna.

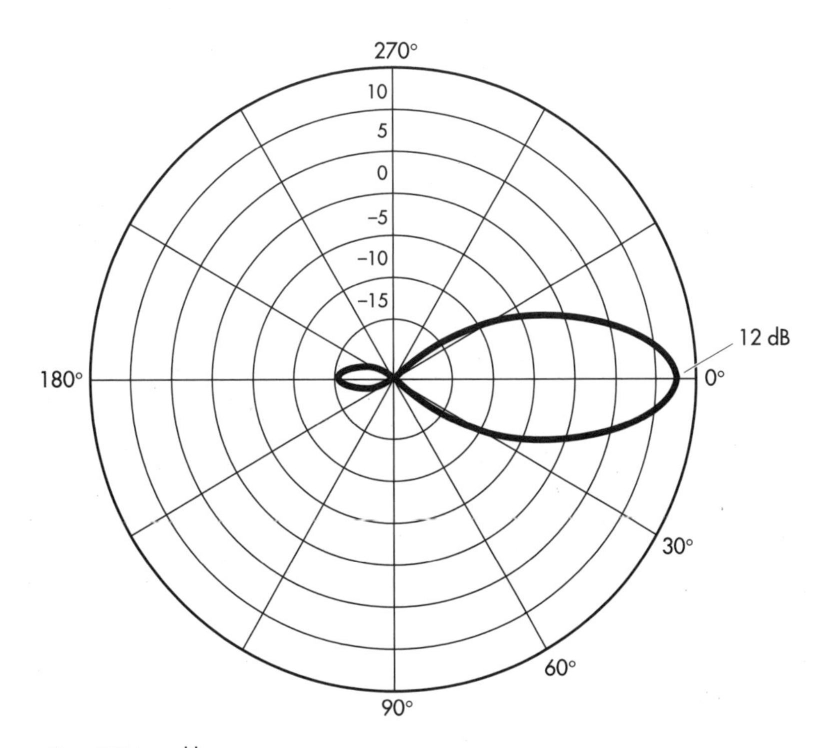

Figure 9.23 Antenna lobe.

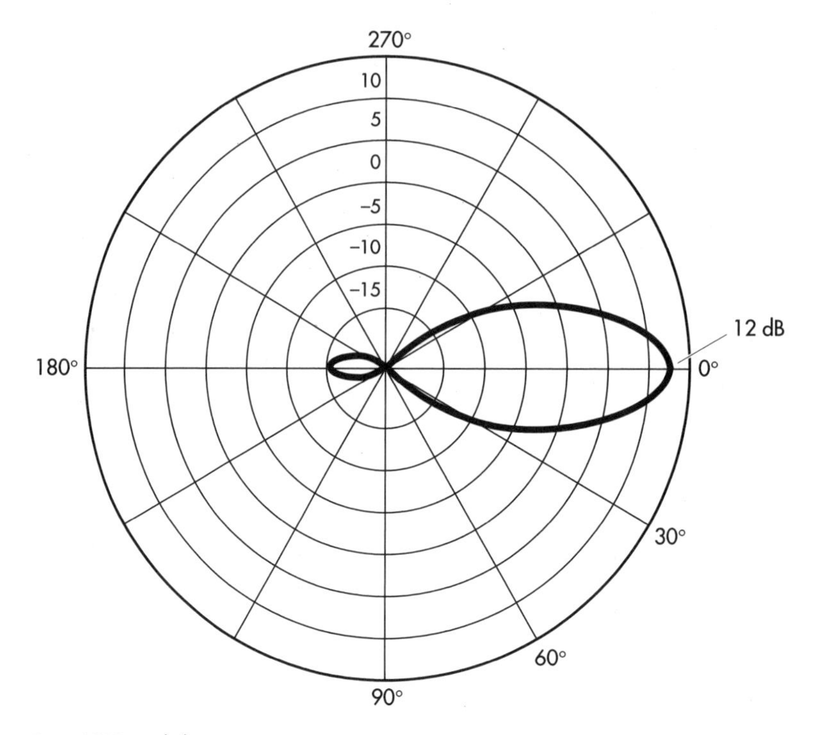

Figure 9.24 Front-to-back ratio.

Q. *What is the front-to-back ratio?*

A. The *front-to-back ratio* is the ratio of the amount of radiation energy that is directed in the exact opposite direction of the energy in the main lobe of the antenna. The front-to-back ratio is an approximation and is applicable only for a directional antenna. A high front-to-back ratio is important for improving the *C/I* levels in a system; however, if the antennas are installed on a building, the front-to-back ratio is not an important FOM. A diagram of a 27-dB front-to-back ratio is shown in Fig. 9.24.

Q. *What is passive intermodulation?*

A. *Passive intermodulation* (PIM) is the term used to describe the intermodulation products generated by an antenna or any passive element in a radio system network. System engineers focus on reducing passive intermodulation products generated by antennas as a way to improve network system performance. The following is an example of a PIM calculation:

$$\text{PIM value} = 150 \text{ dBc}$$

$$\text{Tx power} = +40 \text{ dBm} \qquad \text{(raw power into antenna)}$$

$$\text{PIM} = +40 \text{ dBm} - 150 \text{ dBc}$$

$$= -110 \text{ dBm}$$

For the above example, a PIM level of −110 dBm for intermodulation products that have the potential to fall in the receive band may be of concern in a duplexed system. In that situation, all the signals below −90 dBm, 20 dB *C/I* or *C/N*, will be degraded.

Q. *Do portable or mobile units use diversity schemes?*

A. Neither portable nor mobile units use diversity schemes because they have only one antenna.

Q. *What is the most common diversity scheme used?*

A. The most common diversity scheme used for mobile communication is spacial diversity, which is achieved by horizontally separating the base station antennas. The determination of how much separation there should be between antennas is accomplished through the following equation (see also Fig. 9.25):

$$n = \frac{h}{d} = 11$$

where h = the antenna height, ft, and d = the distance between antennas, ft.

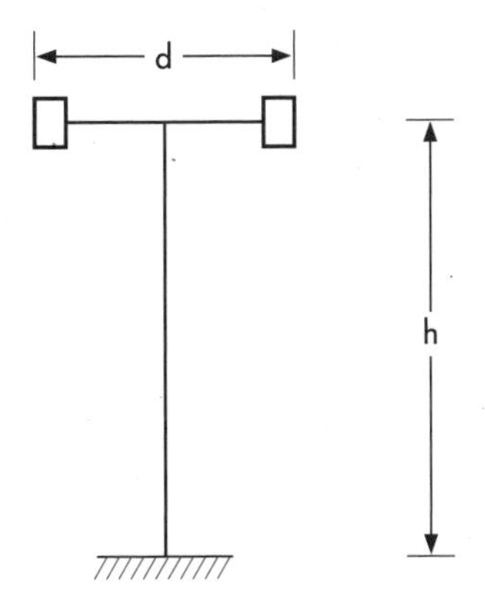

Figure 9.25 Horizontal separation. h = height, d = distance.

Q. *What amount of horizontal separation between antennas is needed?*

A. The objective is to have the antennas spaced from 10 to 20 wavelengths apart within the operating frequency.

Q. *What is vertical antenna isolation?*

A. *Vertical antenna isolation* (VI) is a measure in decibels of how much a signal is attenuated between two vertical antennas. The amount of vertical isolation is a critical design issue for colocation situations. Vertically separating antennas achieves the highest single level of attenuation.

Q. *How is the vertical isolation calculated?*

A. The amount of vertical isolation, attenuation, that is achieved is dependent upon the operating frequency and the spacing between the antennas. Figure 9.26 and the following equation illustrate how to calculate vertical isolation.

$$\mathrm{VI} = 28 + 40 \log_{10}\left(\frac{y}{\lambda}\right) \quad \mathrm{dB}$$

where $y > \lambda$

$$\therefore \text{ if } f = 890 \text{ mHz} \qquad y = 10 \text{ ft} = 120 \text{ in}$$

$$\mathrm{VI} = 28 + 40 \log_{10}\left(\frac{120}{13.267}\right)$$

$$= 28 + 38.256$$

$$= 66.25 \text{ dB}$$

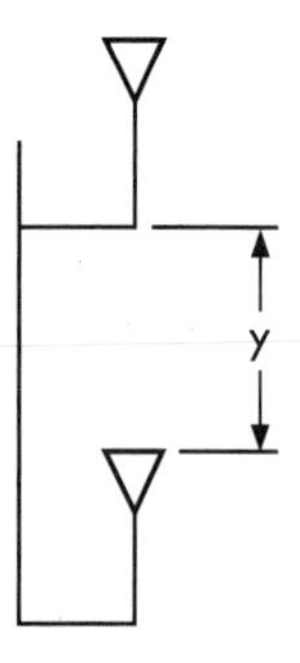

Figure 9.26 Vertical isolation. y = distance between ends of antennas.

Q. *What is horizontal antenna isolation?*

A. *Horizontal antenna isolation* (HI) is a measure in decibels of how much a signal is attenuated between two antennas that are horizontally separated. The amount of horizontal isolation is a critical design issue for colocation situations. Horizontally separating antennas, however, does not achieve the level of attenuation attained with vertical isolation. Isolation is normally expressed in terms of the decibels of attenuation the signal from the transmitter will undergo as it travels toward the receiver antenna. Horizontal separation can also be used successfully to achieve isolation between communication systems. However, the amount of physical separation that is needed to achieve the same level of isolation as that attained with vertical separation is much greater.

Q. *How is the horizontal isolation calculated?*

A. Figure 9.27 and the following equation illustrate how to calculate horizontal isolation in an antenna system.

$$\text{Decibels} \approx 22 + 20 \log_{10}\left(\frac{x}{\lambda}\right) = \text{HI}$$

where $\frac{x}{\lambda} > 10$

$$\therefore \text{ if Tx} = 10 \text{ dB} \qquad f = 890 \text{ mHz}$$

$$\text{Rx} = 10 \text{ dB} \qquad x = 10 \text{ ft}$$

$$\text{VI} = 22 + 20 \log_{10}\left(\frac{10 \times 12}{13.267}\right)$$

$$= 22 + 19$$

$$= 41 \text{ dB}$$

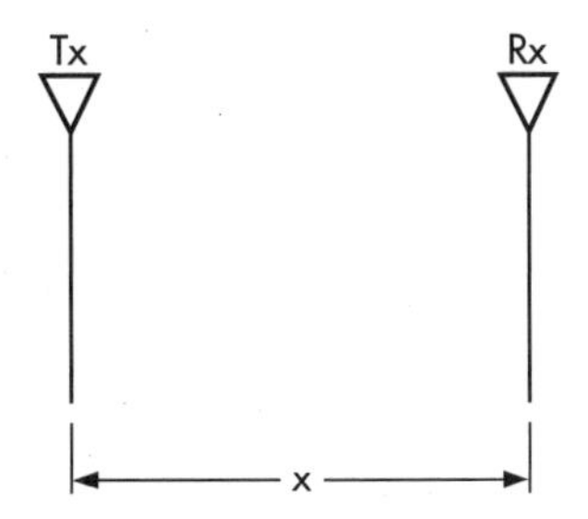

Figure 9.27 Horizontal isolation. x = distance between antennas Tx and Rx.

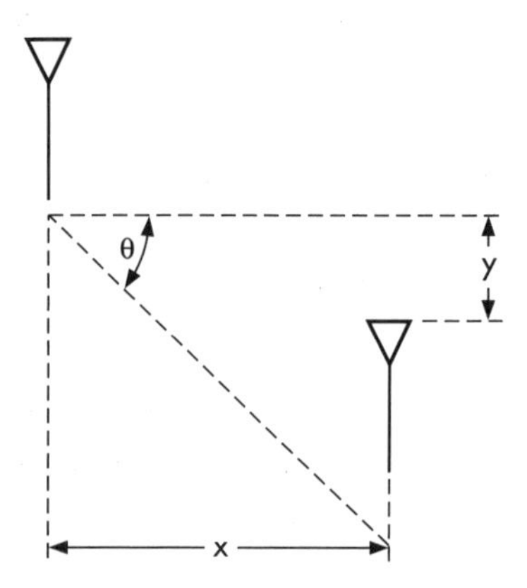

Figure 9.28 Slant isolation. x = horizontal distance, y = vertical distance, θ = slant angle relative to horizontal plane.

Q. *What is slant-angle isolation?*

A. The isolation that is attained in a slant-angle antenna installation is shown in Fig. 9.28 and in the following equation.

$$(\text{VI} - \text{HI}) \times \frac{\theta}{90} + \text{HI}$$

$$\therefore \text{ if } f = 890 \text{ mHz}$$

$$x = 10 \text{ ft}$$

$$y = 10 \text{ ft}$$

$$\theta = 45°$$

$$\text{SI} = (66.25 - 41) \times \frac{45}{90} + 41$$

$$= 53.625 \text{ dB}$$

Q. *How is isolation calculated when neither pure vertical nor horizontal separation is exhibited?*

A. To calculate isolation when neither the vertical nor horizontal separation is pure, the preceding slant-angle isolation equation may be followed. Most installations do not exhibit pure vertical or horizontal separation.

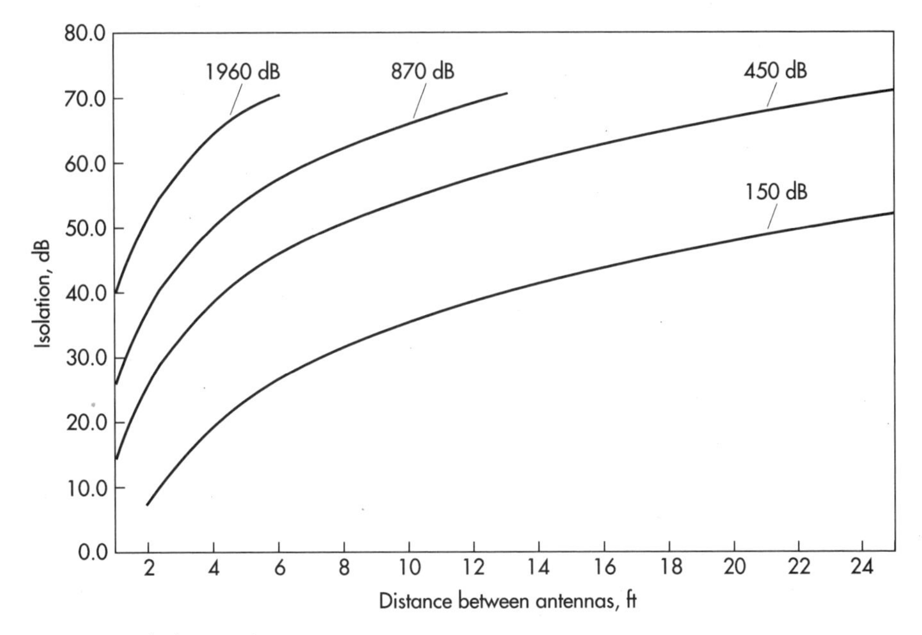

Figure 9.29 Vertical isolation versus frequency.

Q. *How much isolation can be expected?*

A. The isolation that can be expected between antennas for both vertical and horizontal installations is shown in Figs. 9.29 and 9.30. These charts provide the relative isolation for similar distances for both vertical and horizontal installations as a function of both distance and operating frequency.

Q. *What types of connectors are used in wireless systems?*

A. In wireless communication systems, many types of connectors are utilized. Often at a wireless communication site, one system will employ multiple types of connectors depending on the application at hand. The type of connector is chosen for its electrical compatibility with the coaxial cable utilized. The following is a list of connectors that are commonly used:

1. N type
2. 7/16 DIN
3. EIA flang

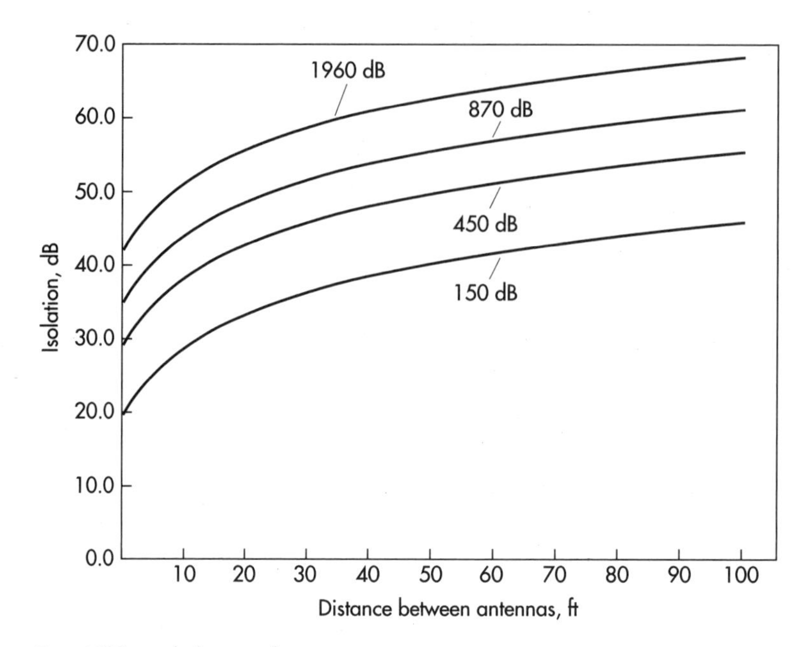

Figure 9.30 Horizontal isolation versus frequency.

4. SMA
5. TNC
6. Mini UHF
7. UHF

Q. *Why use 7/16-DIN versus N-type connectors?*

A. N-type connectors have enjoyed wide use, but 7/16-DIN connectors have gained popularity because of their better intermodulation performance. The intermodulation performance is a direct result of the larger contact area that the 7/16-DIN connector has over the N-type connector. The larger the contact surface, the better the intermodulation suppression the connector attains.

Q. *How is the right antenna selected for use in a base station?*

A. The antenna selected for the application should meet minimally the following criteria:

1. The elevation and azimuth patterns must meet the design requirements.

2. The antenna exhibits the desired gain.

3. The antenna is available from common stock within company inventory.

4. The antenna can be mounted properly at the location—that is, it can be physically mounted at the desired location.

5. The antenna will not adversely affect the tower, wind, and ice loading capacities for the installation.

6. Any negative visual impact of the antenna has been minimized in the antenna design and selection phase.

7. The antenna meets the desired system performance specifications.

Q. *What is return loss?*

A. *Return loss* is the amount of energy that is absorbed by the load that is connected to the transmitter. Specifically, return loss is a measurement of the energy reflected, or SWR, from a device or antenna system. In wireless communications the return-loss measurement is made to ensure the integrity of the antenna system including the jumpers, feed line, and of course, the antenna itself.

An SWR of 1:1.5 results in a return loss of about 14 dB. Therefore, depending on the amount of the return loss originating in the system, the relative integrity of the system and its components can be computed. For example:

SWR = 1:1.5:	14 dB	
Feed-line loss:	3 dB	(to antenna)
Feed-line loss:	3 dB	(from antenna)
	20 dB	Return loss expected for a marginal antenna with 3 dB of line loss

Q. *What is a through test?*

A. A *through test,* also called an *S21 test,* is a measure of the amount of gain or loss in the device being tested. The S21 test is used to measure the gain of amplifiers and the loss for filters in a wireless system. The S21 test will enable the engineer to understand not only the loss or gain of a device but also its relative gain or loss in decibels over different operating frequencies.

Q. *What is a time domain relfectometer?*

A. A *time domain reflectometer* (TDR) is a device that is used to determine cable or feed-line electrical integrity. A TDR is used to determine if a coaxial cable has a sharp bend in it or has a failure, like and open. By reading the results, the TDR can reveal the relative position of the fault in addition to the type of the fault. The TDR is a valuable piece of field test equipment. If a TDR is not available, a network analyzer, if the proper options have been selected, can also be used to evaluate cable or feed-line integrity.

Design Guidelines

The topic "design guidelines" encompasses a wide range of issues associated with a wireless system. This chapter addresses many of the common questions, and even some of the uncommon questions, that are asked about design issues. The design issues for a system are vast, and often a network issue also has RF components that need to be factored into the topics. The same can be said for most RF issues in that they have some traffic or network element that also needs to be accounted for. Of course, there is the underlying commonality in that construction, implementation, real estate, and the all-encompassing operations group also are involved with the design process.

Q. *What are services?*

A. The *services* for a wireless system are the various services offered to the subscribers who utilize the system. Just what constitutes a service and what does not is largely driven by who you talk with. However, in general, services are functions that are offered to the subscribers, and the functions have features as well. Normally, a service offering has a specific price plan associated with it. Some examples of services offered are the following:

Weekend calling

Digital

Voice mail

Caller ID

Short-message service

CDPD

Wireless data transmission

Group 3 fax

Q. *What are features?*

A. Features are capabilities built into the services offered the subscribers. Features are therefore a subset of the services offered. It is important to note that the list of features available for use within a network is usually greater than those that are actually offered. The reason for the discrepancy is that some are yet potential features that may or may not be used due to cost, technological issues, or other reasons. Some examples of features offered by a wireless operator are the following:

Caller ID

Group voice mail

Prepaid calling

Four-digit dialing

Virtual private network (VPN)

Roaming

Three-party conferencing

Call forwarding

Call waiting

Call barring

Q. *How do I find out what features and services are offered?*

A. One of the best locations from which to obtain a list of features and services a particular company offers is the company's Web page. This information is also available from its telemarketing group and, of course, its sales and marketing department.

Q. *Why is it important to know what services and features are offered?*

A. It is important to understand what services are offered to the subscribers because the inclusion or exclusion of a feature or service will have a direct impact on the dimensioning of the system and its subcomponents. For example, if voice mail service is offered to all new subscribers, the voice mail system should be properly dimensioned to support that service. Also, if the features offered with voice mail are unlimited storage and group distributions, then the dimensioning of the voice mail system would need to be set up accordingly. Without knowing or understanding the features and services offered from both a financial and engineering aspect, there is a high probability that miscommunication will result, leading to network platform problems.

Q. *How is the technology chosen for the system?*

A. The determination of the type of technology to use for the system is driven largely by what the adjacent markets utilize. There is, of course, the holy war between the IS-136, GSM, and CDMA platforms, to mention the key combatants. However, most if not all support similar features and services. The voice-quality issue that once drove decision making is no longer a critical differentiator due to improvements in vocoder technology. Therefore, the decision is now based on the current infrastructure in place and the business and marketing goals for increasing the revenue from the system.

Q. *Is CDMA better than GSM and IS-136?*

A. There is no single answer to this question as opinions vary considerably. However, the basic difference is that CDMA has some technological advantages over GSM and IS-136, which result in its having a higher capacity, or erlangs per square kilometers. But CDMA has some implementation drawbacks in its bandwidth and infrastructure size that give an advantage to IS-136 and GSM technologies. Implementing CDMA is more advantageous for a new system due to its spectrum clearing and capacity.

Q. *Is GSM better than IS-136 and CDMA?*

A. There is no single answer to this question. GSM is a well-established technology and offers many advantages in services and features that are also supported by a variety of other platforms. However, the GSM deployed in the United States is different from the GSM deployed in the rest of the world in the frequency band used. This difference in frequency band prevents the GSM system from being a truly universal service. It is interesting to note that CDMA and IS-136 derived many of their feature offerings from GSM technology.

Q. *Is IS-136 better than CDMA and GSM?*

A. There is no single answer to this question. IS-136 is preferable to CDMA and GSM when working with an existing network or when dealing with limited spectrum availability in a new system due to the requirement for 30-kHz channels. The IS-136 technology platform is well understood and is available in both the cellular and PCS bands in the United States.

Q. *What is downbanded IS-136?*

A. *Downbanded IS-136* is IS-136, TDMA, which utilizes the IS-136 specification with the exception that it operates in the SMR band. Downbanded IS-136 has the advantage of offering traditional cellular digital service in the SMR band, opening up the potential for unique roaming agreements.

Q. *What is iDEN?*

A. *iDEN* is a technology, developed by Motorola, that is used primarily by Nextel at this time. iDEN uses a 25-kHz channel with 16-QAM signaling. iDEN has many of the same features as GSM with the exception that it also offers dispatch and uses only 25 kHz of spectrum versus 200 kHz. An iDEN radio channel can handle three or six interconnect calls or six dispatch conversations at any time. The configuration that is used is determined by the system requirements for that area.

Q. *How wide is the IS-136 channel?*

A. The IS-136 channel uses a bandwidth of 30 kHz.

Q. *What is the bandwidth of an IS-136 channel?*

A. The bandwidth of an IS-136 channel is 30 kHz.

Q. *How wide is the CDMA channel?*

A. The CDMA channel uses a bandwidth of 1.25 MHz.

Q. *What is the bandwidth of a CDMA channel?*

A. The bandwidth of a CDMA channel is 1.25 MHz.

Q. *How wide is the GSM channel?*

A. The GSM channel uses a bandwidth of 200 kHz.

Q. *What is the bandwidth of a GSM channel?*

A. The bandwidth of a GSM channel is 200 kHz.

Q. *How wide is the iDEN channel?*

A. The iDEN channel uses a bandwidth of 25 kHz.

Q. *What is the bandwidth of an iDEN channel?*

A. The bandwidth of an iDEN channel is 25 kHz.

Q. *How wide is an AMPS channel?*

A. The AMPS channel uses a bandwidth of 30 kHz.

Q. *What is the bandwidth of an AMPS channel?*

A. The bandwidth of an AMPS channel is 30 kHz.

Q. *What is the theoretical sensitivity of an iDEN channel?*

A. The theoretical sensitivity for an iDEN channel is −130.02 dBm.

Q. *What is the theoretical sensitivity of a GSM channel?*

A. The theoretical sensitivity for a GSM channel is −120.98 dBm.

Q. *What is the theoretical sensitivity of an IS-136 channel?*

A. The theoretical sensitivity for an IS-136 channel is −129.23 dBm.

Q. *What is the theoretical sensitivity of an AMPS channel?*

A. The theoretical sensitivity for an AMPS channel is −129.23 dBm.

Q. *What is the theoretical sensitivity of a CDMA channel?*

A. The theoretical sensitivity for a CDMA channel is −113.03 dBm.

Q. *What is the system design process?*

A. The *system design process* for a wireless network is an ongoing process of refining and adjusting a design as a multitude of variables change, most of which are not under the control of the engineering department. The system design process involves both RF and network engineering efforts with implementation, operations, customer care, marketing, and of course operations.

Q. *What is the RF system design process?*

A. The *RF system design process* that should be followed is listed below. This process can, and should, be used for both existing and new systems.

- Assess marketing requirements.
- Create a methodology.
- Decide on a technology platform.
- Define the types of cell sites.

- Establish a link budget.
- Define the coverage requirements.
- Define the capacity requirements.
- Complete an RF system design.
- Issue search area requirements.
- Conduct *site qualification test* (SQT).
- Reach agreement on whether to accept or reject the candidate sites.
- Acquire the real estate for the site (buy or lease).
- Integrate the system.
- Hand over the cell site to operations.

Q. *What information is needed for a system design?*

A. The information needed for a system design varies from market to market. However, some information is needed for all markets and technology platforms. The following is a brief list of the most important pieces of information needed for a system design:

1. Time frames for design completion
2. Subscriber growth projections (current and future by quarter)
3. Subscriber usage projections (current and forecasted by quarter)
4. Subscriber types (mobile, portable, blend)
5. New features and services to be offered
6. Design criteria (technology-specific issues)
7. Baseline system numbers for building on the growth study
8. Cell site construction schedule expectations (ideal and with land-use entitlement issues factored in)
9. Fixed network equipment (FNE) ordering intervals
10. New-technology deployment and time frames

11. Budget constraints
12. Due date for design report
13. Maximum and minimum offloading for cell sites when new cell is added to design
14. Utilization factors for *central processing unit* (CPU) and network equipment

Q. *What information is needed from marketing for a system design?*

A. The marketing department supplies the following basic input parameters for the RF and network design. This information is critical regardless of whether the system will be new or already exists.

- The projected subscriber growth for the system
- The projected millierlangs per subscriber expected at discrete time intervals
- The dilution rates for the subscriber usage
- The types of subscriber equipment used in the network and percent distribution of CPE projections (i.e., portable or mobile units in use and their percent distribution)
- The special promotion plans to be implemented like local calling or free weekend use
- The projected number of mobile unit data users
- The top 10 customer complaint areas in the network requiring coverage improvements
- The identification of key coverage areas in the network that need to be included for system turnon

Q. *What are RF design criteria?*

A. The *RF design criteria* are the rules, or parameters, that are used by the RF engineering department not only to design the network and the new components that are added—that is, cell sites—but also to improve the performance of the network. The values that are included for each of the design criteria topics are driven by the desire to offer the best service possible within existing monetary and technological constraints.

Q. *What is included in the RF design criteria?*

A. There are many items that can be included in the RF design criteria. The establishment of the RF design criteria directly affects the quality of service offered the subscribers for the cost. The following is a basic list of RF design criteria that every engineer and engineering manager should have readily available:

1. Marketing input
2. RF spectrum available for the RF plan
3. Type of grade-of-service table to be used for the plan—that is, Erlang B P02
4. Minimum and maximum offloading factors for new cells
5. Coverage requirements
6. Identification of coverage sites
7. Busy hours (BH) peak traffic, 10-day high average for month (existing system)
8. Infrastructure equipment constraints
9. Digital and/or analog radio growth
10. New-technology considerations
11. Cell site configurations used for new cells
12. Baseline system numbers
13. Cell site deployment considerations

Q. *What is the purpose of a link budget?*

A. The objective of defining the link budget is to arrive at the size of the cell sites needed for the network design. The link budget will determine if the system is limited in the uplink or downlink direction. Based on the link budget, the *D/R* ratio chosen for the system will determine the radius of the site and also the distance between the cells themselves. Also the link budget can be used to determine if the system can support in-car or in-building portable units. The link budget has a direct impact on the number of cell sites required for a network and will determine the coverage requirements in terms of decibels, milliwatts. For example, if the link budget indicates that a −85-dBm value is needed on the street for in-building coverage, this will require fewer cell sites to achieve than a −75-dBm value.

Q. *What is a balanced system, or balanced path?*

A. A *balanced system,* also called a *balanced path,* is a situation in which the downlink and the uplink path losses for the system are the same. A balanced system is also referred to as a *balanced path.* The balanced system is the design objective for determining the number of cell sites and how much area they are expected to cover. However, depending on the technology, a conscious decision can be made to make the downlink path stronger than the uplink due to rescan issues associated with the phone.

Downlink path loss, dB = uplink path loss, dB

Q. *What is an uplink-limited system?*

A. An *uplink-limited system* is a situation in which the cell site's coverage is less along the uplink than along the downlink, as shown below:

Downlink path	Uplink path
Cell site ERP = +50 dBm	Subscriber max ERP = +28 dBm
Subscriber sensitivity = −102 dBm	Cell site receive antenna = 12 dBd
	Cell site cable loss = 2 dB
	Cell site receiver sensitivity = −105 dBm
Max downlink path loss = 152 dB	Max uplink path loss = 143 dB

The maximum downlink path loss is greater than the maximum uplink path loss. Therefore, the system is uplink limited. One solution would be to operate the cell site at a lower power (−8 dB) or 16 W. This issue, of course, is independent of other system performance issues.

Q. *What is a downlink-limited system?*

A. A downlink-limited system is a situation in which the cell site's coverage is less along the downlink than along the uplink, as shown below:

Downlink path	Uplink path
Cell site ERP = +10 dBm	Subscriber max ERP = +28 dBm
Subscriber sensitivity = −102 dBm	Cell site receive antenna = 12 dBd
	Cell site cable loss = 2 dB
	Cell site receiver sensitivity = −105 dBm
Max downlink path loss = 112 dB	Max uplink path loss = 143 dB

The maximum uplink path loss is greater than the maximum downlink path loss. Therefore, the system is downlink limited. One solution would be to operate the cell site at a higher power (32 dB) or 16 W. This issue, of course, is independent of other system performance issues.

Q. *What is a cell site?*

A. Basically, a *cell site* is a physical location where there is an antenna that is connected to a radio for the purpose of establishing a radio link between the cell site itself and a subscriber unit. Cell sites come in many shapes and sizes.

Q. *What is included in the RF design guidelines?*

A. The proposed RF design guidelines are shown in Fig. 10.1. The guidelines can easily be crafted to reflect the particular design guidelines utilized for the market where it will be applied to.

RF DESIGN GUIDELINES

System name:

Date:

	RSSI	ERP	Cell area	Antenna type
Urban	−80 dBm	16 W	3.14 km^2	12 dBd 90H/14E
Suburban	−85 dBm	40 W	19.5 km^2	12 dBd 90H/14E
Rural	−90 dBm	100 W	78.5 km^2	10 dBd 110H/18E

Voice channel	*C/I*	17 dB (90th percentile)
Frequency reuse		$N = 7$
Maximum no. channels per sector		19
Antenna system		
Sector cell orientation		0, 120, 240
Antenna height		100 ft, or 30 m
Antenna pass band		825–894 MHz
Antenna feed-line loss		2 dB
Antenna system return loss		20–25 dB
Diversity spacing		$d = h/11$ (d = receive antenna spacing, h = antenna AGL)

Receive antennas per sector	2 (3 for iDEN)
Transmit antennas per sector	1
Roof height offset	$h = x/5$ (h = height of antenna from roof; x = distance from roof edge)
Performance criteria	
Lost-call rate	<2%
Attempt failure (access denied)	<2%
RF blocking	1%><2%
FER or BER	<1%
System level	
% area in-car portable (ICP) units	95%
% area in-building portable (IBP) units	75%
% area > 17 dB *C/I*	95%

Figure 10.1 RF design guidelines.

Q. *What are the steps in a cell site design?*

A. There are numerous steps in a cell site design. The following items are some of the more basic issues that need to be addressed regardless of the technology platform used:

- Search area
- Site qualification test (SQT)
- Site acceptance
- Site rejection
- FCC guidelines
- FAA guidelines
- Planning and zoning boards
- EMF compliance
- Microwave clearance (PCS)

Q. *What is a search area?*

A. A *search area* is a geographic area in which it would be desirable to locate a cell site.

Q. *What is a search area information form?*

A. A *search area information form* (SAIF) is a document that defines all the search area parameters such as the size of the room, the height of the room, the ideal location for the cell site, and tolerances from that ideal location. The search area form issued needs to follow the design objectives for the area according to the RF system design objectives. The search area information form should be put together by the RF engineer responsible for the site's design. The final paper needs to be reviewed and signed by the appropriate company authorities, usually including the department manager, to ensure that there is a check and balance in the process. The specifications for the search area document need not only to meet the RF engineering department's requirements but also the real estate, construction, and operations groups' needs. It is imperative that the search area request form undergo a design review prior to its issuance. A suitable format for the SAIF is shown in Fig. 10.2.

RF Engineering

Search area code:________________ Capital funding code:________________
Target-on-air date:________________
Search area type: (capacity, coverage, frequency plan, competitive, new technology)

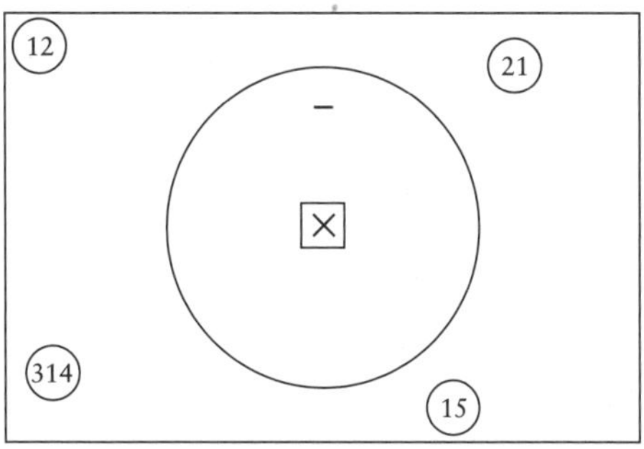

12
21
314
15

Search area map

Cell site configuration: (omni, 3 sector, 6 sector, other)
Type of infrastructure:________________
Physical size of equipment room:________________(ft^2)
Antenna info:

1. Number of antennas:__________
2. Type of antennas: (attach manufacturer's specifications)
3. Antenna height
   AGL:________
   AMSL:________

Maximum cable length:____________

Comments: ________________________________

Search area request
Document:________ Date:________
Design engineer:____________
Review by:____________
Revision

Figure 10.2 Search area request.

Q. *What is a transmitter test?*

A. A *transmitter test* is another name for a *site qualification test.*

Q. *What is a drive test?*

A. A *drive test* is any test that involves using field measurement equipment that collects signal strength data where the measurement equipment is located in a vehicle. One type of drive test is also referred to as a *transmitter test* or *site qualification test.* Another type of drive test involves performance analysis and validation.

Q. *What is a site qualification test?*

A. A *site qualification test* (SQT), also called a *transmitter test,* is a physical test where a transmitter is used for the purpose of simulating what the expected coverage could be from a proposed site location. The SQT is a critical test and should be conducted for every potential cell site that meets a minimum set of acceptance criteria.

Q. *What is a site qualification test form?*

A. A *site qualification test form* is used to specify the requirements that the testing team needs to accomplish when conducting the site qualification test. There are many SQT forms in use, and the form shown in Fig. 10.3 is one example. However, the format of the SQT form needs to directly match the input requirements of the RF engineers, the SQT measurement team, and the group responsible for postprocessing the data.

Site Qualification Test

Date:__________
Search area code:_____ Capital funding code:_____
Test site address: __________
____
____
Site contact: Name:__________
Phone #:__________
Requested test date:__________
Test parameters:
1. 7.5-minute map: (name or # of map(s) that encompass SQT)
2. Test antenna: (make, model #)
3. Test antenna height
   AGL:_____
   AMSL:_____
4. Test ERP:_____watts
5. Antenna orientation:_____
6. Test frequency/channel:_____
7. Clearance/coordination:_____

Test implementation:
1. Antenna mounting info: (roof, tower, crane, water tank, misc.)
2. Rigger required: (Y/N)
3. Antenna test location sketch attached: (Y/N)
4. Test routes attached: (Y/N)
5. SQT team leader:__________

Postprocessing:
1. Map scale:__________
2. Color code:__________
3. Data reduction method:__________

Test equipment:
1. SQT test equipment cal date__________
2. SQT transmitter cal date__________

SQT	
Document #:_____	Date:_____
Design engineer:_____	Reviewed by:____

Figure 10.3 Site qualification test.

Q. *Why is a drawing needed for an SQT?*

A. The SQT needs to include a sketch of the test location to define where to place the actual test transmitter antenna. For example, when using a crane to place the transmitter antenna, it is important to define where the crane should be parked and the test antenna height actually used. If the test location is not properly defined in advance, an error could occur in the test transmitter placement significant enough to pass or fail the SQT requirements.

Q. *What is site acceptance?*

A. *Site acceptance* (SA) refers to the process followed by the RF engineering department in approving a location as a valid candidate to become a cell site. The site acceptance is usually an RF engineering issue, but there are numerous other factors that can preclude the candidate from becoming a cell site. The site acceptance must, however, meet the engineering design requirements.

Q. *What is a site acceptance form?*

A. A *site acceptance form* (SAF) is a document that is used not only for engineering control but also for communicating to external departments engineering's approval of a location for use as a cell site. The site acceptance form will also need to be sourced with a document control number to ensure that changes in personnel during the project's life are as transparent as possible (Fig. 10.4).

RF Engineering

Search area code:_____ Capital funding code:_____
SAF document #:_____
Site address:__________
____
____

Latitude:_____ AGL:_____
Longitude:_____ AMSL:_____

Regulatory issues:
1. FAA analysis attached: (Y/N)
2. FAA lighting/marking required: (Y/N)
3. FCC contour extension required: (Y/N)

Site-specific information:
1. Antenna configuration attached: (Y/N)
2. Radio equipment location defined: (Y/N)
3. Radio equipment location sketch attached: (Y/N)
4. Radio equipment type: ______________
5. Antenna structure: (roof, tower monopole, water tank)
6. Equipment room: (prefab, interior fitup, exterior)
7. Approximate cable length:_____

Type and quantity of antenna

Sector	Type	Quantity	Orientation	ERP

8. Existing transmitters on structure: (Y/N)
9. If (yes), state frequency, ERP, call signal, and physical location for each

Qualification information:
1. SQT document #:_____
2. Design propagation plot attached: (Y/N)
3. SQT drive test plot attached: (Y/N)
4. Site type
   Coverage (Y/N)
   Capacity (Y/N)
5. IMD study completed: (Y/N)
6. Site particular comments:____
____
____

SAF
Document #:_____ Date:_____
Design engineer:_____ Reviewed by:_____

Figure 10.4 Site acceptance form.

Q. *What is a site rejection?*

A. A *site rejection* is a process through which engineering notifies real estate and other pertinent departments that a cell site candidate was tested or evaluated and did not meet the requirements for the design. Note that a site rejection may be reversed later under different circumstances.

Q. *What is a site rejection form?*

A. A *site rejection form* is shown in Fig. 10.5 for reference. There should be a site rejection form prepared for every site not selected for use.

SITE REJECTION FORM

RF Engineering

Search area code

The (*name of test location*) was tested on (*date of test*) and did not meet the design criteria for the search area defined.

The test location did not meet the design criteria for the following reasons (*state reasons*).

RF engineer ____________________

Engineering manager _____________

Figure 10.5 A site rejection form.

Q. *What is site activation?*

A. *Site activation* refers to the process of integrating a cell site into a system network for the purpose of handling traffic by paying subscribers.

Q. *What is the CFR #47?*

A. The Code of Federal Regulations Title 47 (CFR 47) is a set of rules and regulations pertinent to wireless operators. There are numerous chapters within the CFR 47, covering all the types of services and radio frequency spectrums.

Q. *What material is needed for zoning board and planning board meetings?*

A. The material that is needed for a zoning and/or a planning board meeting will vary from municipality to municipality. However, the following are some items commonly needed:

1. Statement about why the site is needed
2. Statement about how the site will improve the network
3. Drawing of what the site will look like
4. Views of the cell site from local residences
5. EMF compliance chart
6. EMF information sheets and handouts for the audience (rare)

Q. *What is radiation in the context of wireless communications systems?*

A. *Radiation* is the energy that is emitted from a cell site's antennas. The term *radiation* refers to the electromagnetic radio wave that is nonionizing.

Q. *What is nonionizing radiation in the context of wireless communications systems?*

A. *Nonionizing radiation* refers to the energy emitted from a cell site or transmitter. The term *nonionizing* refers specifically to energy that is not ionizing and therefore does not affect the DNA of people or animals.

Q. *What is radiation compliance?*

A. *Radiation compliance* is another term—although not exactly synonymous—that is used for *EMF compliance.*

Q. *What is RF exposure?*

A. *RF exposure* refers to a situation in which a person is exposed to RF energy. Note that RF energy is pervasive and that people are constantly exposed to RF energy from a multitude of sources.

In terms of wireless communications, *RF exposure* is commonly used to describe the power or energy emitted from a cell site to which people near the site may have been exposed.

Q. *What is compliance?*

A. In wireless communications, the term *compliance* is often used loosely to mean *EMF compliance* or *RF exposure compliance.*

Q. *What is frequency planning?*

A. *Frequency planning,* or rather *frequency management,* is an integral and critical part of the system design. The extent of frequency planning for a given network is determined largely by the technology platform that is chosen by the operator. There are many steps in frequency planning ranging from coordinating a single transmit channel to orchestrating the manipulation of hundreds of radio channels. Within the cellular and PCS arenas, the amount of frequency planning can range from segmentation of the available spectrum to defining the different PN short codes for CDMA.

Q. *What is the spectrum for cellular systems in the United States?*

A. In the United States an operator can operate only in either the A or B blocks, as shown in Fig. 10.6.

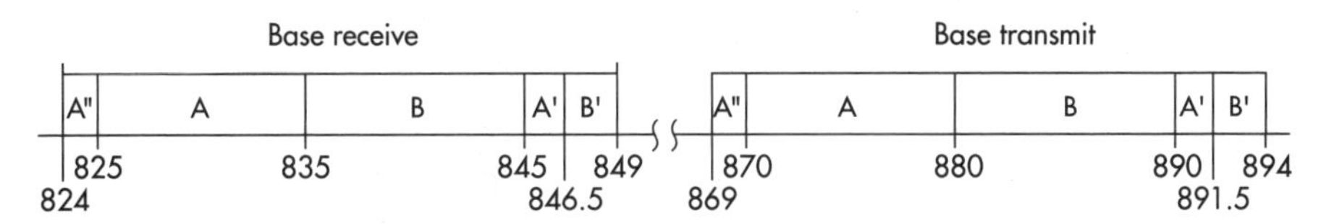

Figure 10.6 U.S. cellular spectrum.

Q. *What is the spectrum for PCS operators in the United States?*

A. The spectrum that is allocated for PCS operation in the United States is shown in Fig. 10.7.

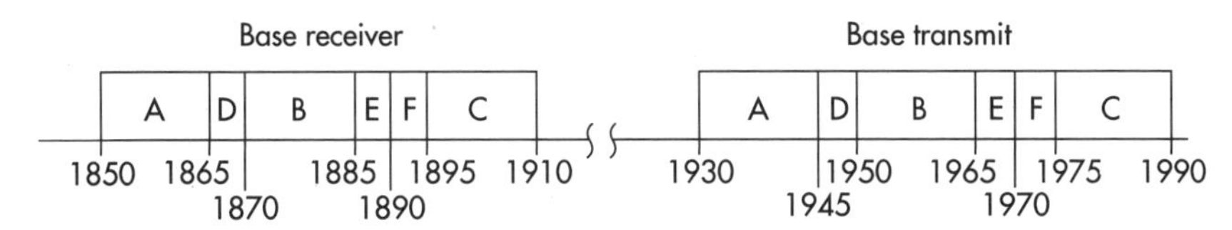

Figure 10.7 U.S. PCS spectrum.

Q. *What is a method of procedure?*

A. *A method of procedure* (MOP) is an outline of the steps that need to be taken to ensure a smooth operation of a wireless communications company. As such, it is an integral part of all aspects associated with successful work in the company.

Q. *What is the MOP for integrating a new cell site?*

A. An MOP integrating new cell sites into a network is shown below. Obviously the MOP will vary according to the peculiarities of a given integration situation.

Method of Procedure for New Cell Site Integration

Rev: *Date*

Preactivation process

DATE

X-X-XX New cell sites to be activated defined

X-X-XX Project leader(s) named and time tables specified, as well as the scope of work associated with the project

X-X-XX Phase 1 design review (frequency planning only and RF engineer for site)

X-X-XX Phase 2 design review (all engineering)

X-X-XX Phase 3 design review (operations and engineering)

X-X-XX Phase 4 design review (adjacent markets if applicable)

X-X-XX Frequency assignment and handoff topology sheets given to operations

X-X-XX New cell site integration procedure meeting

X-X-XX Performance evaluation test completed

X-X-XX Executive decision to proceed with new cell site integration

X-X-XX Adjacent markets contacted and informed of decision

X-X-XX Secure post-cell site activation war room area

X-X-XX Briefing meeting with drive test teams

X-X-XX MIS support group confirms readiness for post-processing efforts

X-X-XX Customer care and sales notified of impending actions

New Cell Site Activation Process (begins X-X-XX at time XXXX)

X-X-XX

- Operations informs key personnel of new cell site activation results
- Operations personnel conduct brief post-turn-on test to ensure that call processing is working on every channel and that handoff and handins are occurring with the new cell site
- Operations manager notified of key personnel of testing results

Post-Turn-On Process (begins X-X-XX at time XXXX)

Voice mail message left from engineering indicating status of new cell sites (time)

Begin post-turn-on drive testing, phase 1 (time)

Database check

Statistics analysis

Voice mail message left from RF engineering indicating status of post-retune effort (time)

Phase 2 of post-turn-on drive testing begins

Commit decision made with directors for new cell site (time)

Phase 3 of post-turn-on drive testing begins

X-X-XX

- Continue drive testing areas affected
- Statistics analysis
- Conduct post-turn-on analysis and corrections where required

X-X-XX

- Post-turn-on closure report produced
- New site files updated and all relevant information about the site transferred to performance engineering

Q. *Whom should be notified when new cells are turned on or major system changes are made?*

A. The primary groups to keep informed are the following:

- Sales
- Marketing
- Customer service
- Operations, real estate, and engineering
- Corporate communications
- Legal and regulatory

Q. *Whom should be notified in the event of a major service outage?*

A. At the very least, the following groups need to be notified in the event that a major service outage occurs. The definition of a major service outage should be established in the escalation procedure and provided to operations.

- Engineering
- Sales
- Marketing
- Customer service
- Operations, real estate, and engineering
- Corporate communications
- Legal and regulatory

Q. *Who gets a new cell site report?*

A. A new cell site report is issued by RF design and given to the performance and optimization group within either engineering or operations.

Q. *What needs to be included in a new cell site report?*

A. A new cell site performance report needs to have, at the very least, the following items included in it:

- Search area request form
- Site acceptance report
- New cell site integration MOP
- Cell site configuration drawing
- Frequency plan for site
- Handoff and cell site parameters
- System performance report indicating the following parameters one week after site activation:
  - Lost calls
  - Blocking
  - Access failures
  - Customer complaints
  - Usage/RF loss
  - BER/FER
  - Soft handoff percentage
  - Handoff failures
  - RF call completion ratio
- Radios out of service
- Cell site span outage
- Technician trouble reports
- FCC site information
- FAA clearance analysis
- EMF power budget
- Copy of lease
- Copy of any special planning board or zoning board requirements

Q. *What is an RF system design report?*

A. An *RF system design report* is a report that reviews and defines the RF design goals and objectives for the wireless system. This report should be prepared every 6 or 12 months.

Q. *What is included in a RF system design report?*

A. The following outline is suitable for a standard RF system design report. This report format is structured for an existing system. When crafting a report for a new system, the particular issues associated with that system should be incorporated into it.

1.0 Executive summary

2.0 Introduction

3.0 Design criteria

4.0 Cell site analysis

- 4.1 Analysis
- 4.2 Link budget
- 4.3 Coverage maps
- 4.4 Summary of requirements

5.0 Network capital requirements

Q. *What is the methodology for establishing a system design?*

A. The methodology for establishing a system design consists of the following seven-step process:

1. Define the objective.
2. Identify the variables.
3. Isolate the system components.
4. Design the system.
5. Present the report (design review).
6. Communicate the design to respective departments.
7. Implement the design.

Q. *What is the difference between a design for a new versus an existing system?*

A. The primary difference between an existing system design and a new system design is that the existing system obviously has some traffic-loading and performance parameters that are associated with it.

Q. *Under what circumstances are system designs prepared?*

A. There are fundamentally three situations that require system designs for wireless communication systems:

- Existing system expansion
- New system
- Introduction of new technology platform to existing system

Q. *What is the process for RF design for a new wireless system?*

A. The RF design process for a new wireless system is as follows:

1. Obtain marketing plan and objectives.
2. Establish system coverage area.
3. Establish system on-air projections.
4. Establish technology platform.
5. Determine maximum radius per cell (link budget).
6. Establish environmental corrections.
7. Determine desired signal level.
8. Establish the maximum number of cells to cover area.
9. Generate coverage propagation plot for system.
10. Determine subscriber usage.
11. Determine usage per square kilometer.
12. Determine maximum number of cells for capacity.
13. Determine if system is capacity or coverage driven.
14. Establish total number of cells required for coverage and capacity.

15. Generate coverage plot incorporating coverage and capacity cell sites (if different).
16. Reevaluate results and make assumption corrections.
17. Determine revised (if applicable) number of cells required for coverage and capacity.
18. Check number of sites against budget objective; if number of sites is exceeded, reevaluate design.
19. Using known database of sites, overlay on the system design and check off matches or close matches ($<0.2R$).
20. Adjust system design using site-specific parameters from known database matches.
21. Generate propagation and usage plots for system design.
22. Evaluate design objective with time frame and budgetary constraints, and readjust design if necessary.
23. Issue search rings.

Q. *What is the process for RF design for an existing wireless system?*

A. The RF design process for an existing wireless system is as follows:

1. Obtain marketing plan.
2. Identify coverage problem areas.
3. Establish technology platform decisions (for example, CDPD).
4. Determine maximum radius per cell (link budget).
5. Establish environmental corrections.
6. Determine desired signal level.
7. Establish the maximum number of cells to cover area(s).
8. Generate coverage propagation plot for system and areas showing before and after coverage.
9. Determine subscriber usage.
10. Allocate percentage of system usage to each cell.
11. Determine maximum number of cells for capacity (technology dependent).

12. Establish which cells need capacity relief.
13. Determine where new cells are needed for capacity relief.
14. Establish total number of cells required for coverage and capacity.
15. Generate coverage plot incorporating coverage and capacity cell sites (if different)
16. Reevaluate results and make assumption corrections.
17. Determine revised (if applicable) number of cells required for coverage and capacity.
18. Check number of sites against budget objective; if number of sites exceeds budget, reevaluate design.
19. Using known database of sites, overlay on system design and check off matches or close matches ($<0.2R$).
20. Adjust system design using site-specific parameters from known database matches.
21. Generate propagation and usage plots for system design.
22. Evaluate design objective with time frame and budgetary constraints and readjust design if necessary.
23. Issue search rings.

Q. *What is the process for a new technology platform introduction to an existing wireless system?*

A. The process for a new technology platform introduction to an existing wireless system is as follows:

1. Obtain marketing plan.
2. Establish technology platform introduction time table.
3. Determine new technology implementation tradeoffs.
4. Determine new technology implementation methodology.
5. Identify coverage problem areas.
6. Determine maximum radius per cell (link budget for each technology platform).
7. Establish environmental corrections.

8. Determine desired signal level (for each technology platform).
9. Establish the maximum number of cells to cover area(s).
10. Generate coverage propagation plot for system and areas showing before and after coverage.
11. Determine subscriber usage.
12. Allocate percentage of system usage to each cell.
13. Adjust cell maximum capacity by spectrum reallocation method (if applicable).
14. Determine maximum number of cells for capacity (technology dependent).
15. Establish which cells need capacity relief.
16. Determine number of new cells needed for capacity relief.
17. Establish total number of cells required for coverage and capacity.
18. Generate coverage plot incorporating coverage and capacity cell sites (if different).
19. Reevaluate results and make assumption corrections.
20. Determine revised (if applicable) number of cells required for coverage and capacity.
21. Check number of sites against budget objective; if number of sites exceeds budget, reevaluate the design.
22. Using known database of sites, overlay on system design and check off matches or close matches ($<0.2R$).
23. Adjust system design using site-specific parameters from known database matches.
24. Generate propagation and usage plots for system design.
25. Evaluate design objective with time frame and budgetary constraints, and readjust design if necessary.
26. Issue search rings.

Q. *What are the AMPS channel assignments?*

A. Table 10.1 on pp. 402 and 403 is a chart of the common channel assignments for an $N = 7$.

Q. *What is IS-95?*

A. *IS-95* is the specification that is used by CDMA technology.

Q. *What is code division multiple access?*

A. *Code division multiple access* (CDMA) is a wideband spread-spectrum technology platform that enables multiple users to occupy the same radio channel, or frequency spectrum, at the same time. CDMA is based on the principle of *direct sequence* (DS). The CDMA channel utilized is reused in every cell of the system and is differentiated by the *pseudo-random number* (PN) *code* that it utilizes. Whether the system is to be deployed in an existing AMPS or new PCS band system, the design concepts are fundamentally the same.

Utilizing the IS-95 and J-008 standards, the same functionality, with the exception of frequency band particulars, applies to both the 800- and 1900-MHz bands. The primary difference between the 800- and 1900-MHz bands is that CDMA has a few nuances that are directly applicable to the channel assignments in an existing cellular band.

Q. *What is the CDMA spectrum allocation for cellular (AMPS)?*

A. The spectrum in which the IS-95 CDMA technology is designed to operate is shown below:

	A band	B band
Primary	283	384
Secondary	691	777

Cellular operation for CDMA is shown in Fig. 10.8.

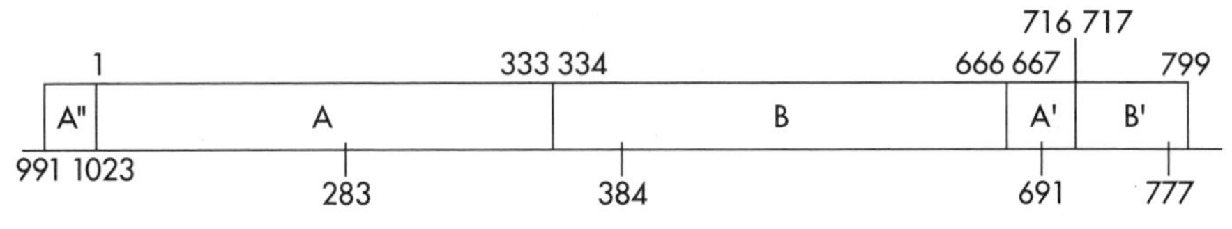

Figure 10.8 Cellular CDMA assignment.

TABLE 10.1 FCC CHANNEL CHART FOR $N = 7$

Channel group:	A1	B1	C1	D1	E1	F1	G1	A2	B2	C2	D2	E2	F2	G2	A3	B3	C3	D3	E3	F3	G3
											Wire-line B-band channels										
Control channel:	334	335	336	337	338	339	340	341	324	343	344	345	346	347	348	349	350	351	352	353	354
	335	356	357	358	359	360	361	362	345	364	365	366	367	368	369	370	371	372	373	374	375
	376	377	378	379	380	381	382	383	366	385	386	387	388	389	390	391	392	393	394	395	396
	397	398	399	400	401	402	403	404	387	406	407	408	409	410	411	412	413	414	415	416	417
	418	419	420	421	422	423	424	425	408	427	428	429	430	431	432	433	434	435	436	467	438
	439	440	441	442	443	444	445	446	429	448	449	450	451	452	453	454	455	456	457	458	459
	460	461	462	463	464	465	466	467	450	469	470	471	472	473	474	475	476	477	478	479	480
	481	482	483	484	485	486	487	488	471	490	491	492	493	494	495	496	497	498	499	500	501
	502	503	504	505	506	507	508	509	492	511	512	513	514	515	516	517	518	519	520	521	522
	523	524	525	526	527	528	529	530	513	532	533	534	535	536	537	538	539	540	541	542	543
	544	545	546	547	548	549	550	551	534	553	554	555	556	557	558	559	560	561	562	563	564
	565	566	567	568	569	570	571	572	555	574	575	576	577	578	579	580	581	582	583	584	585
	586	587	588	589	590	591	592	593	576	595	596	597	598	599	600	601	602	603	604	605	606
	607	608	609	610	611	612	613	614	597	616	617	618	619	620	621	622	623	624	625	626	627
	628	629	630	631	632	633	634	635	618	637	638	639	640	641	642	643	644	645	646	647	648
	649	650	651	652	653	654	655	656	639	658	659	660	661	662	663	664	665	666			
	717	718	719	720	721	722	723	724	725	726	727	728	729	730	731	732	733	734	735	736	737
	738	739	740	741	742	743	744	745	746	747	748	749	750	751	752	753	754	755	756	757	758
	759	760	761	762	763	764	765	766	767	768	769	770	771	772	773	774	775	776	777	778	779
	780	781	782	783	784	785	786	787	788	789	790	791	792	793	794	795	796	797	798	799	

Non-wire-line A-band channels																					
Control channel:	333	332	331	330	329	328	327	326	325	324	323	322	321	320	319	318	317	316	315	314	313
	312	311	310	309	308	307	306	305	304	303	302	301	300	299	298	297	296	295	294	293	292
	291	290	289	288	287	286	285	284	283	282	281	280	279	278	277	276	275	274	273	272	271
	270	269	268	267	266	265	264	263	262	261	260	259	258	257	256	255	254	253	252	251	250
	249	248	247	246	245	244	243	242	241	240	239	238	237	236	235	234	233	232	231	230	229
	228	227	226	225	224	223	222	221	220	219	218	217	216	215	214	213	212	211	210	209	208
	207	206	205	204	203	202	201	200	199	198	197	196	195	194	193	192	191	190	189	188	187
	186	185	184	183	182	181	180	179	178	177	176	175	174	173	172	171	170	169	168	167	166
	165	164	163	162	161	160	159	158	157	156	155	154	153	152	151	150	149	148	147	146	145
	144	143	142	141	140	139	138	137	136	135	134	133	132	131	130	129	128	127	126	125	124
	123	122	121	120	119	118	117	116	115	114	113	112	111	110	109	108	107	106	105	104	103
	102	101	100	99	98	97	96	95	94	93	92	91	90	89	88	87	86	85	84	83	82
	81	80	79	78	77	76	75	74	73	72	71	70	69	68	67	66	65	64	63	62	61
	60	59	58	57	56	55	54	53	52	51	50	49	48	47	46	45	44	43	42	41	40
	39	38	37	36	35	34	33	32	31	30	29	28	27	26	25	24	23	22	21	20	19
	18	17	16	15	14	13	12	11	10	9	8	7	6	5	4	3	2	1			
	1020	1019	1018	1017	1016	1015	1014	1013	1012	1011									1023	1022	1021
	725	724	723	722	721	720	719	718	717	716	1010	1009	1008	1007	1006	1005	1004	1003	1002	1001	1000
	704	703	702	701	700	699	698	697	696	695	715	714	713	712	711	710	709	708	707	706	705
	683	682	681	680	679	678	677	675	674	694	693	692	691	690	689	688	687	686	685	684	
										673	672	671	670	669	668	667					

Q. *What is a primary CDMA channel?*

A. A *primary CDMA channel* is a CDMA channel that is allocated for use in the cellular A or B band and that is part of the initialization process used for a cellular dual-mode phone. If a dual-mode subscriber unit does not find a pilot channel on either the primary or secondary channel, then it reverts to an analog mode. Therefore, the primary or secondary CDMA channel location is a fixed location.

Q. *What is a secondary CDMA channel?*

A. A *secondary CDMA channel* is a CDMA channel that is allocated for use in the cellular A or B band and that is part of the initialization process used for a cellular dual-mode phone. If the dual-mode subscriber unit does not find a pilot channel on either the primary or secondary channel, then it reverts to an analog mode. Therefore, the primary or secondary CDMA channel location is a fixed location.

Q. *What is a preferred channel?*

A. A *preferred channel* is a channel that is recommended for CDMA channel assignments.

Q. *What are the PCS CDMA channels?*

A. The PCS CDMA channels that are preferred are shown in Table 10.2. The initialization algorithm is that when the subscriber powers up, it will search in its preferred block, A through F, for a pilot channel using the preferred channel.

Q. *What are the CDMA power classes?*

A. The different CDMA power classes are shown in Table 10.3.

Q. *What is standard CDMA cell site configuration?*

A. There are several general types of CDMA cell site configurations that are being deployed at this time. The radio equipment for both cellular and PCS is fundamentally the same. The difference between the two is that for PCS the frequency for transmit and receive has been upbanded. The configurations in Fig. 10.9 are meant for only PCS and cellular CDMA cells.

Table 10.2 PCS CDMA Preferred Pilot Channel

PCS block	CDMA channel no.	Valid CDMA assignment	Preferred set channel numbers
A (15 MHz)	0–24	NV	25, 50, 75, 100, 125, 150, 175, 200, 225, 250, 275
	25–275	V	
	276–299	CV	
D	300–324	CV	325, 350, 375
	325–375	V	
	376–399	CV	
B	400–424	CV	425, 450, 475, 500, 525, 550, 575, 600, 625, 650, 675
	425–675	V	
	676–699	CV	
E	700–724	CV	725, 750, 775
	725–775	V	
	776–799	CV	
F	800–824	CV	825, 850, 875
	825–875	V	
	876–899	CV	
C	900–924	CV	925, 950, 975, 1000, 1025, 1050, 1075, 1100, 1125, 1150, 1175
	925–1175	V	
	1176–1199	NV	

NV = not valid.
V = valid.
CV = conditionally valid.

Table 10.3 CDMA Subscriber Power Levels

Station class	EIRP (max), dBw
I	3
II	0
III	−3
IV	−6
V	−9

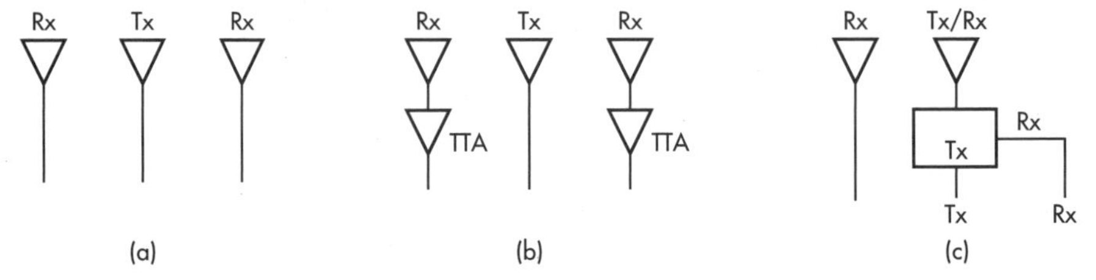

Figure 10.9 PCS and cellular CDMA only installations. (*a*) Optimal sector configuration. (*b*) Utilizing TTAs in receive path. (*c*) Duplex configuration.

Q. *How are the Walsh codes assigned in CDMA?*

A. The Walsh codes for CDMA are assigned, or rather allocated, following the method shown in the table below. From the assignments, there are a total of 64 Walsh codes from which to choose.

Channel type	No. Walsh codes
Pilot	1
Sync	1
Paging	1–7
Traffic channels	55

Q. *What is the CDMA pole capacity?*

A. The *pole capacity* for CDMA is the theoretical maximum number of simultaneous users that can coexist on a single CDMA carrier. However, at the pole, the system will become unstable, and therefore operating at less than 100 percent of the pole capacity is the desired method of operation.

Q. *How is the number of traffic channels determined for CDMA systems?*

A. The actual number of traffic channels for a CDMA cell site is determined using the following equation:

$$\text{Actual traffic channels} = \text{effective traffic channels} + \text{soft handoff channels}$$

The maximum capacity for a CDMA cell site should be 75 percent of the pole.

Q. *What types of CDMA handoffs are available?*

A. There are several types of handoffs available with CDMA technology—namely, soft, softer, and hard. The difference between the types is dependent upon what is trying to be accomplished.

Q. *What is a search window?*

A. A *search window* is defined as the amount of time, in terms of chips, that the CDMA subscriber's receiver will hunt for a pilot channel. There is a slight difference in how the receiver hunts for pilots depending on its type. The search windows are active, neighbor, and remaining.

Each of the search windows has its own role in the process, and it is not uncommon to have different search window sizes for each of the windows for a particular cell site. Additionally, the search window for each site needs to be set based on actual system conditions; however, there are several system startup values available.

Q. *What are soft handoffs?*

A. *Soft handoffs,* an integral part of CDMA, involve the use of two or more cell sites or sectors handling the same conversation at the same time. The determination of which pilots will be used in the soft handoff process has a direct impact on the quality of the call and the capacity for the system. Therefore, setting the soft handoff parameters is a key element in a CDMA system design.

The parameters associated with soft handoffs involve the determination of which pilots are in the active, candidate, neighbor, and remaining sets. The list of neighbor pilots is sent to the subscriber unit when it acquires the cell site or is assigned a traffic channel.

Q. *What is the soft handoff region?*

A. The *soft handoff region* is an area that exists between two or more CDMA cell sites. The soft handoff region in Fig. 10.10 is an area between cells A and B. Naturally as the subscriber unit travels farther away from cell A, cell B, in this example, increases in signal strength for the pilot. When the pilot from cell B reaches a certain threshold, it is added to the active pilot list.

Q. *How does the CDMA neighbor list change?*

A. The CDMA neighbor list, or, rather, candidate list, changes based on the relative strength of the pilot channel.

Example 10.1

In Fig. 10.11 is a brief description of how a pilot channel moves from the neighbor to the candidate to the active and then to the neighbor sets.

1. Pilot exceeds T_ADD and subscriber unit sends a PSMM and transfer pilot to candidate set.
2. Base station sends an extended handoff direction message.
3. Subscriber unit transfers pilot to active set and acknowledges this with a handoff completion message.

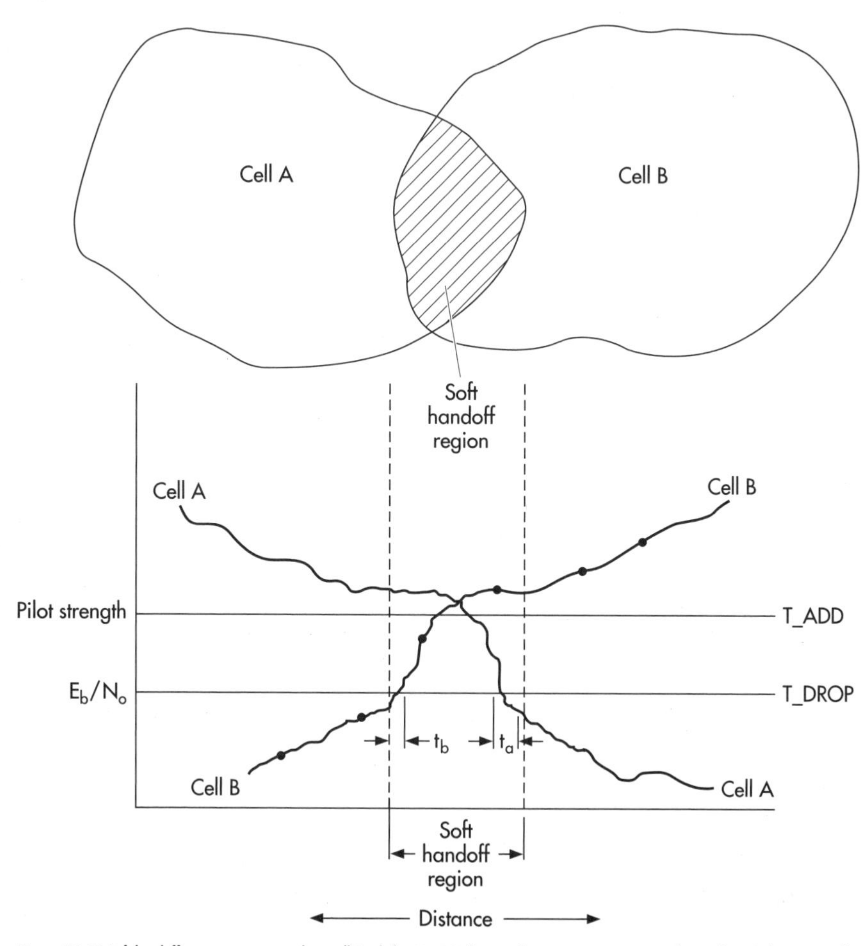

Figure 10.10 Soft handoff region. t_a = area where cell A is below T_DROP but is still in active set. t_b = area where cell B is below T_DROP but is still in active set.

4. Pilot strength drops below T_DROP and subscriber unit begins handoff drop time.
5. Pilot strength goes above T_DROP prior to handoff drop time expiring and T_DROP sequences topping.
6. Pilot strength drops below T_DROP and subscriber unit begins the handoff drop times.
7. Handoff drop timer expires and subscriber unit sends a PSMM.

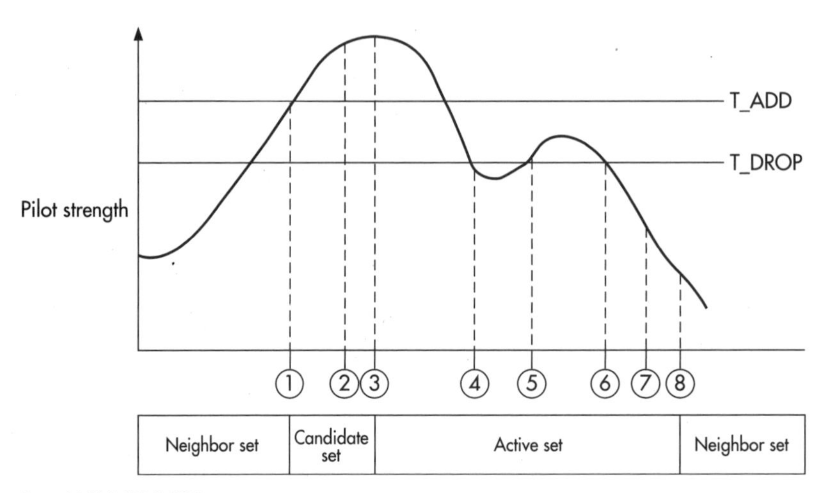

Figure 10.11 T_ADD, T_DROP.

8. Base station sends an extended handoff direction message.
9. Subscriber unit transfers the pilot from the active set to the neighbor set and acknowledges this with a handoff completion message.

Q. *What is a pilot channel?*

A. A *pilot channel* is the beacon signal for the CDMA cell site. The pilot channel does not carry data, but it is used by the subscriber unit to acquire the system and assist in the process of soft handoffs, synchronization, and channel estimation. A separate pilot channel is transmitted for each sector of the cell site. The pilot channel is uniquely identified by its PN offset, or rather PN short code.

Q. *What is a PN sequence?*

A. A *PN sequence* is also referred to as a *PN code.* The PN sequence has some 32,768 chips that when divided by 64, results in a total of 512 possible PN codes that are available for use. The fact that there are 512 PN short codes available ensures that there will be few, if any, problems associated with the assignment of these PN codes. However, there are some simple rules that must be followed to ensure that there are no problems encountered with the selection of the PN codes for the cell and its surrounding cell sites. The example below shows the relationship between the chips and the distance:

$$\frac{32.768}{64} = 512 \qquad \text{possible PN offsets}$$

$$\text{Chip frequency} = 1.228 \times 10^6 \qquad \text{chips/per second}$$

Therefore,

$$\text{Time} = \frac{1}{\text{chip frequency}} = 0.8144\ \mu\text{s/chip}$$

$$\text{Distance} = 244\ \text{m/chip}$$

Q. *How are PN codes selected?*

A. The following table can be used to establish the PN codes for any cell site in the network. To do so, first determine whether you wish to have a 4, 7, 9, 19, or other reuse pattern for the PN codes.

Sector	PN code
Alpha	$3 \cdot P \cdot N - 2P$
Beta	$3 \cdot P \cdot N$
Gamma	$3 \cdot P \cdot N - P$
Omni	$3 \cdot P \cdot N$

N = reuse cell.
P = PN code increment.

Q. *What PN code reuse pattern should be used for a new system?*

A. The suggested PN reuse pattern is a $N = 19$ pattern for a new CDMA system.

Q. *What PN code reuse pattern should be used for a cellular system?*

A. The suggested PN reuse pattern for overlaying the CDMA system onto a cellular system is a $N = 14$ pattern when the analog system utilizes a $N = 7$ voice channel reuse pattern.

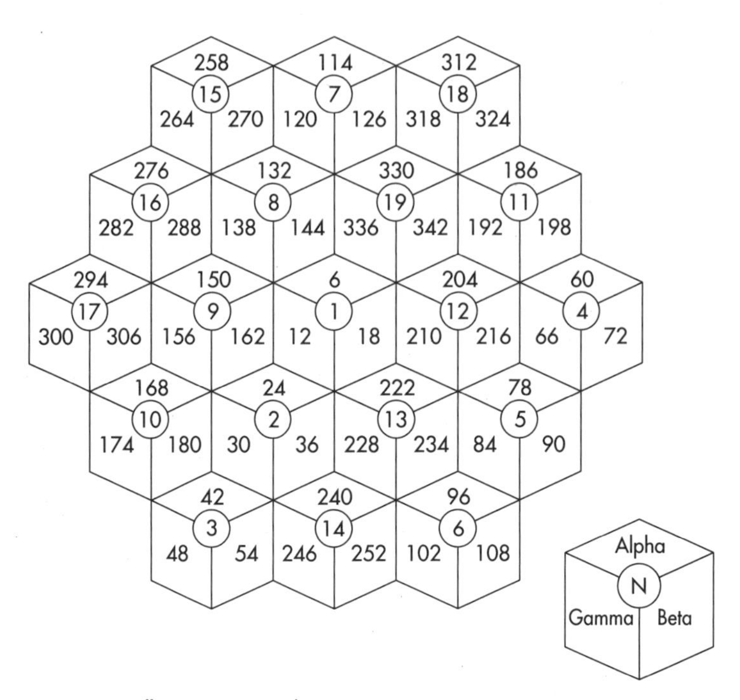

Figure 10.12 PN offset pattern. $N = 19$, Pilot_INC = 6.

Q. *What does an N = 19 PN code reuse pattern look like?*

A. An $N = 19$ PN code reuse pattern is shown in Fig. 10.12. Note that not all the codes have been utilized in the $N = 19$ pattern. The remaining codes should be left in reserve for use when there is a PN code problem that arises. In addition, a PN_INC value of 6 is also recommended for use.

Q. *At what value should the Pilot_INC be set?*

A. The Pilot_INC has a valid range of 0 to 15. The Pilot_INC is the PN sequence offset index, and it is a multiple of 64 chips. The subscriber unit uses the Pilot_INC to determine which are the valid pilots to be scanned. A simple table that can be used to determine the Pilot_INC as a function of the distance between reusing sites is shown in Table 10.4.

TABLE 10.4 PILOT OFFSETS

R, km	R, chips	S	C/I	m, chips	Pilot_INC	No. of offsets
25	103	14	24	622	10	50
20	82	12	24	499	8	64
15	61	12	24	390	6	85
12.5	51	10	24	325	5	102
10	41	10	24	271	4	128
7	29	10	24	207	4	128
5	21	10	24	165	3	170
3	12	10	24	117	2	256
2.5	10	10	24	106	2	256
2	8	10	24	96	2	256

Q. *What is the CDMA link budget?*

A. There are two links, forward and reverse, and they utilize different coding and modulation formats. The first step in the link budget process is to determine the forward, then the reverse link's maximum path losses. The link budget calculations utilized influence directly the performance of the CDMA system since the budget is used to determine power setting and capacity limits for the network.

Q. *What is a guardzone?*

A. A *guardzone* is the physical area outside the CDMA coverage area that can no longer utilize the AMPS channels once they are occupied by the CDMA system. The establishment and size of the guardzone is dependent upon the traffic load expected by the CDMA system. The guardzone is usually defined in terms of a signal strength level from which analog cell sites operating with the CDMA channel sets cannot contribute to the overall interference level of the system (Fig. 10.13).

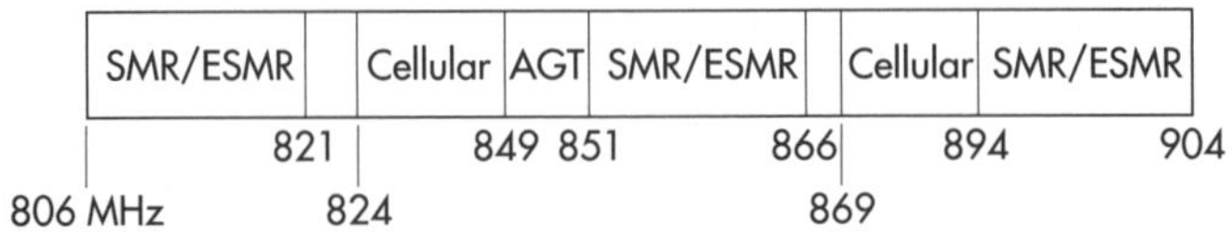

SMR/ESMR | Cellular | AGT | SMR/ESMR | Cellular | SMR/ESMR

806 MHz 821 824 849 851 866 869 894 904

Figure 10.13 Guardzone.

Q. *What CDMA system deployment options exist?*

A. There are two distinct CDMA system deployment options, or methods, that can be used. The first method is to deploy CDMA in every cell site, for the defined service area, on a one-to-one basis. The other method available is to deploy CDMA on an *N*-to-one basis. Both one-to-one and *N*-to-one deployment strategies have advantages and disadvantages, as listed in Table 10.5.

TABLE 10.5 TWO CDMA SYSTEM DEPLOYMENT OPTIONS

	Advantages	Disadvantages
One to one	Consistent coverage	Higher cost
	Facilitates gradual growth	Guardzone requirements
	Integrates into existing system	Digital-to-analog boundary handoff
	Easier to engineer	Slower deployment
	Larger initial capacity gain	
N to one	Lower cost	Harder to engineer properly
	Faster to implement	Lower capacity gain

Q. *What does an N:1 CDMA system look like?*

A. The *N*:1 CDMA system has the system configuration shown in Fig. 10.14.

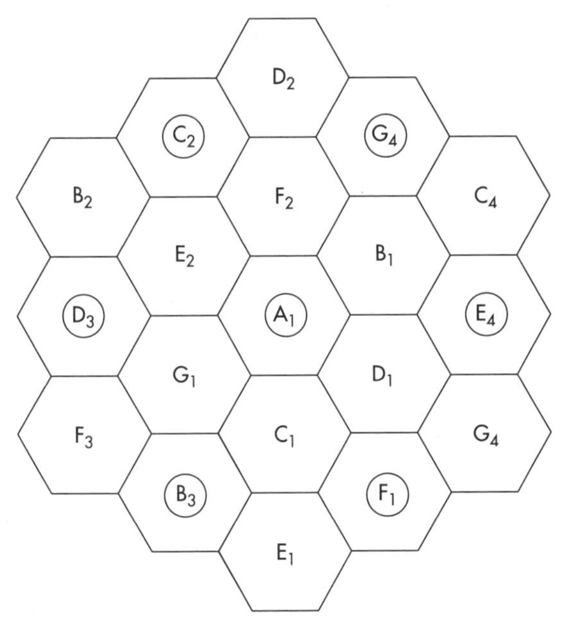

Figure 10.14 *N*:1 layout. The cells with the circled numbers are CDMA and analog.

Q. *What are the initial system parameters for an N:1 CDMA system?*

A. The initial system parameters for an *N*:1 CDMA system are shown below:

Pilot PN offset, 14

Pilot_INC, 6

T_Add, −16

T_Drop, −20

T_Comp, 3

T_Tdrop, 5

Search Window_Active, 8

Search Window_Neighbor, 11

Search Window_Remaining, 10

Pilot power = 20%

Q. *What does a 1:1 CDMA system look like?*

A. The 1:1 CDMA system has the system configuration shown in Fig. 10.15.

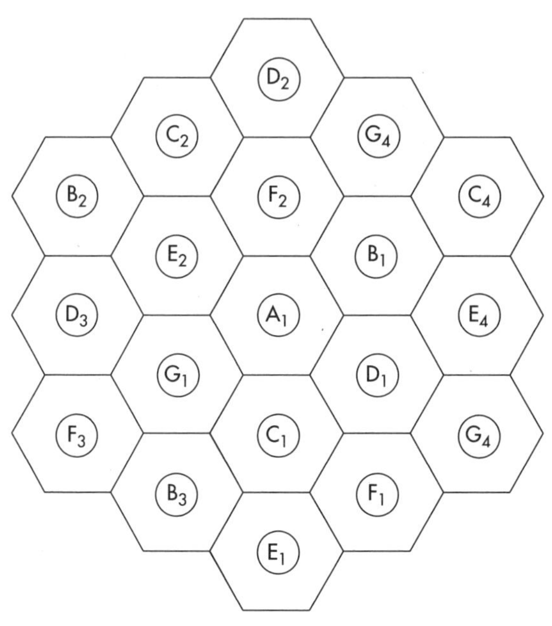

Figure 10.15 CDMA. 1:1 layout.

Q. *What are the initial system parameters for a 1:1 CDMA system?*

A. The initial system parameters for a 1:1 CDMA system are shown below:

Pilot PN offset, 14

Pilot_INC, 6

T_Add, −16

T_Drop, −20

T_Comp, 3

T_Tdrop, 5

Search Window_Active, 6

Search Window_Neighbor, 11

Search Window_Remaining, 10

Pilot power = 20%

Q. *What does a core-only CDMA system look like?*

A. The core-only CDMA system has the system configuration shown in Fig. 10.16.

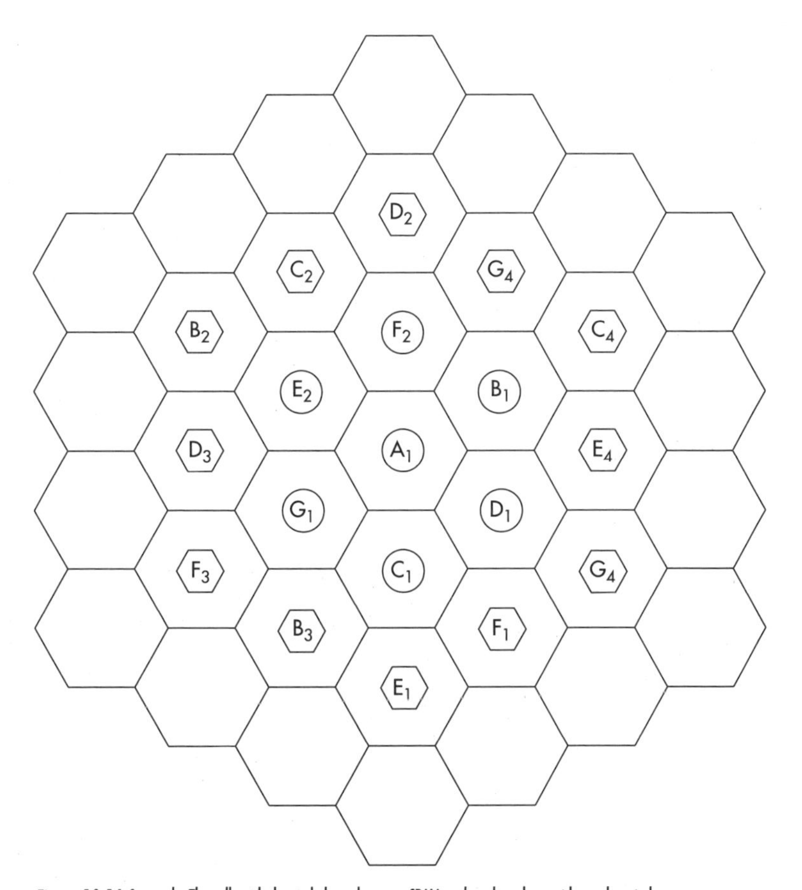

Figure 10.16 Core only. The cells with the circled numbers are CDMA and analog; those with numbers in hexagons are guardzone; the blank cells are full analog.

Q. *What are the initial system parameters for a core-only CDMA system?*

A. The initial system parameters for a core-only CDMA system are shown below:

Pilot PN offset, 14

Pilot_INC, 6

T_Add, −16

T_Drop, −20

T_Comp, 3

T_Tdrop, 5

Search Window_Active, 6

Search Window_Neighbor, 11

Search Window_Remaining, 10

Pilot power = 20%

Q. *What is IS-136?*

A. IS-136 is the NADC standard that is the next evolution of IS-54. IS-136 is currently being deployed in cellular and for new PCS systems throughout the United States. IS-136 is very similar to IS-54 with a few exceptions that tie into the digital control channel and short-message services.

Q. *Is there a difference in the channels assigned for IS-136 and IS-54?*

A. The allocation of channels for IS-136 is the same as that for IS-54 in that both use the same modulation format and occupy 30 kHz of spectrum for each physical radio channel. Each of the radio channels are divided into six time slots of which two are used per call—that is, three subscribers per physical radio.

IS-136 brings to the table the *digital control channel* (DCCH), and it enables the delivery of adjunct features that in cellular were not really possible. The DCCH occupies two of the six time slots, and therefore, if a physical radio also has a DCCH assigned to it, only two subscribers can use the physical radio for communication purposes.

Q. *What is the PCS spectrum allocation for IS-136 channels?*

A. The spectrum allocation for IS-136 in PCS bands is shown in Table 10.6.

Q. *What are the DCCH channel assignments for cellular IS-136?*

A. The DCCH channel assignments for cellular preferred channel sets are broken down into 16 relative probability blocks for each frequency band of operation (Table 10.7). The relative probability block 1 is the first group of channels the subscriber unit uses to find the DCCH for the system and cell. The subscriber unit will then scan through the entire frequency band going through channel sets according to the relative probability blocks until it finds a DCCH.

If no DCCH is found, the subscriber unit reverts to the control channel for a dual-mode phone and then acquires the system either through the control channel or is directed to a specific channel that has the DCCH.

Q. *What is the DCCH channel assignment for PCS IS-136?*

A. The DCCH channel assignments for PCS preferred channel sets are broken down into 16 relative probability blocks for each frequency band of operation (Table 10.8). The relative probability block #1 is the first group of channels the subscriber unit uses to find the DCCH for the system and cell. The subscriber unit will then scan through the entire frequency band, going through channel sets according to the relative probability blocks until it finds a DCCH.

Q. *What are the channel assignments for PCS IS-136?*

A. The channel assignments for PCS IS-136 are shown in Table 10.6 in the various columns associated with each PCS block that conform to an $N = 7$ three-sector channel chart. Glancing at the table, it is obvious that the D, E, and F blocks do not have enough channels in the preferred locations for these groups. Therefore, in the figures listed for channel assignments, the remaining channels that fill out the channels that should make up the DCCH list include those in probability block 2.

Table 10.6 Spectrum Allocation for IS-136 Channels

A1	B1	C1	D1	E1	F1	G1	A2	B2	C2	D2	E2	F2	G2	A3	B3	C3	D3	E3	F3	G3
										A block										
498	497	496	495	494	493	492	491	490	489	488	487	486	485	484	483	482	481	480	479	478
477	476	475	474	473	472	471	470	469	468	467	466	465	464	463	462	461	460	459	458	457
456	455	454	453	452	451	450	449	448	447	446	445	444	443	442	441	440	439	438	437	436
435	434	433	432	431	430	429	428	427	426	425	424	423	422	421	420	419	418	417	416	415
414	413	412	411	410	409	408	407	406	405	404	403	402	401	400	399	398	397	396	395	394
393	392	391	390	389	388	387	386	385	384	383	382	381	380	379	378	377	376	375	374	373
372	371	370	369	368	367	366	365	364	363	362	361	360	359	358	357	356	355	354	353	352
351	350	349	348	347	346	345	344	343	342	341	340	339	338	337	336	335	334	333	332	331
330	329	328	327	326	325	324	323	322	321	320	319	318	317	316	315	314	313	312	311	310
309	308	307	306	305	304	303	302	301	300	299	298	297	296	295	294	293	292	291	290	289
288	287	286	285	284	283	282	281	280	279	278	277	276	275	274	273	272	271	270	269	268
267	266	265	264	263	262	261	260	259	258	257	256	255	254	253	252	251	250	249	248	247
246	245	244	243	242	241	240	239	238	237	236	235	234	233	232	231	230	229	228	227	226
225	224	223	222	221	220	219	218	217	216	215	214	213	212	211	210	209	208	207	206	205
204	203	202	201	200	199	198	197	196	195	194	193	192	191	190	189	188	187	186	185	184
183	182	181	180	179	178	177	176	175	174	173	172	171	170	169	168	167	166	165	164	163
162	161	160	159	158	157	156	155	154	153	152	151	150	149	148	147	146	145	144	143	142
141	140	139	138	137	136	135	134	133	132	131	130	129	128	127	126	125	124	123	122	121
120	119	118	117	116	115	114	113	112	111	110	109	108	107	106	105	104	103	102	101	100
99	98	97	96	95	94	93	92	91	90	89	88	87	86	85	84	83	82	81	80	79
78	77	76	75	74	73	72	71	70	69	68	67	66	65	64	63	62	61	60	59	58
57	56	55	54	53	52	51	50	49	48	47	46	45	44	43	42	41	40	39	38	37
36	35	34	33	32	31	30	29	28	27	26	25	24	23	22	21	20	19	18	17	16
15	14	13	12	11	10	9	8	7	6	5	4	3	2							

B block

1165	1164	1163	1162	1161	1160	1159	1158	1157	1156	1155	1154	1153	1152	1151	1150	1149	1148	1147	1146	1145
1144	1143	1142	1141	1140	1139	1138	1137	1136	1135	1134	1133	1132	1131	1130	1129	1128	1127	1126	1125	1124
1123	1122	1121	1120	1119	1118	1117	1116	1115	1114	1113	1112	1111	1110	1109	1108	1107	1106	1105	1104	1103
1102	1101	1100	1099	1098	1097	1096	1095	1094	1093	1092	1091	1090	1089	1088	1087	1086	1085	1084	1083	1082
1081	1080	1079	1078	1077	1076	1075	1074	1073	1072	1071	1070	1069	1068	1067	1066	1065	1064	1063	1062	1061
1060	1059	1058	1057	1056	1055	1054	1053	1052	1051	1050	1049	1048	1047	1046	1045	1044	1043	1042	1041	1040
1039	1038	1037	1036	1035	1034	1033	1032	1031	1030	1029	1028	1027	1026	1025	1024	1023	1022	1021	1020	1019
1018	1017	1016	1015	1014	1013	1012	1011	1010	1009	1008	1007	1006	1005	1004	1003	1002	1001	1000	999	998
997	996	995	994	993	992	991	990	989	988	987	986	985	984	983	982	981	980	979	978	977
976	975	974	973	972	971	970	969	968	967	966	965	964	963	962	961	960	959	958	957	956
955	954	953	952	951	950	949	948	947	946	945	945	943	942	941	940	939	938	937	936	935
934	933	932	931	930	929	928	927	926	925	924	923	922	921	920	919	918	917	916	915	914
913	912	911	910	909	908	907	906	905	904	903	902	901	900	899	898	897	896	895	894	893
892	891	890	889	888	887	886	885	884	883	882	881	880	879	878	877	876	875	874	873	872
871	870	869	868	867	866	865	864	863	862	861	860	859	858	857	856	855	854	853	852	851
850	849	848	847	846	845	844	843	842	841	840	839	838	837	836	835	834	833	832	831	830
829	828	827	826	825	824	823	822	821	820	819	818	817	816	815	814	813	812	811	810	809
808	807	806	805	804	803	802	801	800	799	798	797	796	795	794	793	792	791	790	789	788
787	786	785	784	783	782	781	780	779	778	777	776	775	774	773	772	771	770	769	768	767
766	765	764	763	762	761	760	759	758	757	756	755	754	753	752	751	750	749	748	747	746
745	744	743	742	741	740	739	738	737	736	735	734	733	732	731	730	729	728	727	726	725
724	723	722	721	720	719	718	717	716	715	714	713	712	711	710	709	708	707	706	705	704
703	702	701	700	699	698	697	696	695	694	693	692	691	690	689	688	687	686	685	684	683
682	681	680	679	678	677	676	675	674	673	672	671	670	669	668						

Table 10.6 Spectrum Allocation for IS-136 Channels (*Continued*)

A1	B1	C1	D1	E1	F1	G1	A2	B2	C2	D2	E2	F2	G2	A3	B3	C3	D3	E3	F3	G3
										C block										
1998	1997	1996	1995	1994	1993	1992	1991	1990	1989	1988	1987	1986	1985	1984	1983	1982	1981	1980	1979	1978
1977	1976	1975	1974	1973	1972	1971	1970	1969	1968	1967	1966	1965	1964	1963	1962	1961	1960	1959	1958	1957
1956	1955	1954	1953	1952	1951	1950	1949	1948	1947	1946	1945	1944	1943	1942	1941	1940	1939	1938	1937	1936
1935	1934	1933	1932	1931	1930	1929	1928	1927	1926	1925	1924	1923	1922	1921	1920	1919	1918	1917	1916	1915
1914	1913	1912	1911	1910	1909	1908	1907	1906	1905	1904	1903	1902	1901	1900	1899	1898	1897	1896	1895	1894
1893	1892	1891	1890	1889	1888	1887	1886	1885	1884	1883	1882	1881	1880	1879	1878	1877	1876	1875	1874	1873
1872	1871	1870	1869	1868	1867	1866	1865	1864	1863	1862	1861	1860	1859	1858	1857	1856	1855	1854	1853	1852
1851	1850	1849	1848	1847	1846	1845	1844	1843	1842	1841	1840	1839	1838	1837	1836	1835	1834	1833	1832	1831
1830	1829	1828	1827	1826	1825	1824	1823	1822	1821	1820	1819	1818	1817	1816	1815	1814	1813	1812	1811	1810
1809	1808	1807	1806	1805	1804	1803	1802	1801	1800	1799	1798	1797	1796	1795	1794	1793	1792	1791	1790	1789
1788	1787	1786	1785	1784	1783	1782	1781	1780	1779	1778	1777	1776	1775	1774	1773	1772	1771	1770	1769	1768
1767	1766	1765	1764	1763	1762	1761	1760	1759	1758	1757	1756	1755	1754	1753	1752	1751	1750	1749	1748	1747
1746	1745	1744	1743	1742	1741	1740	1739	1738	1737	1736	1735	1734	1733	1732	1731	1730	1729	1728	1727	1726
1725	1724	1723	1722	1721	1720	1719	1718	1717	1716	1715	1714	1713	1712	1711	1710	1709	1708	1707	1706	1705
1704	1703	1702	1701	1700	1699	1698	1697	1696	1695	1694	1693	1692	1691	1690	1689	1688	1687	1686	1685	1684
1683	1682	1681	1680	1679	1678	1677	1676	1675	1674	1673	1672	1671	1670	1669	1668	1667	1666	1665	1664	1663
1662	1661	1660	1659	1658	1657	1656	1655	1654	1653	1652	1651	1650	1649	1648	1647	1646	1645	1644	1643	1642
1641	1640	1639	1638	1637	1636	1635	1634	1633	1632	1631	1630	1629	1628	1627	1626	1625	1624	1623	1622	1621
1620	1619	1618	1617	1616	1615	1614	1613	1612	1611	1610	1609	1608	1607	1606	1605	1604	1603	1602	1601	1600
1599	1598	1597	1596	1595	1594	1593	1592	1591	1590	1589	1588	1587	1586	1585	1584	1583	1582	1581	1580	1579
1578	1577	1576	1575	1574	1573	1572	1571	1570	1569	1568	1567	1566	1565	1564	1563	1562	1561	1560	1559	1558
1557	1556	1555	1554	1553	1552	1551	1550	1549	1548	1547	1546	1545	1544	1543	1542	1541	1540	1539	1538	1537
1536	1535	1534	1533	1532	1531	1530	1529	1528	1527	1526	1525	1524	1523	1522	1521	1520	1519	1518	1517	1516
1515	1514	1513	1512	1511	1510	1509	1508	1507	1506	1505	1504	1503	1502	1501						

D block

665	664	663	662	661	660	659	658	657	656	655	654	653	652	651	650	649	648	647	646	645
644	643	642	641	640	639	638	637	636	635	634	633	632	631	630	629	628	627	626	625	624
623	622	621	620	619	618	617	616	615	614	613	612	611	610	609	608	607	606	605	604	603
602	601	600	599	598	597	596	595	594	593	592	591	590	589	588	587	586	585	584	583	582
581	580	579	578	577	576	575	574	573	572	571	570	569	568	567	566	565	564	563	562	561
560	559	558	557	556	555	554	553	552	551	550	549	548	547	546	545	544	543	542	541	540
539	538	537	536	535	534	533	532	531	530	529	528	527	526	525	524	523	522	521	520	519
518	517	516	515	514	513	512	511	510	509	508	507	506	505	504	503	502				

E block

1332	1331	1330	1329	1328	1327	1326	1325	1324	1323	1322	1321	1320	1319	1338	1337	1336	1335	1334	1333	1332
1331	1330	1329	1328	1327	1326	1325	1324	1323	1322	1321	1320	1319	1318	1317	1316	1315	1314	1313	1312	1311
1310	1309	1308	1307	1306	1305	1304	1303	1302	1301	1300	1299	1298	1297	1296	1295	1294	1293	1292	1291	1290
1289	1288	1287	1286	1285	1284	1283	1282	1281	1280	1279	1278	1277	1276	1275	1274	1273	1272	1271	1270	1269
1268	1267	1266	1265	1264	1263	1262	1261	1260	1259	1258	1257	1256	1255	1254	1253	1252	1251	1250	1249	1248
1247	1246	1245	1244	1243	1242	1241	1240	1239	1238	1237	1236	1235	1234	1233	1232	1231	1230	1229	1228	1227
1226	1225	1224	1223	1222	1221	1220	1219	1218	1217	1216	1215	1214	1213	1212	1211	1210	1209	1208	1207	1206
1205	1204	1203	1202	1201	1200	1199	1198	1197	1196	1195	1194	1193	1192	1191	1190	1189	1188	1187	1186	1185
1184	1183	1182	1181	1180	1179	1178	1177	1176	1175	1174	1173	1172	1171	1170	1160	1168				

TABLE 10.6 SPECTRUM ALLOCATION FOR IS-136 CHANNELS (CONTINUED)

F block																				
1498	1497	1496	1495	1494	1493	1492	1491	1490	1489	1488	1487	1486	1485	1484	1483	1482	1481	1480	1479	1478
1477	1476	1475	1474	1473	1472	1471	1470	1469	1468	1467	1466	1465	1464	1463	1462	1461	1460	1459	1458	1457
1456	1455	1454	1453	1452	1451	1450	1449	1448	1447	1446	1445	1444	1443	1442	1441	1440	1439	1438	1437	1436
1435	1434	1433	1432	1431	1430	1429	1428	1427	1426	1425	1424	1423	1422	1421	1420	1419	1418	1417	1416	1415
1414	1413	1412	1411	1410	1409	1408	1407	1406	1405	1404	1403	1402	1401	1400	1399	1398	1397	1396	1395	1394
1393	1392	1391	1390	1389	1388	1387	1386	1385	1384	1383	1382	1381	1380	1379	1378	1377	1376	1375	1374	1373
1372	1371	1370	1369	1368	1367	1366	1365	1364	1363	1362	1361	1360	1359	1358	1357	1356	1355	1354	1353	1352
1351	1350	1349	1348	1347	1346	1345	1344	1343	1342	1341	1340	1339	1338	1337	1336	1335	1334			

Table 10.7 Cellular DCCH Preferred Assignments

Block number	Channel number	Band	Number of channels	Relative probability
		A block		
1	1–26	A	26	4
2	27–52	A	26	5
3	53–78	A	26	6
4	79–104	A	26	7
5	105–130	A	26	8
6	131–156	A	26	9
7	157–182	A	26	10
8	183–208	A	26	11
9	209–234	A	26	12
10	235–260	A	26	13
11	261–286	A	26	14
12	287–312	A	26	15
13	313–333	A	21	16 (lowest)
14	667–691	A′	25	3
15	692–716	A′	25	2
16	991–1023	A″	33	1 (highest)
		B block		
1	334–354	B	21	16 (lowest)
2	355–380	B	26	15
3	381–406	B	26	14
4	407–432	B	26	13
5	433–458	B	26	12
6	459–484	B	26	11
7	485–510	B	26	10
8	511–536	B	26	9
9	537–562	B	26	8
10	563–588	B	26	7
11	589–614	B	26	6
12	615–640	B	26	5
13	641–666	B	26	4
14	717–741	B′	25	3
15	742–766	B′	25	2
16	767–799	B′	33	1 (highest)

TABLE 10.8 DCCH PREFERRED ASSIGNMENTS

Block number	Channel number	Band	Number of channels	Relative probability
		A block		
1	2–31	A	30	16 (lowest)
2	32–62	A	31	15
3	63–93	A	31	14
4	94–124	A	31	13
5	125–155	A	31	12
6	156–186	A	31	11
7	187–217	A	31	10
8	218–248	A	31	9
9	249–279	A	31	8
10	280–310	A	31	7
11	311–341	A	31	6
12	342–372	A	31	5
13	373–403	A	31	4
14	404–434	A	31	3
15	435–465	A	31	2
16	466–498	A	33	1 (highest)
		B block		
1	668–698	B	31	16 (lowest)
2	699–729	B	31	15
3	730–760	B	31	14
4	761–791	B	31	13
5	792–822	B	31	12
6	823–853	B	31	11
7	854–884	B	31	10
8	885–915	B	31	9
9	916–946	B	31	8
10	947–977	B	31	7
11	978–1008	B	31	6
12	1009–1039	B	31	5
13	1040–1070	B	31	4
14	1071–1101	B	31	3
15	1102–1132	B	31	2
16	1133–1165	B	33	1 (highest)

Table 10.8 DCCH Preferred Assignments (Continued)

Block number	Channel number	Band	Number of channels	Relative probability
		C block		
1	1501–1531	C	31	16 (lowest)
2	1532–1562	C	31	15
3	1563–1593	C	31	14
4	1594–1624	C	31	13
5	1625–1655	C	31	12
6	1656–1686	C	31	11
7	1687–1717	C	31	10
8	1718–1748	C	31	9
9	1749–1779	C	31	8
10	1780–1810	C	31	7
11	1811–1841	C	31	6
12	1842–1872	C	31	5
13	1873–1903	C	31	4
14	1904–1934	C	31	3
15	1935–1965	C	31	2
16	1966–1998	C	33	1 (highest)
		D block		
1	502–511	D	10	16 (lowest)
2	512–521	D	10	15
3	522–531	D	10	14
4	532–541	D	10	13
5	542–551	D	10	12
6	552–561	D	10	11
7	562–571	D	10	10
8	572–581	D	10	9
9	582–591	D	10	8
10	592–601	D	10	7
11	602–611	D	10	6
12	612–621	D	10	5
13	622–631	D	10	4
14	632–641	D	10	3
15	642–651	D	10	2
16	652–665	D	14	1 (highest)

Table 10.8 DCCH Preferred Assignments (Continued)

Block number	Channel number	Band	Number of channels	Relative probability
		E block		
1	1168–1177	E	10	16 (lowest)
2	1178–1187	E	10	15
3	1188–1197	E	10	14
4	1198–1207	E	10	13
5	1208–1217	E	10	12
6	1218–1227	E	10	11
7	1228–1237	E	10	10
8	1238–1247	E	10	9
9	1248–1257	E	10	8
10	1258–1267	E	10	7
11	1268–1277	E	10	6
12	1278–1287	E	10	5
13	1288–1297	E	10	4
14	1298–1307	E	10	3
15	1308–1317	E	10	2
16	1318–1332	E	15	1 (highest)
		F block		
1	1335–1344	F	10	16 (lowest)
2	1345–1354	F	10	15
3	1355–1364	F	10	14
4	1365–1374	F	10	13
5	1375–1384	F	10	12
6	1385–1394	F	10	11
7	1395–1404	F	10	10
8	1405–1414	F	10	9
9	1415–1424	F	10	8
10	1425–1434	F	10	7
11	1435–1444	F	10	6
12	1445–1454	F	10	5
13	1455–1464	F	10	4
14	1465–1474	F	10	3
15	1475–1484	F	10	2
16	1485–1498	F	14	1 (highest)

Q. *Is the frequency reuse the same for IS-136 and IS-54?*

A. The frequency reuse for IS-136 is the same as that for IS-54.

Q. *What is the C/I level used for IS-136?*

A. The *C/I* level desired is 17 dB, and it is the same for DCCH and the *digital traffic channel* (DTC).

Q. *What are the valid DCC values for IS-136?*

A. The valid DCC values for IS-136 range from 0 to 3.

Q. *What are the SDCC values for IS-136?*

A. The SDCC values for IS-136 range from 0 to 15.

Q. *What are the DVCC values for IS-136?*

A. The DVCC values for IS-136 range from 1 to 255.

Q. *What is the reuse pattern for DCC and SDCC assignments with IS-136?*

A. The reuse pattern for DCC and SDCC assignments with IS-136 is shown in Fig. 10.17.

Q. *What is the DVCC reuse pattern for IS-136?*

A. The reuse pattern for DVCC assignments with IS-136 is shown in Fig. 10.18.

Q. *Where are the DVCCs used in IS-136?*

A. DVCCs are used in IS-136 for defining the identity of the radio channel. The same DVCC should be assigned to *all* the channels in a sector in the same fashion that an SAT is assigned to a frequency group in analog. However, in the event that more than one frequency group is assigned to a particular sector, then it is recommended that the same DVCC be used for all the channels in the sector.

Q. *What is a BER?*

A. A BER is an abbreviation for *bit error rate.*

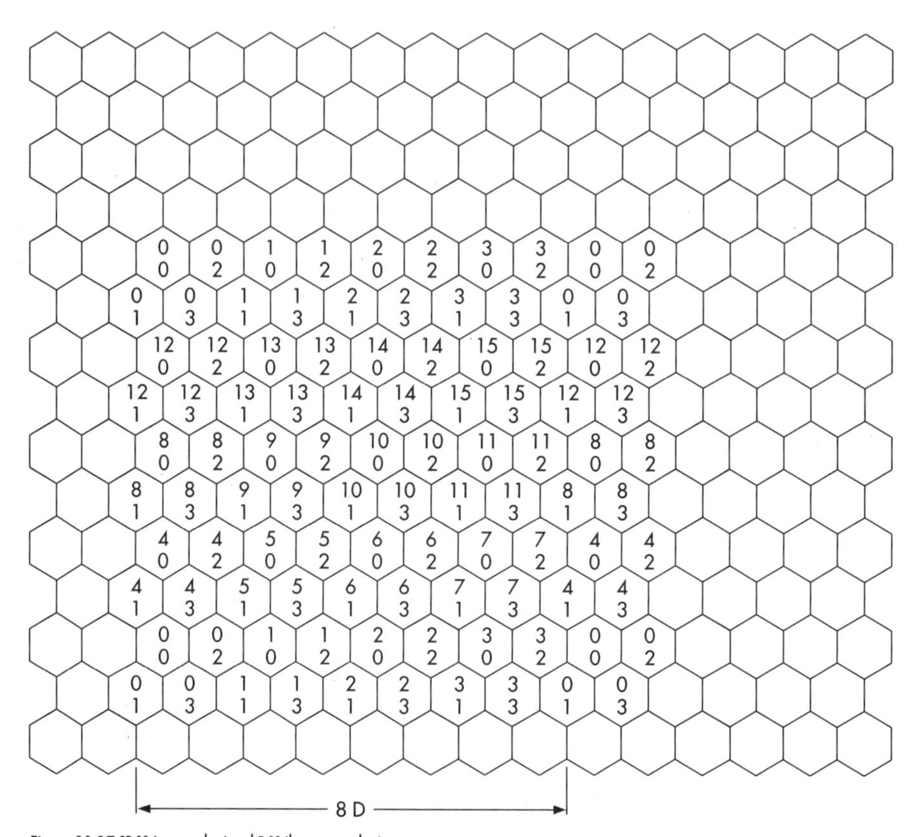

Figure 10.17 SDCC (top number) and DCC (bottom number) reuse patterns.

Q. *What are the bit-error-rate classes?*

A. The BER relationship to BER class is shown in Table 10.9. A brief relationship between BER and voice quality is given below for reference:

BER, %	**<1%**	**1–3%**	**>3%**
Voice quality	Good	Marginal	Bad

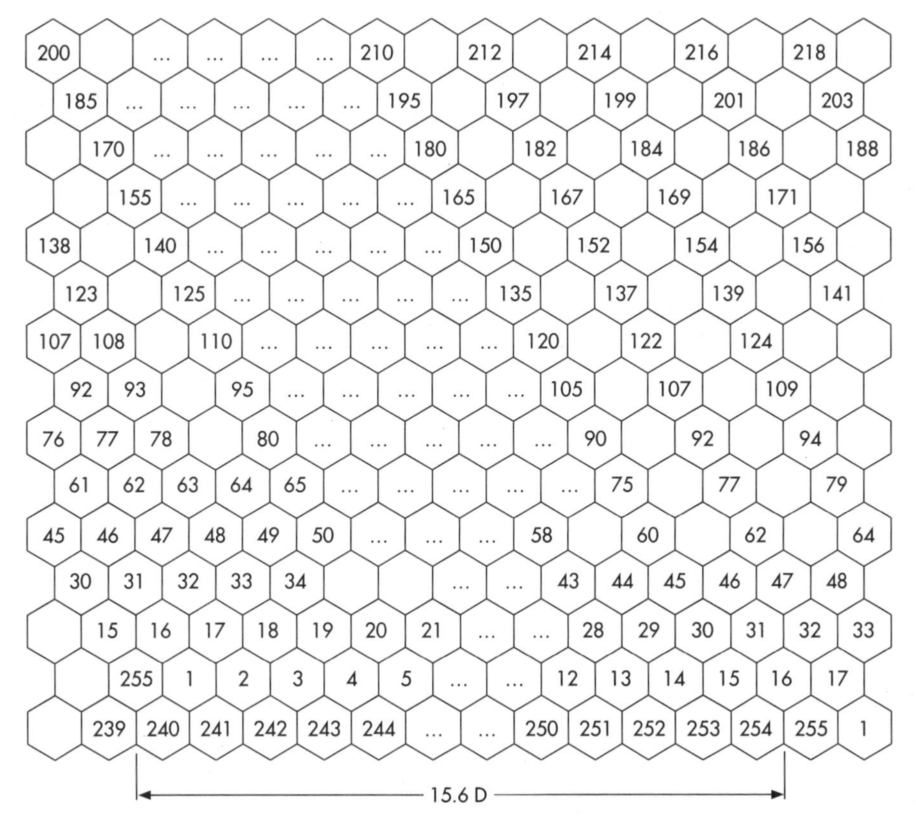

Figure 10.18 DVCC and traffic channel reuse.

TABLE 10.9 BER CLASSES

BER class	BER, %
0	<0.1
1	0.01–0.10
2	0.10–0.50
3	0.50–1.0
4	1.0–2.0
5	2.0–4.0
6	4.0–8.0
7	8.0<

TABLE 10.10 CELLULAR SYSTEM CAPACITY WITH IS-136 CHANNELS

Channels				**Erlangs, 29% erlang B**		
		Digital				
No. of analog	**No. of**	**DTC**	**DCCH**	**Analog**	**Digital**	**Total***
19	0			12.3	0	12.3
18	1	3	0	11.5	0.602	14
	1	2	1		0.223	13.2
17	2	6	0	10.7	2.28	15.8
	2	5	1		1.66	16.6
16	3	9	0	9.83	4.34	17.5
	3	8	1		3.63	16.6
15	4	12	0	9.01	6.61	19.3
	4	11	1		5.84	18.4
14	5	15	0	8.2	9.01	19.3
	5	14	1		8.2	20.2

*Assumes trunking efficiency.

Q. *What is the relationship between IS-136 channels and the capacity for a cellular system?*

A. The relationship between IS-136 channels and the capacity for a cellular wireless network is best illustrated in Table 10.10.

Q. *What is GSM?*

A. GSM is a technology platform that has gained worldwide acceptance and use. The GSM spectrum occupies several frequency bands. The particular band within which the GSM operates is dependent upon the country that it is in. For instance, GSM operates in the 900-MHz range, DCS-1800, the 1800-MHz range, and DCS-1900 or PCS-1900, the 1900-MHz range. All of the names GSM DCS-1800, DCS-1900, or PCS-1900 are GSM systems. The features and functionality of GSM is the same for each of the bands mentioned.

Q. *What are the GSM channels for PCS-1900?*

A. The channel chart for PCS-1900 is shown in Table 10.11 using an $N = 3$ reuse pattern. If another reuse pattern is chosen for the system, then a reallocation of the channels is easily achieved for PCS blocks A, B, and C. However, for PCS blocks D, E, and F, the flexibility to expand or relax the reuse distance does not really exist because of the allotted spectrum.

TABLE 10.11 CHANNEL CHART FOR PCS-1900

A1	B1	C1	A2	B2	C2	A3	B3	C3
				A block				
512	513	514	515	516	517	518	519	520
521	522	523	524	525	526	527	528	529
530	531	532	533	534	535	536	537	538
539	540	541	542	543	544	545	546	547
548	549	550	551	552	553	554	555	556
557	558	559	560	561	562	563	564	565
566	567	568	569	570	571	572	573	574
575	576	577	578	579	580	581	582	583
584	585							
				B block				
611	612	613	614	615	616	617	618	619
620	621	622	623	624	625	626	627	628
629	630	631	632	633	634	635	636	637
638	639	640	641	642	643	644	645	646
647	648	649	650	651	652	653	654	655
656	657	658	659	660	661	662	663	664
665	666	667	668	669	670	671	672	673
674	675	676	677	678	679	680	681	682
683	684	685						

Table 10.11 Channel Chart for PCS-1900 (*Continued*)

A1	B1	C1	A2	B2	C2	A3	B3	C3
				C block				
736	737	738	739	740	741	742	743	744
745	746	747	748	749	750	751	752	753
754	755	756	757	758	759	760	761	762
763	764	765	766	767	768	769	770	771
772	773	774	775	776	777	778	779	780
781	782	783	784	785	786	787	788	789
790	791	792	793	794	795	796	797	798
799	800	801	802	803	804	805	806	807
808	809	810						
				D block				
586	587	588	589	590	591	592	593	594
595	596	597	598	599	600	601	602	603
604	605	606	607	608	609	610		
				E block				
686	687	688	689	690	691	692	693	694
695	696	697	698	699	700	701	702	703
704	705	706	707	708	709	710		
				F block				
711	712	713	714	715	716	717	718	719
720	721	722	723	724	725	726	727	728
729	730	731	732	733	734	735		

Q. *How does call processing work for GSM?*

A. The flowchart in Fig. 10.19 best describes call processing for a GSM system.

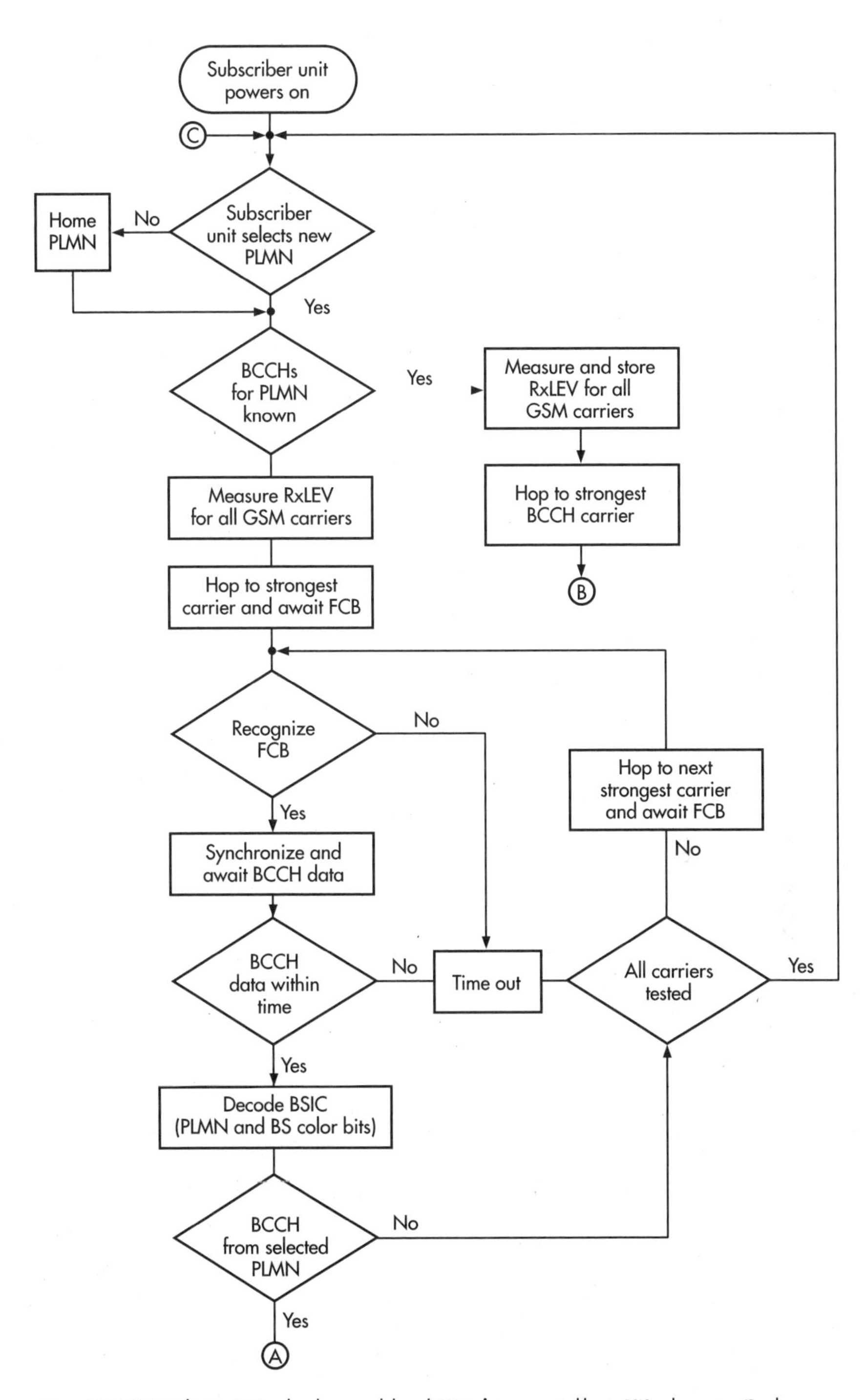

Figure 10.19 GSM initialization. BCCH = broadcast control channel, FCB = frequency control bursts, BSIC = base system ID code.

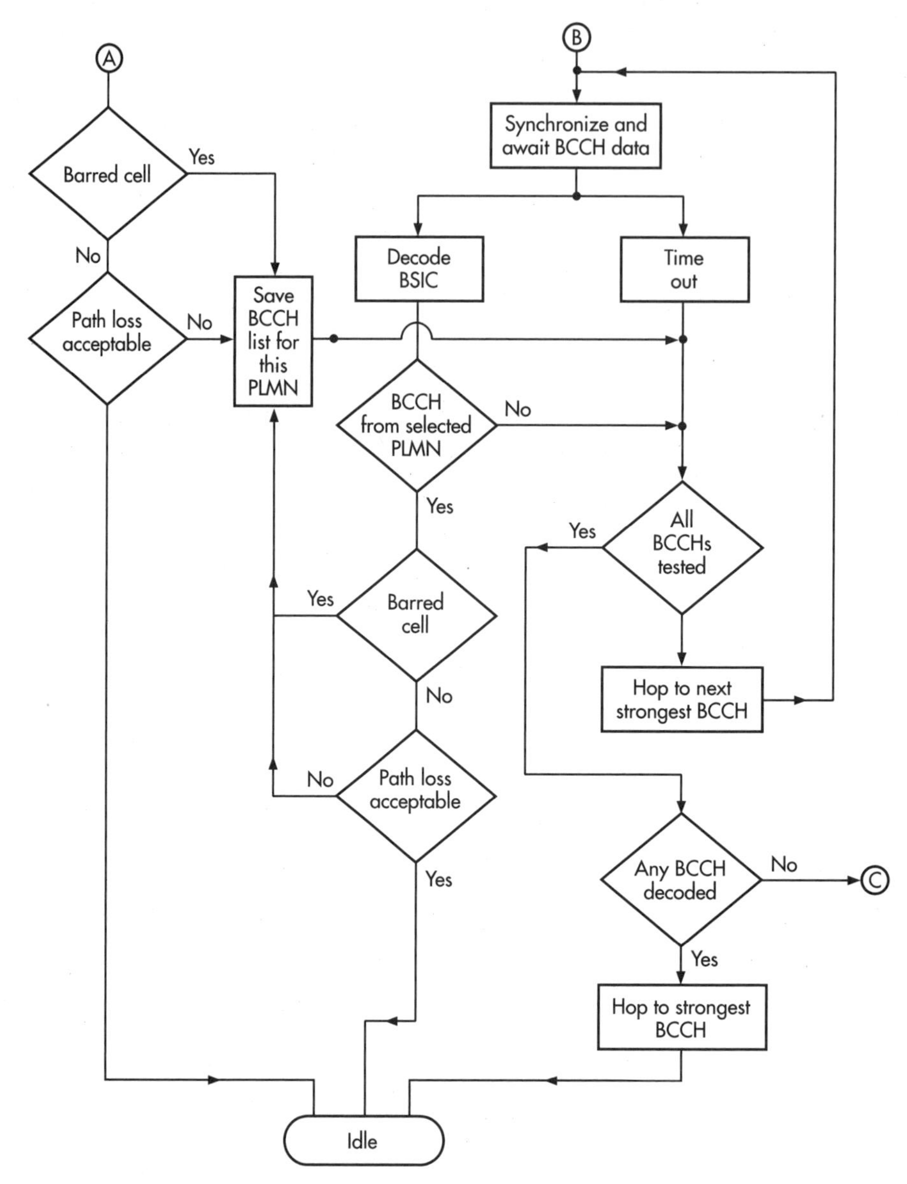

Figure 10.19 *(Continued)*

Q. *How does a GSM handoff work?*

A. The flowchart in Fig. 10.20 best describes the handoff process that takes place in a GSM system.

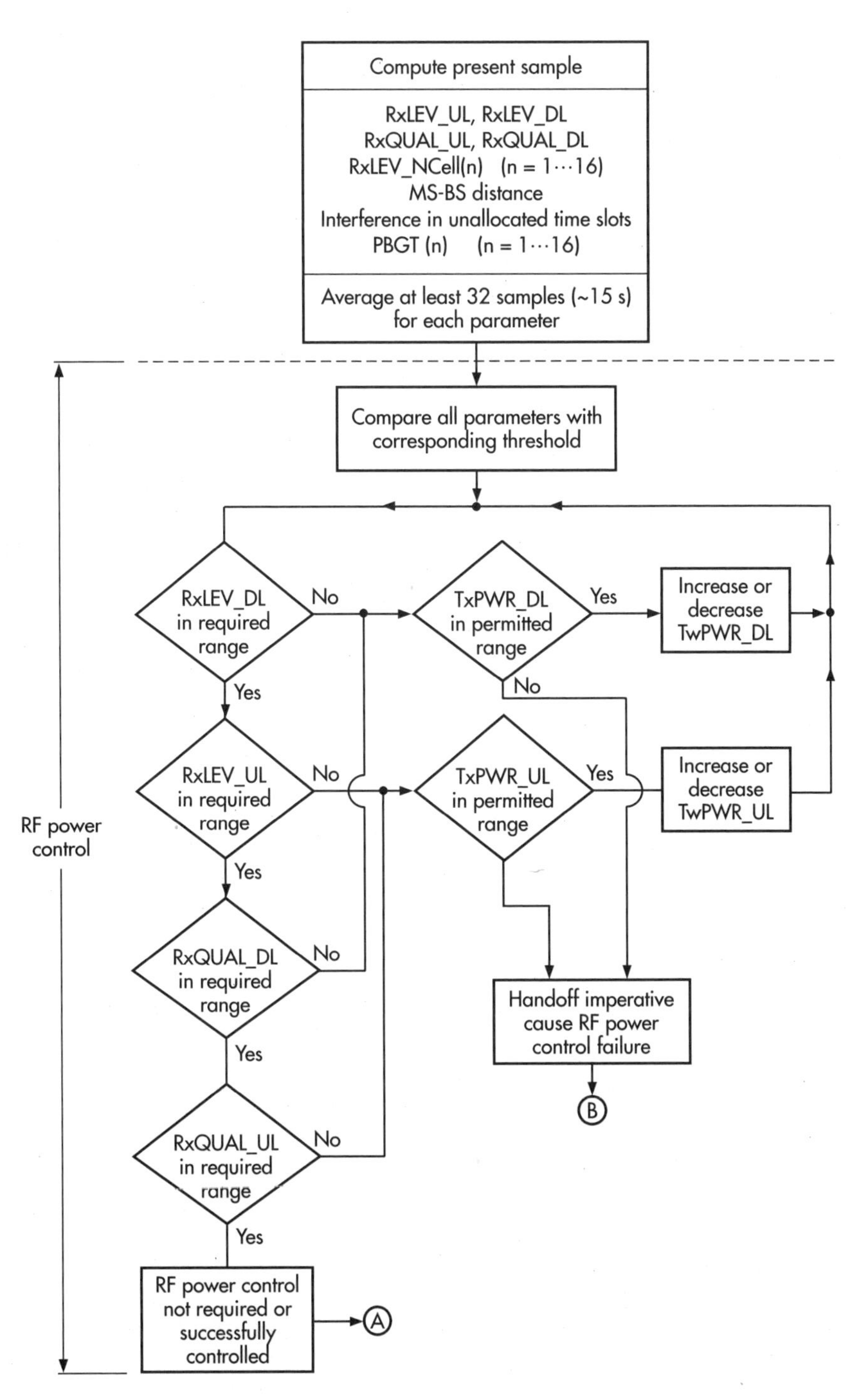

Figure 10.20 Flow diagram of handover process.

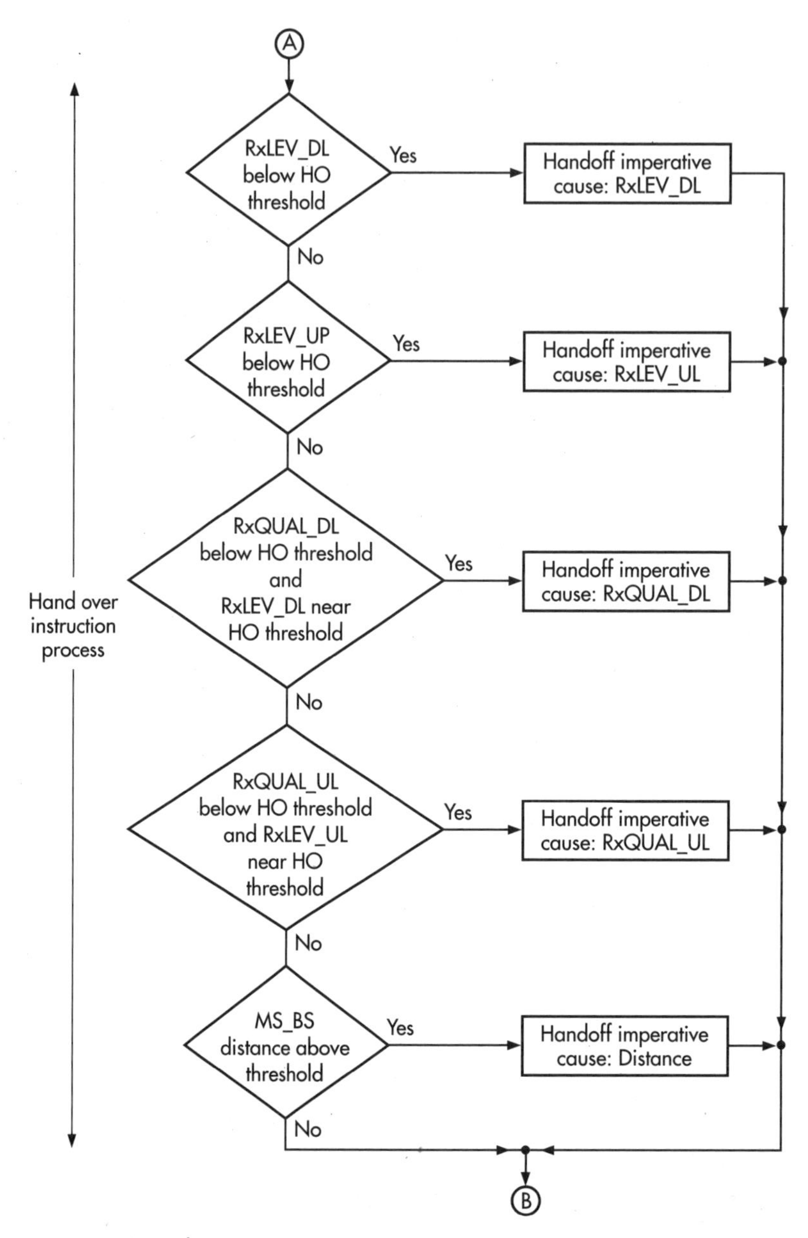

Figure 10.20 (*Continued*)

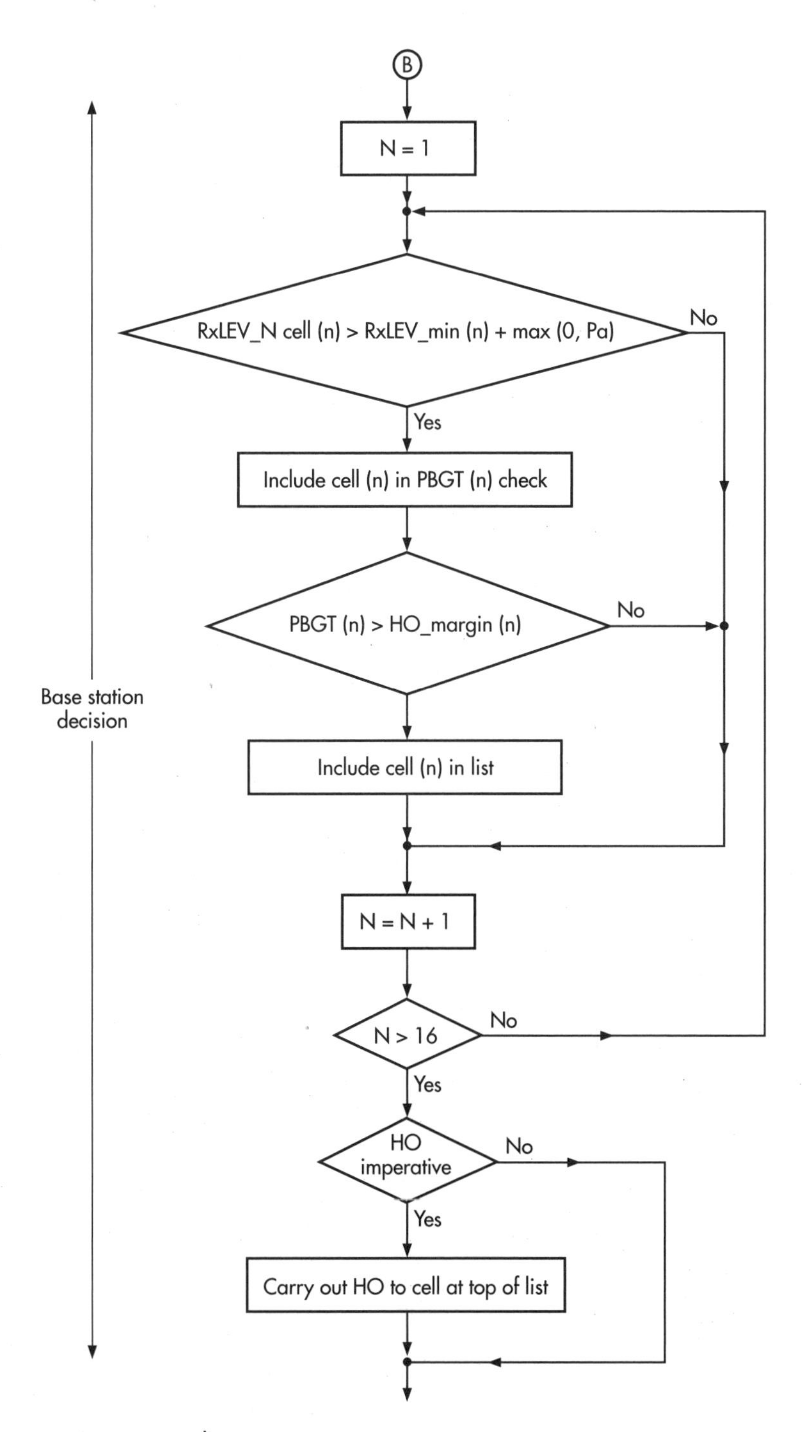

Figure 10.20 (*Continued*)

Q. *What is RxQUAL?*

A. *RxQUAL* is a quality metric used in GSM that directly relates to the BER for the channel used.

Q. *What is the relationship between RxQUAL and BER for GSM?*

A. The relationship between RxQUAL and BER for GSM is shown in Table 10.12.

TABLE 10.12 RxQUAL versus BER

RxQUAL	BER, %
0	<0.2
1	0.2–0.4
2	0.4–0.8
3	0.8–1.6
4	1.6–3.2
5	3.2–6.4
6	6.4–12.8
7	12.8>

Q. *What are the system design parameters required for a cellular network?*

A. The following is a list of the parameters that are valuable in the design and operation of a cellular system. These data should be updated on a regular basis as input to the quarterly network design review and should be compared to the previous quarter's data to detect large changes in any specific parameter and to determine if the modeling and prediction efforts by the network group are improving or remaining consistent.

- Size of initial subscriber base
- Projected growth of subscriber base over a two-year period
- Estimated usage per subscriber (millierlangs per subscriber)

- Estimated calls per subscriber
- Estimated calls per second
- Estimated average call-holding time (data from other known systems in the area are used or a typical industry value is used)
- Estimated switch initial traffic (erlangs)
- Projected switch traffic growth over a two-year period
- Estimated switch calls processed per second
- Projected switch calls processed per second over a two-year period
- Estimated number of cell sites in service at the time of system cut
- Projected cell site growth over a two-year period
- Estimated number of PSTN trunks in service at the time of the system cut
- Projected number of PSTN trunks required over a two-year period
- Estimated number of IMT trunks in service at the time of the system cut
- Estimated number of data links (SS7) in service at the time of the system cut
- Projected number of data links (SS7) required over a two-year period
- Auxiliary systems in service at the time of system cut:
  - Voice mail system (initial and projected required capacity)
  - Switch manager
  - Validation systems (precall and/or postcall)
  - Fraud prevention systems
  - Billing systems
  - Network management system

Q. *What are some key system switching design issues?*

A. Some of the key system switching design issues pertain to the capacity of the switching platform itself. The capacity is driven by the traffic load expected, the busy-hour call attempts, inbound and outbound traffic volume, the amount of ports on the platform, the storage or memory, and of course the CPU utilization factors.

Q. *What are some main factors for choosing a switch product?*

A. Some of the main factors to consider when choosing a switch product are the following:

- Compatibility with industry standard telecommunication protocols
- Standardized billing formats
- Industry- and market-specific switch functions
- Switch reliability and maintenance
- Switch vendor software, hardware, and operations support

Q. *What are the standard switching functions?*

A. The standard switching functions for a wireless system include the following:

- Subscriber features (three-party conferencing, call waiting, call forwarding, etc.)
- Paging area definitions
- Mobile autonomous registration
- Subscriber validation and denial functionality

Q. *What is network diversity?*

A. *Network diversity* is the alternate routing capacity provided in the network. Typically a route in the network will have at least one alternate or backup route to serve as standby in the event that the first route is out of service. The type of diversity routing used is dependent upon many different factors ranging from the type of facilities, equipment available for use, location, and distances of the area serving central offices—and of course, the budgetary constraints.

Q. *What is involved with a system interconnect design (voice)?*

A. The following is a brief list of some items associated with a system interconnect design:

1. Determine the types of interconnection facilities needed from the proposed service offering and the switch interface chosen.
2. Determine the grade of service to be used in the network design for landline facilities. Typical value is $P = 0.001$.
3. Determine the amount of diversity required in the network alternate routing design. A typical routing design will have at least one backup or secondary route for every primary route.
4. Collect and assess data for the network design including the type of interconnection facilities available for use, and the number, location, and distance of the area serving central offices for PSTN interconnections.
5. Formulate an interconnection plan using the information from the previous steps to develop the recommended system tables and diagrams.

Q. *What does the SS7 network do?*

A. The SS7 data network of a cellular system does more than just carry messages for performing mobile unit call delivery. This network also provides the needed messaging for interswitch handoffs, mobile unit validations, and feature updates. A cell site will go off the air from time to time, but problems with the SS7 network will tend to have a much larger impact on the service to the customer base. It is for this reason that this network should be properly designed from the beginning and should be constantly maintained separately from the voice network described above.

Q. *What are the SS7 network basic design parameters?*

A. The following is a list of basic design parameters for use in designing, building, monitoring, and expanding an SS7 data network:

1. The maximum recommended utilization for an SS7 data link is 40 percent.
2. Follow the ANSI specifications for SS7 network designs.
3. When designing the network facilities at each of the system MTSOs, be sure to incorporate patch points in each data link.
4. Separate, as much as possible, the voice interconnection facilities from the SS7 data network facilities.

Q. *How do you measure network reliability?*

A. The reliability of the network as a whole should be reviewed at every quarterly engineering design review. Some of the metrics used to measure the network reliability are given below:

1. Description and size of the network outages including duration of the outage and causes for the outage
2. Number of system swaps taken place (use of backup or redundant systems or facilities)
3. Life-cycle performance tracking data for the network equipment and facilities
4. Plans to improve current network reliability performance

Q. *What are demographics?*

A. The term *demographics* when applied to wireless applications refers to the segmentation of different subscriber groups or potential subscriber groups for differing levels of radio and system usage. The demographics information will determine which areas within a network are used for portable or inbuilding applications. Also the demographic information will be used to determine the amount of usage in terms of millierlangs per sub expected from different regions within the network.

Q. *What are capital constraints?*

A. *Capital constraints* are limits on the amount of capital dollars from which to accomplish all the engineering activities for the year or several years. In reality, every system has capital constraints.

Q. *What is a macrocell?*

A. A *macrocell* is a traditional cellular cell site whose coverage area is over 1 km in diameter.

Q. *What is a three-sector cell?*

A. A *three-sector cell* is any cell site that employs three different sectors for providing coverage and capacity for a given geographic area. Usually each sector is responsible for one-third of the cell's coverage area, and this represents 120 degrees coverage per sector.

Q. *What is an omnicell?*

A. An *omnicell* is any cell site that is not a sectorized cell site and that employs omnidirectional antennas for covering 360 degrees of an area.

Q. *What is a six-sector cell?*

A. A *six-sector cell* is any cell site that employs six different sectors for providing coverage and capacity for a given geographic area. Usually each sector is responsible for one-sixth of the cell's coverage area, which represents 60 degrees per sector.

Q. *What is a microcell?*

A. The term *microcell* has many different meanings to operators and vendors. Often a *microcell* means any cell site equipment that is much smaller than that used in the past; however, this definition can and is misleading. A microcell can use any technology platforms or size of equipment. A *microcell* is actually any cell site that has a radius of less than 0.5 km.

Q. *What are the types of microcells available?*

A. There are currently several types of microcells on the market, and they are available in a variety of technology platforms:

- Fiber-fed microcell
- T-1 microcell
- Microwave microcell

- High-power rerad
- Low-power rerad
- Bidirectional amplifier

Q. *What is a fiber-fed microcell?*

A. The *fiber-fed microcell* can and is being deployed in cellular systems. There are many advantages and disadvantages in using a fiber-fed microcell technology platform. Currently two distinct types of fiber-fed microcells are available, analog and digital. Both of these microcells require the use of dark fiber to make them operate in a cellular network. Dark fiber is often not readily available in many of the areas where a microcell could be deployed and requires that the pair of fibers utilized for the microcell be dedicated only to that microcell. However, a fiber-fed microcell can be utilized with any of the cellular infrastructure vendors' equipment, making it vendor transparent.

Q. *What is a microwave microcell?*

A. *Microwave-fed microcells* have some distinct advantages in that they are backhaul and infrastructure vendor independent. This type of microcell is also directly compatible with different technology platforms. The microwave microcell utilizes microwave point-to-point communication to connect the feeding cell site with the remote locations. The advantage with utilizing a point-to-point method for connecting the microcell to the donor cell is the fact it doesn't require the use of landline facilities, T-1, or fiber, to connect the donor cell and the microcell. The disadvantages of using this type of microcell are difficulty in frequency coordination of the microwave path and maintaining path clearance and the necessity for an additional antenna and feed line for the donor cell.

Q. *What is a bidirectional amplifier?*

A. The *bidirectional amplifier* is very similar to that of the low-power rerad. The bidirectional amplifier does not translate voice or control channel frequency. The bidirectional amplifier has the same isolation and path clearance requirements as does a low-power rerad. The bidirectional amplifier can be utilized for in-

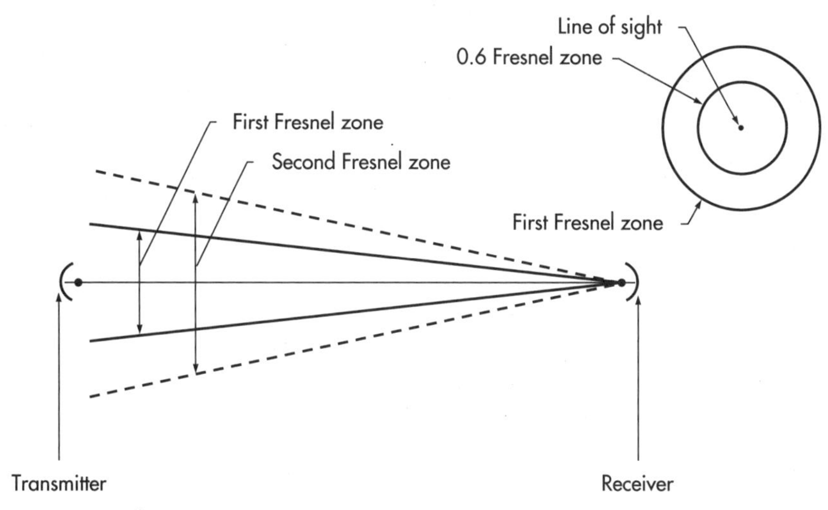

Figure 10.21 Fresnel zone.

building applications either by itself or adjunct to microcell deployments. The bidirectional amplifier offers a cost-effective method of providing coverage in an in-building environment.

Q. *What is the Fresnel zone?*

A. The *Fresnel zone* is shown in Fig. 10.21. There are effectively an infinite number of Fresnel zones for any communication link. The Fresnel zone is a function of the frequency of operation for the communication link. The primary energy of the propagation wave is contained within the first Fresnel zone. The Fresnel zone is important for the path clearance analysis because it determines the effect of the wave bending on the path above the earth and the reflections caused by the earth's surface itself. The odd-numbered Fresnel zones will reinforce the direct wave while the even-numbered Fresnel zones will cancel.

In a point-to-point communication system, it is desirable to have at least a 0.6 first Fresnel zone clearance to achieve path attenuation approaching free-space loss between the two antennas. The clearance criteria apply to all sides of the radio beam, not just the top and bottom portions represented by the drawing in Fig. 10.21.

Q. *How is a point-to-point path clearance study performed?*

A. A quick example for use in determining the path clearance required for a point-to-point communication site is shown below:

Example

Refer to Fig. 10.22.

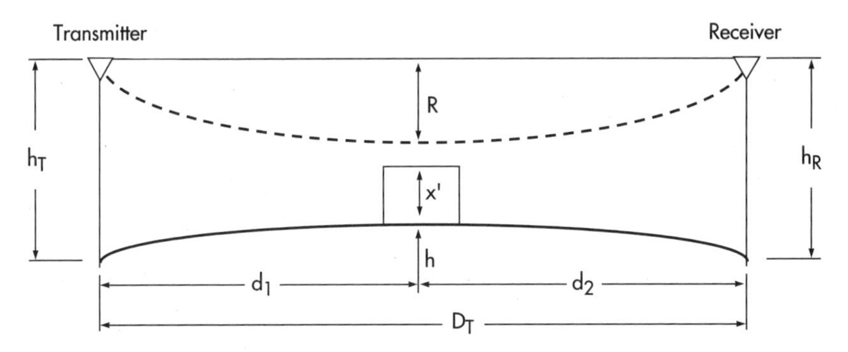

Figure 10.22 Path clearance.

First Fresnel zone: $$R = 72\sqrt{\frac{d_1 d_2}{D_T f}}$$

Earth curvature: $$h = \frac{d_1 d_2}{1.5k}$$

with f = frequency in GHz; d_1, d_2, D_T in miles; x, h, h_T, h_R in feet; and where d_1 = 1.6 mi, d_2 = 2.1 mi, D_T = 3.7 mi, and f = 0.88 GHz (or 880 MHz).

$$R = 72\sqrt{\frac{(1.6)\,(2.1)}{(3.7)\,(.88)}} = \sqrt{\frac{3.36}{3.256}} = 73.14 \text{ ft}$$

$$R' = (0.6) \quad R = 43.884$$

$$h = \frac{(1.6)\,(2.1)}{(1.5)\,(4/3)} = 1.68 \text{ ft} \qquad k = 4/3$$

Assume that the transmitter, receiver, and obstruction have the same ASML.

Earth curvature	1.68
0.6 Fresnel zone	43.88
Obstruction height	100
	145.56 ft

The minimum Tx and Rx heights for the system are $h_T = h_R =$ 145.56 ft

Q. *What are the design considerations for an in-building system?*

A. Some of the design considerations that need to be factored into an in-building design are the following:

1. Base to mobile unit power
2. Mobile unit to base power
3. Link budget
4. Coverage area
5. Antenna system type and placement
6. Frequency planning

Q. *What is microwave clearance?*

A. *Microwave clearance* is also called OSF clearance, and it refers to the requirement that PCS operators in the United States move existing microwave point-to-point systems from the existing PCS band to another band. The PCS operators need to work with the existing microwave users to achieve an orderly and financially viable solution to this issue.

Q. *What is lightning protection?*

A. *Lightning protection* refers to the process of providing protection for personnel and equipment at the communication site from a lightning strike. The strike can come from the antenna system, the telco system, or the power system itself. Different types of surge protectors are used for lightning protection in addition to the establishment of a good earth grounding program.

Q. *What is datafill?*

A. *Datafill* refers to the process of populating the cell site and/or switching database with the necessary parameter required to either have the cell site process calls properly or have an auxiliary platform as backup. Examples of datafill for a cell site would be the frequencies used at the site, the neighbor list, or even the radio slot assignments within the equipment shelf itself.

Q. *What are parameter settings?*

A. *Parameter settings* are the actual parameter values that define how a cell site will operate. An example of parameter settings, or rather cell site parameters, are those on the neighbor list for the cell site.

Q. *What is post-turn-on testing?*

A. *Post-turn-on testing* is the process whereby engineering and operations personnel ensure that a cell site, switch, auxiliary platform, or even a radio is integrated into a network properly so that they do not degrade the performance of the network with their introduction.

Q. *What is spectrum aggregation?*

A. *Spectrum aggregation* is the assignment of an allotted services spectrum to a wireless operator. There are numerous uses of spectrum aggregation such as the allocation of spectrum for CDMA, which is separate from that of the analog system for cellular, and the allocation of one PCS block of spectrum for GSM and another block for CDMA.

Q. *What is a handoff border?*

A. A *handoff border* is the region where handoffs occur either between cell sites or between different services within the same system. Handoff borders typically follow some geographic boundary like a switch border or even a river. A handoff border is actually a transition point from one part of a system to another.

Q. *What intersystem design issues should be addressed?*

A. There are numerous intersystem design issues that engineering departments for a wireless system need to factor into their designs. Some simple intersystem design issues could be an incompatible technology platform such as TDMA or CDMA being used by a neighboring system. Other examples of inter-system design issues could be IS-41 border issues, switch vendor software loads, or, of course, a frequency plan used by either system.

Q. *What are isolation requirements?*

A. *Isolation requirements* are the amount in decibels required to maintain radio frequency isolation between different systems when colocating or locating near another operator. There are several excellent isolation requirements documents available in the public domain to assist an operator in determining optimal isolation amounts. The isolation requirements defined in terms of decibels can then be converted into feet or meter separation requirements based on vertical or horizontal path-loss predictions.

Q. *What are MSC borders?*

A. *MSC borders* are the borders between switching areas. Typically the MSC borders are located where there is minimal traffic activity in comparison to other potential locations. The MSC borders tend to follow some geographic definition but are then designed to minimize the amount of inter-MSC traffic that would result from the selection of the border itself.

Q. *What are intersystem borders?*

A. An *intersystem border* is the border between different wireless operators. In cellular networks, the intersystem border is usually defined as the CGSA border for either the MSA or RSA for either the A- or B-band operator since there is normally a roaming agreement with all wire-line or non-wire-line networks. For PCS operators, the intersystem border could be between any other operator's system that borders the PCS operator's MTA or BTA border.

Q. *What is the iDEN channel list?*

A. The *iDEN channel list* is rather unusual in that the channels available for use can be any of the available 601 channels. The almost random channel assignments, or rather availability, leads to many operational issues. There is not a specific channel used for assignment for control, which results in the subscriber unit's having to scan and reacquire the control channel for use when it either turns on or has to rescan due to a call processing problem.

Q. *What is the PCCH?*

A. The PCCH is the *primary control channel* for an iDEN system.

Q. *What is the DCCH?*

A. The DCCH is the *dedicated control channel* that is used by many TDMA-based technology platforms for conveying various control material to and from the subscriber's unit.

Q. *What is a beacon channel?*

A. A *beacon channel* is a dedicated channel that identifies a sector of a given cell site for use in an MAHO handoff. The beacon channel is usually the digital channel that has the control information on it. The beacon channel is also the channel that is entered into the database for identifying the potential neighbor list for the subscriber unit to scan.

Q. *What is I and Q?*

A. The abbreviation *I and Q* refers to the in phase, I, and quadrature portion of a digitally modulated signal. The relative difference in phase and amplitude for the I and Q portions of the modulated wave determine the particular information content that it carries.

Q. *What is a meger?*

A. A *meger* is a piece of test equipment that is used to measure the resistivity of a grounding system.

Q. *What is a biddle test?*

A. A *biddle test* measures the effectiveness of the connection that usually exists between the ground bus bar and the ground itself.

Q. *What is frequency planning?*

A. *Frequency planning* is the process of assigning specific frequencies or PN codes to an individual cell site or sector.

Q. *What is a frequency plan?*

A. A *frequency plan* is the actual frequency or PN code assignment that is given to an individual cell site or sector. The frequency plan for a wireless network directly impacts the performance of the system in either a positive or negative fashion.

Q. *What is the dynamic channel allocation?*

A. The *dynamic channel allocation* (DCA) is a software parameter that is available, at least with Lucent, to enable the assignment of the same frequencies at adjacent cell sites or sectors within a system. The DCA effectively removes from operation the offending cochannel radio when the other one is in use, allowing a quick fix to a frequency planning problem or traffic relief. However, DCA is extremely inefficient for spectrum allocation as well as for capital equipment.

Q. *What is frequency reuse?*

A. *Frequency reuse* is the use of the same radio frequency multiple times within a given market. For example, channel 256 may be reused 15 times within the same system in order to obtain the required capacity needed to support the subscriber base.

Q. *What is power control?*

A. *Power control* is a process by which power levels are adjusted so that as a subscriber unit moves closer to a cell site, the amount of power it transmits is reduced, and the amount of power it transmits is increased as it moves away from the cell site. Depending on the technology platform used, the number of power steps—that is, the increments in decibels of added or removed attenuation—will be different, although the concept is still the same. Power control is exerted at the cell site.

Q. *What is a signaling tone?*

A. A *signaling tone* (ST) is a 10-kHz tone that is used for control functions, flash hooks, handoff processes, and so on in a cellular system.

Q. *What is a handoff?*

A. A *handoff* is the process by which the subscriber unit is passed from one frequency to another for the purpose of maintaining a communication link as the subscriber traverses through a network.

Q. *What is an S11?*

A. *S11* is the S-parameter test, which is also commonly called a *sweep test,* for an antenna system. The S11 test result is the SWR, or return loss, of a system, and it is used in a wireless system to test the integrity of the antenna system without having to disconnect any of the components for the test itself. There are numerous devices capable of conducting S11 tests, but the most common is a network analyzer.

Q. *What is the difference between vector and scalar network analyzers?*

A. The difference between vector and scalar network analyzers is that the scalar network analyzer provides information relative to amplitude while the vector network analyzer can also provide phase information. For a wireless system, either a vector or scalar analyzer is sufficient for conducting an antenna sweep test, or S11 test.

Q. *What is an S21?*

A. An *S21* is an S-parameter test, also called a *through test,* and it is used to measure, by frequency, the gain or loss of a device under test. The usual devices assessed by an S21 test are filters, duplexers, cables, and of course, amplifiers.

Q. *What is a sniffer port?*

A. A *sniffer port* is a port from a directional coupler used for diagnostic purposes. The sniffer port can have either 20 or 40 dB of attenuation from the primary signal. The sniffer port is usually associated with the transmit portion of the cell site where the sniffer port is usually placed after the transmit combining takes place.

Q. *What are the classes of mobile units?*

A. There are numerous classes of mobile units available for a wireless system to match the technology platform utilized. However, there are four general classes of subscriber units, as follows:

Class	Power, W	
1	4	Mobile unit
2	1.2	Bag phone
3	0.6	Portable unit
4	0.6	Dual-band, portable unit

Q. *What are the key elements in a scope of work?*

A. The key elements in a *scope of work* (SOW) involve defining what needs to be done followed by who is to do each part and of course the project due date. For an internal scope of work, the cost factor is typically computed as the amount of time needed to complete the project plus any equipment that needs to be purchased, including software. However, for a scope of work involving external resources, the responsibilities as well as the deliverables need to be well defined. The cost for a scope of work for external resources does not have to be a fixed cost depending on the nature of the work involved. Sometimes it is best to have the work performed on a T&M basis using tight control rather than demanding a fixed cost. For construction issues, a fixed cost is most relevant; however, for engineering services a T&M basis proves to be the most cost-effective method for containing the cost of the project, provided it is managed properly.

The following is a brief outline for a scope of work that can be easily populated and modified to meet the specific requirements at hand:

Name

Project number

Overview

Subprojects

Project duration

Starting time

Responsibilities

Input required

Due date

Output (closure requirements)

Cost

- Hourly (T&M)
- Fixed
- Hybrid (hourly and fixed)

Q. *What does an N = 12 reuse pattern look like?*

A. Figure 10.23 depicts an N = 12 reuse pattern.

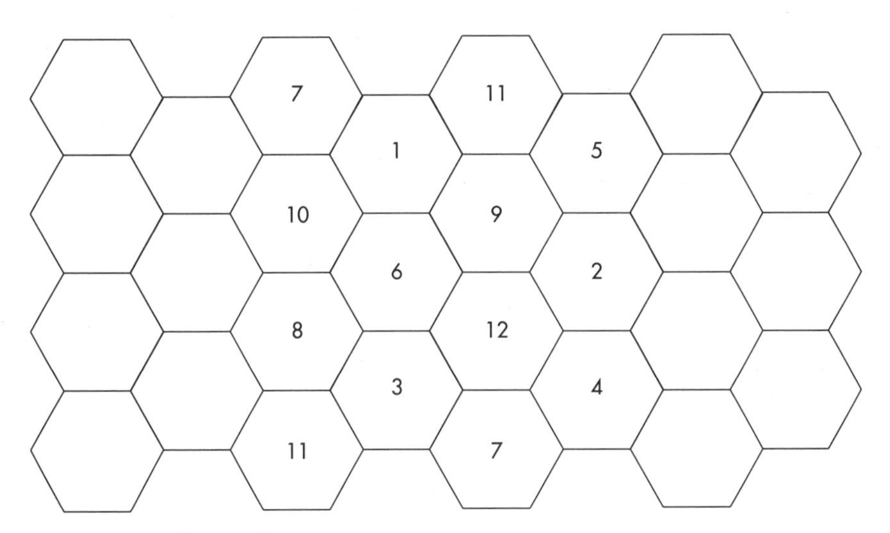

Figure 10.23 $N = 12$ frequency grid.

Q. *What does an N = 7 reuse pattern look like?*

A. Figure 10.24 depicts an $N = 7$ reuse pattern.

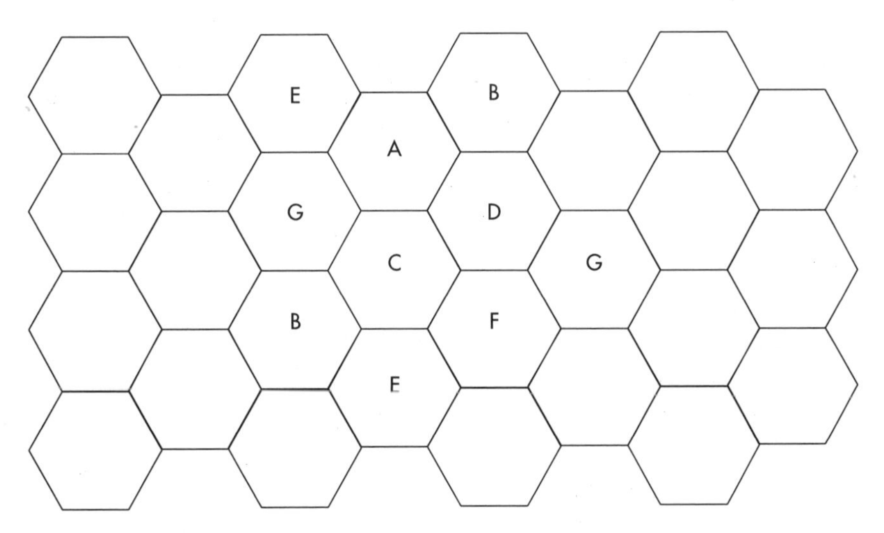

Figure 10.24 $N = 7$ frequency grid.

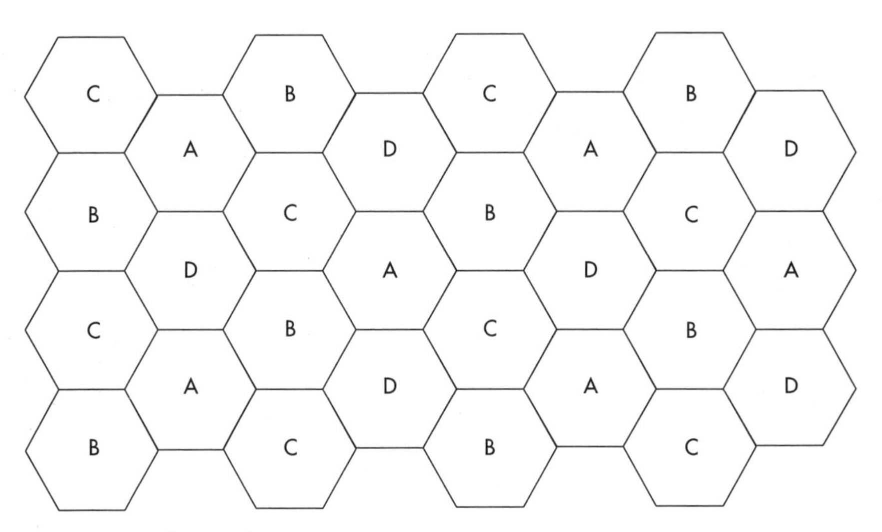

Figure 10.25 $N = 4$ frequency grid.

Q. *What does an N = 4 reuse pattern look like?*

A. Figure 10.25 depicts an $N = 4$ reuse pattern.

Traffic Engineering

Traffic engineering is a critical function within a wireless company since the traffic engineering directly determines the growth requirements needed for the success of the network. The traffic engineer has to make numerous tradeoffs that directly affect the grade of service as well as the capital requirements for a wireless system.

The following are many of the common questions asked that pertain to traffic engineering.

Q. *What is an erlang?*

A. An *erlang* is a basic unit for measuring traffic volume. It is important to note that an erlang is dimensionless. However, the industry norm is to reference 1 erlang of usage to one hour or 60 minutes of use for the circuit or facility. Depending on the grade of service and the probability model used—Erlang B, Erlang C, or Poisson—the number of facilities or circuits can be determined from knowing the number of erlangs used.

Q. *What is a lost, or dropped, call?*

A. A *lost,* or *dropped, call* with respect to traffic engineering is another name for the Erlang B model for mobility traffic. Specifically, a call is lost from a traffic perspective when it tries to gain access to the network and is denied. For an Erlang B, or rather lost call, the subscriber does not enter into a queue for potential assignment.

Q. *What is a queue?*

A. A *queue* is another name for *waiting* in a telephony system. When someone refers to a *queue* or to "a subscriber's entering a queue," he or she means the subscriber who requested a facility —that is, a radio channel—and is denied, he or she is put on a waiting list for assignment when a facility becomes available. Wireless systems that utilize queuing use Erlang C for their traffic modeling. An example of a queue is the dispatch system used for an iDEN system.

Q. *What is the busy hour?*

A. The *busy hour* refers to the time period, one hour, in which, during the course of a 24-hour day, there is the most amount of traffic either for the system, the cell site, or even a sector of a cell site. The busy hour is normally a fixed time of day, and in the United States the busy hour is normally between 4:00 and 6:00 P.M. with a secondary busy hour between 10:00 A.M. and 12 noon.

Q. *What are centa call seconds?*

A. *Centa call seconds* (CCS) is a unit of measure that represents 100 call seconds of traffic, or centa call seconds. For example, 90

CCS recorded for a sector over a one-hour period would be equivalent to 9000 seconds of traffic, or 2.5 hours of traffic, or 2.5 erlangs.

Q. *How are CCS converted to erlangs?*

A. To convert from centa call seconds to erlangs, the following equation is used:

$$1 \text{ erlang} = \text{CCS}/36$$

Q. *How are erlangs converted to CCS?*

A. To convert from erlangs to CCS, the following equation is used:

$$\text{CCS} = 36 \cdot X \text{ erlangs}$$

Q. *What is a bouncing busy hour report?*

A. Constructing a *bouncing busy hour report* serves the purpose of determining the peak load that exists for any given hour on the wireless system, cell site, or sector within a network. The bouncing busy hour report is a valuable tool because each cell or sector of a cell has different usage patterns, and these usage patterns may or may not conform to the system load. The bouncing busy hour data are used to evaluate if the system is designed properly on a per cell and sector level to ensure that the appropriate grade of service is achieved.

For example, cell 1, sector A, may have a busy hour between 8:00 and 9:00 A.M. while sector B of the same cell may have a busy hour between 4:00 and 5:00 P.M. If the system busy hour were the only source of data used for traffic engineering, then the potential would exist of having insufficient capacity for sector A.

Q. *How is a fixed hour used to track traffic patterns?*

A. Traffic engineers use fixed hours to monitor project traffic usage trends in a wireless network. The use of a fixed hour has many advantages over a bouncing busy hour in that it enables the designer to view the system as a whole and make design recommendations based on all the cells sites' interacting together over the same period. The fixed hour for most wireless systems in the

United States is either from 4:00 to 5:00 P.M. or from 5:00 to 6:00 P.M.

Q. *What is a system busy hour?*

A. A *system busy hour* is the one-hour period that the wireless system as a whole will have the highest traffic load. The system busy hour is usually a fixed hour, usually 4:00 to 5:00 P.M. or 5:00 to 6:00 P.M.

Q. *What is a cell site busy hour?*

A. A *cell site busy hour* is the time period, one hour, when the cell site experiences the highest amount of traffic in comparison to the remaining hours of the day. For example, cell 1, sector A, may have a busy hour between 8:00 and 9:00 A.M. while sector B of the same cell may have a busy hour between 4:00 and 5:00 P.M. It is common for different sectors of a cell site to have different busy hours; however, on the whole, the system should be designed to support the system busy hour.

Q. *What is subscriber growth?*

A. *Subscriber growth* are the additions, or *net adds,* of subscribers to the wireless system usually tracked over a quarter or a year. For example, subscriber growth for a quarter might be 25,000 new subscribers with 5000 churning. Therefore, the net subscriber growth that occurred was 20,000 subscribers. The rate of *subscriber growth* is also used for planning purposes so that the marketing department might say that they plan "to grow" the subscriber base by 30 percent for the next year. If the base exists at 1 million subscribers, 30 percent growth would result in 1,300,000 total subscribers on the network at the end of the year.

Q. *How are erlangs per subscriber computed?*

A. Erlangs per subscriber is a unit of measurement—that is, it is a dimensioning number—that is used as a foundation for determining the number of radios, cell sites, and of course, switch and landline facilities required. The erlangs per subscriber is often determined from the current traffic load and subscriber levels. For example:

Subscribers = 1,000,000

System busy hour erlangs = 10,000

Therefore, the erlangs per subscriber = 10,000 erlangs/1,000,000 subscribers = 0.01 E/sub, or rather 10 mE/sub.

Q. *How is the unit of measure millierlangs per subscriber used?*

A. The unit millierlangs per subscriber is a traffic dimensioning number that is the foundation for determining the amount of radios, cell sites, and of course, the switch and landline facilities required. The millierlangs per subscriber is often determined from the current traffic load and subscriber levels:

Subscribers = 1,000,000

System busy hour erlangs = 10,000

Therefore, the erlangs per subscriber = 10,000 erlangs/1,000,000 subscribers = 0.01 E/sub, or rather 10 mE/sub. Therefore, if an operator planned to add 100,000 new subscribers to a network, the expected system load would be the following:

New subscriber = 100,000

Millierlangs per subscriber = 10

100,000 · 10 mE = 1000 E of new usage expected for the system busy hour

Q. *What is erlang, or traffic, density?*

A. *Erlang density,* or *traffic density,* is the amount of erlangs per square kilometer or square mile that applies to a wireless system or a subpart of a wireless system. For example, if a system has 100 km^2 of area and has 10 erlangs of usage, then the erlang density = 10/100, or 0.1 E/km^2.

Q. *How is the amount of erlangs per square kilometer computed?*

A. The amount of erlangs per square kilometer is a dimensioning number that is used for wireless traffic engineering purposes. This measure may be computed by two methods, as follows:

Method A

$$\text{System area} = 100\ \text{km}^2$$

$$\text{System busy hour usage} = 10\ \text{erlangs}$$

Thus,

$$\text{Erlangs/km}^2 = 10/100 = 0.1\ \text{erlangs/km}^2$$

Method B

$$\text{System area} = 100\ \text{km}^2$$

$$\text{mE/subscriber} = 10$$

$$\text{Subscribers} = 1000$$

Thus,

$$\frac{\text{mE} \cdot \text{subscribers}}{\text{System area}} = (0.01 \cdot 1000)/100 = 0.1\ \text{erlangs/km}^2$$

Q. *How is the amount of erlangs per square mile computed?*

A. The amount of erlangs per square mile is a dimensioning number that is used for wireless traffic engineering purposes. This measure may be computed by two methods as follows:

Method A

$$\text{System area} = 100\ \text{mi}^2$$

$$\text{System busy hour usage} = 10\ \text{erlangs}$$

Thus,

$$\text{Erlangs/mi}^2 = 10/100 = 0.1\ \text{erlangs/mi}^2$$

Method B

$$\text{System area} = 100\ \text{mi}^2$$

$$\text{mE/subscriber} = 10$$

$$\text{Subscribers} = 1000$$

Thus,

$$\frac{\text{mE} \cdot \text{subscribers}}{\text{System area}} = (0.01 \cdot 1000)/100 = 0.1\ \text{erlangs/km}^2$$

Q. *What is a service feature impact evaluation?*

A. A *service feature impact evaluation* is conducted to ensure that a particular feature is or will not cause adverse effects in a wireless network. For example, if voice mail is given to each subscriber in a network, its impact on the network must be evaluated so that the number of messages left or their storage duration does not deplete the memory available for the voice mail system.

Q. *How is CPU, or processor, utilization evaluated?*

A. Utilization of the central processing unit (or processor)—that is one of several processors that reside in the switch used in a wireless system—must be evaluated. Each switch manufacturer has specific CPU utilization rates at which they suggest the switch operate and also a maximum that should not be exceeded. For example, when a new switch is added to a network, it may have a 40 percent utilization rate, but when it is projected to reach 80 percent another switch would be required to offload that switch. The CPU utilization depends on a number of factors, and each switch has unique variables that are directly tied to the traffic load of the system.

Q. *What is a radio utilization ratio?*

A. A radio utilization ratio is used to express the relationship of the existing traffic load to the offered traffic load at the defined grade of service. The radio utilization ratio is a planning parameter that is used as a trigger for adding new radios to the wireless network.

For example, a sector having eight radios, or time slots, can handle traffic of 3.63 erlangs but it is carrying only 2 erlangs. Therefore, the utilization ratio is 55 percent. If the standard gestation period for adding radios is six months and it is estimated that the system will grow by another 45 percent over the next six months, then the utilization ratio is used to determine when the planning and ordering of radio equipment needs to begin to ensure that there is sufficient capacity available for the subscriber base at the defined grade of service.

It is important to note that the radio utilization ratio should not be used to float the traffic, or rather the radios required, by offering a margin, or fudge factor. Where this has been done, the net result has been the design of a nonblocking system when the stated goal for that operator was a 5 percent grade of service.

Q. *To what relationship does the term trunking efficiency refer?*

A. *Trunking efficiency* refers to the relationship between trunk size and traffic capacity. When the size of the trunk increases—that is, when the number of radios increases—the trunk handles more traffic than several smaller trunks, as seen in the following example:

Trunk group A: 5 radios = 1.66 erlangs offered

Trunk group B: 5 radios = 1.66 erlangs offered

Trunk group C: 10 radios = 5.08 erlangs offered

Therefore, using one larger trunk group C rather than the two smaller trunk groups A and B results in the network's being able to handle more traffic, that is, 1.76 erlangs.

Q. *What is the difference between N = 4 and N = 7 trunking efficiency?*

A. The difference between $N = 4$ and $N = 7$ trunking efficiency is shown below. A typical $N = 4$ system employs a six-sector system while an $N = 7$ uses a three-sector system. Therefore, if a cell is to accommodate 12 erlangs of traffic, assuming uniform distribution, the following comparison can be made:

$N = 4$: Six sectors = 2 erlangs per sector
= 6 radios per sector, or 36 radios per cell site

$N = 7$: Three sectors = 4 erlangs per sector
= 9 radios per sector, or 27 radios per cell site

Therefore, using the example above, an $N = 4$ cell site needs 36 radios while an $N = 7$ cell needs only 27 radios, leaving a savings of 9 radios per cell to handle the same traffic. However, $N = 4$ allows for a closer reuse of cell sites and is more spectrally efficient, allowing for a higher number of erlangs per square kilometer over an $N = 7$ cell site but at the expense of trunking efficiency.

Q. *What is the difference between omni and sector trunking efficiency?*

A. The difference between omni and sector trunking efficiency is easily shown in the example below. As a general rule of thumb, an omni cell site will always be able to handle more traffic than a sectored cell site with the same number of radios due to the trunking efficiency enjoyed by an omni cell site. Therefore, if a cell is to provide 12 erlangs of traffic, assuming uniform distribution, the following comparison can be made:

Omni cell: = 12 erlangs per sector = 20 radios per cell site

$N = 7$: Three sectors = 4 erlangs per sector
= 9 radios per sector, or 27 radios per cell site

Therefore, using the example above, a sector cell site needs 27 radios while an omni cell site needs only 20 radios, leaving a savings of 7 radios per cell site to handle the same traffic. However, this fact does not account for spectral efficiency issues and the need for sectorization for frequency reuse.

Q. *What is a T coder?*

A. A *T coder* is another name for an ADPCM, or a variable-rate ADPCM device, that compresses the signals for a DS-1 to enable one DS-1 to handle twice the traffic load. Specifically, a DS-1 has 24 DS-0s, and with a T coder it is possible to transport 48 DS-0s over the same facility, thereby reducing the overall operating cost of the network.

Q. *What is a T-1?*

A. A *T-1* is a DS-1 in the North American cellular system, and it contains 24 DS-0s. A T-1 is also called a *T carrier* and occupies 1.544 MHz, or 1.544 Mb/s, of throughput with each DS-0 having the maximum throughput of 64 kb/s. The T-1 is a very common circuit in wireless systems.

Q. *What is a T-2?*

A. A *T-2* is a special T carrier that is used when the operator wishes to aggregate its traffic onto a larger pipe. The T-2's bandwidth and traffic-handling capabilities enable it to handle four T-1's or four DS-1's worth of traffic. Another name for the T-2 is the *DS-2*.

Q. *What is a T-3?*

A. A *T-3* is a DS-3 that has the capability of handling 28 DS-1s. The T-3 is chosen by operators when they wish to aggregate traffic onto a very large pipe. The T-3's bandwidth and traffic-handling capabilities enable it to handle 28 T-1's or 28 DS-1's worth of traffic. Another name for a T-3 is a *DS-3*.

Q. *What is a DS-1?*

A. A *DS-1* is the name used for a T-1 or E-1, and it is a single pair gain circuit that enables either 24 DS-0s if a T-1 or 32 DS-0s if it is associated with an E-1 to be carried by the circuit. The DS-1 is one of the more common types of circuits used in wireless systems because their primary function is to connect cell sites to the MSC. The amount of bandwidth associated with the DS-1 is different for a T-1 and an E-1. The T-1 has 1.544 Mb/s while the E-1 has 2.048 Mb/s.

Q. *What is a DS-3?*

A. A *DS-3* is the name used for a T-3 or E-3, and it is a single pair gain circuit. Depending on whether the underlying service is T-1 or E-1, the amount of DS-0s that are associated with the DS-3 will of course be different. For example, 28 T-1s multiplexed together make up a DS-3 while 16 E-1s are required to make a DS-3. The DS-3 is chosen by operators when they wish to aggregate traffic onto a very large pipe. The DS-3's bandwidth and traffic-handling capabilities enable it to handle 28 T-1's or 16 E-1's worth of traffic. The DS-3 is called a *T-3* in North America and an *E-3* in European countries.

Q. *What is the difference between a DS-3 and a T-3?*

A. There is no difference between a DS-3 and a T-3 used in the United States, and the name *T-3* is often interchanged with the name *DS-3*.

Q. *What is an E-1?*

A. An *E-1* is a DS-1 outside the North American system. The E-1 has 32 DS-0s that are contained in it. If the E-1 is associated with voice traffic, typically only 30 DS-0s are utilized; however, if it is used for a PRI, then 31 DS-0s are utilized. The E-1 is a very common circuit in wireless systems outside the United States.

Q. *What is the difference between a T-1 and an E-1?*

A. The difference between a T-1 and an E-1 associated with the wireless industry is the amount of DS-0s that the pair gain can contain. Specifically, a T-1 has 24 DS-0s while an E-1 has 32 DS-0s, 30 for voice traffic or 31 for a PRI. Therefore, an E-1 has typically six more DS-0s to use than a T-1, or rather has 25 percent more capacity when used in a similar configuration. The signaling, however, is different between the two, and they are not directly compatible without an intermediate device in between. Of course, there is also the obvious issue of the other six DS-0s, which is not a problem when going from T-1 to E-1 but only when going from E-1 to T-1.

The issue of T-1/E-1 compatibility is important when reviewing the switch port count in that if the switch is set up for E-1 and there is a T-1 interface, then there is a port utilization problem, which requires a preconversion with a DACS.

Q. *What is an E-3?*

A. An *E-3* is another name for a DS-3 that uses E-1s as its DS-1 format. An E-3 represents a single pair gain circuit that has 16 E-1s multiplexed onto it. The DS-3 is chosen by operators when they wish to aggregate traffic onto a very large pipe. The DS-3's bandwidth and traffic-handling capabilities enable it to handle 16 E-1's worth of traffic. Another name for the E-3 is DS-3.

Q. *What is cell site growth?*

A. *Cell site growth* is the process of expanding the wireless network by adding cell sites.

Q. *What are radio channels?*

A. *Radio channels* are physical radios that are used in transmitting and receiving wireless communications. The radio channels can be of any technology platform. Typically a radio channel will be associated with a specific frequency that is used for transmitting and receiving.

Q. *What is a traffic channel?*

A. The term *traffic channel* refers to time slots used in digital wireless systems. For example, an IS-136 wireless system has six time slots used for conveying voice information. Of those six time slots, two slots are assigned to particular communication and are referred to as a *traffic channel.*

Q. *How is iDEN traffic engineered?*

A. iDEN traffic is engineered in an unusual way in that there are several factors that need to be included when performing a capacity analysis and growth study for an iDEN system. The iDEN system consists of both an interconnect and a dispatch system. The iDEN channel consists of six time slots that occupy a 25-kHz channel. For the interconnect system part, the communication can take place on a 6:1 or 3:1 basis, meaning that for the 6:1 service, one of the six time slots are used while for the 3:1 service two time slots are used. In conjunction with the interconnect service, a dispatch system is also provided, and this is done using one time slot only. Of course, there is also a digital control channel that is broadcast, and this occupies one of the time slots also. For interconnect service, Erlang B is utilized, and for dispatch, Erlang C is used.

Q. *What are busy hour call attempts?*

A. *Busy hour call attempts* (BHCAs) are a measure of traffic that is used by traffic engineers to dimension not only the radio system but primarily the fixed network platforms like the switch itself. The BHCA is a key metric, and it is used for dimensioning purposes for many components of a wireless system. Often there are processor limits that are directly tied to the BHCAs that occur on a wireless system.

Q. *What is registration?*

A. *Registration* is the process used in a wireless system for determining at a periodic interval where the subscriber is within the network. The purpose of registration is to relieve the paging channel from congestion by being more efficient in directing the in-coming call, or page, to where the subscriber was last located instead of paging the entire network.

Q. *What is flood paging?*

A. *Flood paging* is the process of sending a page from all the cell sites within a wireless network for the sole purpose of finding the subscriber to deliver a call. The advantage with flood paging is that if a subscriber is between registration boundaries, the subscriber is paged anyway. However, the disadvantage of flood paging is that its use requires increased capacity by the control or sig channel.

Q. *What is a registration interval?*

A. A *registration interval* is the time that is programmed by the wireless system operator for how often the subscriber units will register with the network. The longer the interval between registrations, the less taxing it is upon the network. However, the disadvantage of increasing the registration interval is the possibility of losing contact with the subscriber unit. Setting the interval too short will tax the system needlessly. Therefore, a compromise needs to be reached for setting the registration interval at an optimal rate.

Q. *What is paging channel capacity?*

A. *Paging channel capacity* is the highest amount of throughput that can be achieved for a cell site and also for the system. With paging areas, the paging channel capacity has been expanded, but if the borders are set incorrectly, a problem with paging channel capacity can occur, resulting in calls not being delivered.

Q. *What is the Erlang B equation?*

A. The *Erlang B equation* is given below for reference. There are numerous books and technical articles regarding Erlang B's statistical nature and accuracy. However, the primary driving element with Erlang B is that it estimates, statistically, the probability that all circuits will be busy when a call is attempted. The problem with the Erlang B calculation is that it assumes that the call, if it is not assigned a circuit, is lost from the network permanently. The equation for Erlang B is as follows:

$$\text{Grade of service} = \sum_{n=0}^{n} \frac{E^n}{n!}$$

where E = erlang traffic and n = number of trunks (voice channels) in group.

Q. *What is the Poisson equation?*

A. The *Poisson equation* is given below for reference, and it should be compared against the Erlang B value arrived at for the same number of circuits and grade of service. The fundamental difference between the Erlang B and Poisson equations is that in the Poisson it is assumed that blocked calls are put into a queue instead of discarded:

$$\text{Grade of service} = e^{-a} \sum_{n=c}^{\infty} \frac{a^n}{n!}$$

where a = erlang traffic, c = required number of servers, and n = index of number of arriving calls.

Q. *What is grade of service?*

A. *Grade of service* (GOS) with respect to telephony refers to the blocking probability that is expected for any given trunk or member size. A typical GOS value used for wireless systems is 1 and 2 percent. This refers to a 1 or 2 percent chance that the caller will be denied a facility, that is, a radio channel, when he or she wants to use the system. The GOS used for a system is different for the radio and switching environments. For example, the radio environment may have a GOS of 2 percent using the Erlang B formula while the landline, PSTN connection, may be designed for a 1 percent GOS Erlang C because there is a facility concentration at the landline side of the wireless system.

Q. *What is an Erlang B table?*

A. Table 11.1 is an Erlang B table that has several GOS values associated with it. The total channel count represented in the Erlang B table is sufficient in size to accommodate any wireless operator's need for dimensioning a radio system.

TABLE 11.1 ERLANG B GRADE OF SERVICE

Channels	1	1.5	2	3	5
1	0.01	0.012	0.0204	0.0309	0.0526
2	0.153	0.19	0.223	0.282	0.381
3	0.455	0.535	0.602	0.715	0.899
4	0.869	0.992	1.09	1.26	1.52
5	1.36	1.52	1.66	1.88	2.22
6	1.91	2.11	2.28	2.54	2.96
7	2.5	2.74	2.94	3.25	3.74
8	3.13	3.4	3.63	3.99	4.54
9	3.78	4.09	4.34	4.75	5.37
10	4.46	4.81	5.08	5.53	6.22
11	5.16	5.54	5.84	6.33	7.08
12	5.88	6.29	6.61	7.14	7.95
13	6.61	7.05	7.4	7.97	8.83
14	7.35	7.82	8.2	8.8	9.73
15	8.11	8.61	9.01	9.65	10.6
16	8.88	9.41	9.83	10.5	11.5
17	9.65	10.2	10.7	11.4	12.5
18	10.4	11	11.5	12.2	13.4
19	11.2	11.8	12.3	13.1	14.3
20	12	12.7	13.2	14	15.2
21	12.8	13.5	14	14.9	16.2
22	13.7	14.3	14.9	15.8	17.1
23	14.5	15.2	15.8	16.7	18.1
24	15.3	16	16.6	17.6	19
25	16.1	16.9	17.5	18.5	20
26	17	17.8	18.4	19.4	20.9
27	17.8	18.6	19.3	20.3	21.9
28	18.6	19.5	20.2	21.2	22.9
29	19.5	20.4	21	22.2	23.8

TABLE 11.1 ERLANG B GRADE OF SERVICE (CONTINUED)

Channels	1	1.5	2	3	5
30	20.3	21.2	21.9	23.1	24.8
31	21.2	22.1	22.8	24	25.8
32	22	23	23.7	24.9	26.7
33	22.9	23.9	24.6	25.8	27.7
34	23.8	24.8	25.5	26.8	28.7
35	24.6	25.6	26.4	27.7	29.7
36	25.5	26.5	27.3	28.6	30.7
37	26.4	27.4	28.3	29.6	31.6
38	27.3	28.3	29.2	30.5	32.6
39	28.1	29.2	30.1	31.5	33.6
40	29	30.1	31	32.4	34.6
41	29.9	31	31.9	33.4	35.6
42	30.8	31.9	32.8	34.3	36.6
43	31.7	32.8	33.8	35.3	37.6
44	32.5	33.7	34.7	36.2	38.6
45	33.4	34.6	35.6	37.2	39.6
46	34.3	35.6	36.5	38.1	40.5
47	35.2	36.5	37.5	39.1	41.5
48	36.1	37.4	38.4	40	42.5
49	37	38.5	39.3	41	43.5
50	37.9	39.2	40.3	41.9	44.5
51	38.8	40.1	41.2	42.9	45.5
52	39.7	41	42.1	43.9	46.5
53	40.6	42	43.1	44.8	47.5
54	41.5	42.9	44	45.8	48.5
55	42.4	43.8	44.9	46.7	49.5
56	43.3	44.7	45.9	47.7	50.5
57	44.2	45.7	46.8	48.7	51.5
58	45.1	46.6	47.8	49.6	52.6

TABLE 11.1 ERLANG B GRADE OF SERVICE (CONTINUED)

Channels	1	1.5	2	3	5
59	46	47.5	48.7	50.6	53.6
60	46.9	48.4	49.6	51.6	54.6
61	47.9	49.4	50.6	52.5	55.6
62	48.8	50.3	51.5	53.5	56.6
63	49.7	51.2	52.5	54.5	57.6
64	50.6	52.2	53.4	55.4	58.6
65	51.5	53.1	54.4	56.4	59.6
66	52.4	54	55.3	57.4	60.6
67	53.4	55	56.3	58.4	61.6
68	54.3	55.9	57.2	59.3	62.6
69	55.2	56.9	58.2	60.3	63.7
70	56.1	57.8	59.1	61.3	54.7

12

Performance

The following series of questions and answers pertains to the all-encompassing area of wireless system performance. Performance issues are so pervasive that they involve all aspects of the wireless system design and operation and technology platforms. There is a common misconception that performance requirements for one type of technology platform are significantly different from others. However, all systems have to deal with capacity, coverage requirements, lost calls, call quality, auxiliary platforms like voice mail, SS7, and various interconnection issues. Every wireless system company also has to wrestle with the issue of land-use acquisition, and those who are involved with this management aspect know that the differences among the available technologies are immaterial in relation to the various necessary approvals.

Q. *What are the key performance metrics for a wireless system?*

A. Most system operators have several key metrics they utilize for monitoring the performance of their networks. They use the metrics for both day-to-day operation and for upper-management reports. The particular system metrics utilized by the operators depend on the actual infrastructure manufacturer they are using and the software loads. However, the following are the key metrics that are independent of the technology platform or vendor used:

1. Lost calls
2. Blocking
3. Access failures
4. BER/FER
5. Customer complaints
6. Usage/RF loss
7. Handoff failures
8. RF call completion ratio
9. Radios out of service
10. Cell site span outage
11. Technician trouble reports

Q. *How is the lost call goal established for a wireless system?*

A. The acceptable percentage of lost calls should be set so that every year the percentage decreases as a function of overall usage and increased subscriber penetration levels. The suggested calculation method is to provide at the end of the third quarter of every year the following year's goal for the lost call rate. Goals should be set so that there is a realistic reduction in lost calls for the coming year based on an estimate of the calls expected to be lost for the current year.

The following questions and answers explore the factors to be considered in setting lost call goals. The lost call goal should not be set in the absence of concrete data by uninformed management. Rather, the goals should be set based on a collaborative effort of system personnel taking into consideration the growth rate expected, budget constraints, personnel, and the overall network build program.

Q. *What issues should be checked if a cell site or sector is performing poorly?*

A. The issues that should be checked if a cell site or sector is performing poorly are the following:

1. What changes were made to the network recently in that area?
2. Have there been any customer complaints regarding this area?
3. Who are the reusers, cochannel and adjacent channel, for the area, two or three rings out?
4. What are the sites' configuration, hardware, antennas, and so on?
5. How well is the topology, handoff, tables for one-way handoffs, either into or out of the site, functioning?
6. Are there any unique cell site parameter settings?
7. Does an access failure problem also exist for the same area?
8. What is the signal level distribution for the mobile units using the site?
9. Is there a maintenance problem with the site?
10. Does one individual radio cause most of the problems?
11. Is there a software problem associated with the cell site load?

Q. *What are attempt failures?*

A. The record of *attempt failures,* also known as *access-denied levels,* is a key metric used to monitor and continuously improve a wireless network. An attempt failure occurs when a subscriber tries to gain access to the network but is denied service when there are sufficient facilities at that time to process the call. Attempt failures can occur as a result of poor coverage, maintenance problems, parameter settings, or software problems. The attempt-failure level is important to act on since it is directly related to revenue. Regardless of the exact cause for an attempt failure, when a customer is denied access on the network due to its received signal level, revenue is lost.

Q. *What are some of the key items to check when troubleshooting access failures?*

A. The items that should be checked when troubleshooting access failure problems are the following:

1. Does a lost call problem also exist for the same area?
2. Who are the co-setup co-DCC sites?
3. What is the signal level distribution for the originating signals on the site?
4. Have there been any customer complaints?
5. Is there a maintenance problem with the site?
6. Does one individual radio cause most of the problems?
7. Is there a software problem associated with the cell site load?
8. Were most of the problems caused by one mobile unit or a class of mobile units?

Q. *What is a lack of a dominant server?*

A. A *lack of a dominant server* is also referred to as a *random-origination location.* The primary problem with random origination is that there is no one setup, or control channel, that dominates the area where the problem occurs. The lack of a dominant server leads to mobile units, originating on distant cells, creating interference problems in both directions, uplink and downlink.

Q. *What is radio blocking or blocking?*

A. *Radio blocking,* or *blocking,* is another term used to describe blocking at the cell site. RF blocking occurs when a subscriber unit tries to gain access to a network but is denied because there are not sufficient channels or time slots available. The nominal range that a system operates within is 1 to 2 percent blocking.

Q. *How should radio additions be planned for a network?*

A. The six-month process below should be used for planning radio additions to a wireless network:

1. Determine, based on growth levels, the system needs for the following 1, 3, 6, and 12 months.
2. Predict the number of sites expected to be available from the build program, accounting for deloading issues.
3. Predict the number of radios that will need to be added or removed from each sector in the network.
4. Modify the existing quarterly plan.
5. Determine the number of facilities and equipment bays that will be needed to support channel expansions.
6. Inform the frequency planners, performance engineers, equipment engineers, and operations of the requirements.
7. Issue a tracking report showing the status of the sites requiring action. This report should have in it as a minimum the following information:
   - Radios currently at the site
   - Net change in the number of radios
   - When exhaustion, 2 percent blocking is expected
   - Radios ordered (if needed)
   - Facilities ordered (if required)
   - Radios secured
   - Radio frequency issues
   - Cell site translations completed
   - Facilities secured (if needed)
   - Radios installed or removed
   - Activation date planned for radios to be added
8. On a quarterly basis conduct a brief one-hour meeting to discuss the provisioning requirements and arrange for personnel.
9. Perform a biweekly traffic analysis report to validate the quarterly plan, and issue the tracking report at the same time.

Q. *What is a retune?*

A. A *retune* is a process used in wireless systems to alter the channel or radio assignments for a particular cell or group of cell sites.

Q. *What are the types of frequency retunes?*

A. There are several types of frequency retunes used, and each has its pros and cons. The retune methods used are the following:

1. Systemwide flash cuts
2. Cell-by-cell retunes
3. Sectional retunes

Q. *What is the retune procedure?*

A. The following is an example of a retune procedure that should be followed when conducting a retune. The procedure outlined here includes the planning as well as the implementation processes.

Method of Procedure for Regional Retune

Revised Date

Pre-Retune Process

X-X-XX Retune area defined.

X-X-XX Project leader(s) named, and time tables specified as well as the scope of work associated with the project.

X-X-XX Traffic engineering provides radio channel count.

X-X-XX Frequency planning begins design.

X-X-XX Phase 1 design review (frequency planning only).

X-X-XX Phase 2 design review (all engineering).

X-X-XX Phase 3 design review (operations and engineering).

X-X-XX Phase 4 design review (adjacent markets if applicable).

X-X-XX Frequency assignment sheets given to operations.

X-X-XX Retune integration procedure meeting.

X-X-XX Executive decision to proceed with retune.

X-X-XX Adjacent markets contacted and informed of decision.

X-X-XX Secure post-retune war room area.

X-X-XX Briefing meeting with implementors of retune.

X-X-XX MIS support group confirms readiness for post-processing efforts.

X-X-XX Customer care and sales notified of impending actions.

RETUNE PROCESS (begins *X-X-XX* at time *XXXX*)

X-X-XX

- Operations informs key personnel of retune results.
- Operations personnel conduct brief post-retune test to ensure that call processing is working on every channel changed.
- Operations manager notifies key personnel of testing results.

POST-RETUNE PROCESS (begins *X-X-XX* at time *XXXX*)

Voice mail message left from engineering indicating status of retune (time).

Begin post-retune drive testing phase 1 (time).

Database check takes place.

Statistics analysis takes place.

Voice mail message left from RF engineering indicating status of post-retune effort (time).

Phase 2 of post-retune drive testing begins.

Commit decision made with directors for retune (time).

Phase 3 of post-retune drive testing begins.

X-X-XX

- Continue drive testing network.
- Statistics analysis.
- Conduct post-retune analysis and corrections where required.

X-X-XX

- Post-retune closure report produced.

Q. *What is the analog retune design review checklist?*

A. The following checklist can be used for a regional or systemwide retune:

- Voice channel assignments
  1. Reason for change.
  2. Number of radio channels predicted for all sites.
  3. New sites expected to be added.
  4. Proposed ERP levels by sector for all sites.
  5. Coverage prediction plots generated.
  6. *C/I* prediction plots generated.
  7. Cochannel reusers identified by channel and SAT.
  8. Adjacent channel cell sites identified by channel and SAT.
  9. SAT assignments checked for cochannel and adjacent channel.
  10. Link budget balance checked.
- Control channel assignments
  1. Reason for change.
  2. Coverage prediction plots generated.
  3. Cochannel *C/I* plots generated.
  4. Proposed ERP levels by sector.
  5. Cocontrol channel reusers identified by channel and DCC.
  6. Adjacent control channel reusers identified by channel.
  7. DCC assignments checked for dual originations.
  8. 333/334 potential conflict checked.

- Frequency design reviews
    1. RF design engineer
    2. Performance engineer
    3. Engineering managers
    4. Adjacent markets (if required)

Q. *What is the IS-136 retune checklist?*

A. The following checklist can be used for a regional or systemwide retune for an IS-136 system. The checklist assumes a dual-mode situation for cellular and can be easily modified to reflect IS-136 for PCS alone:

- Voice channel assignments
    1. Reason for change.
    2. Number of radio channels predicted for all sites for analog.
    3. Number of time slots and therefore radios predicted for all sites for digital.
    4. Digital and analog spectrum allocations.
    5. Guardband defined.
    6. New sites expected to be added.
    7. Proposed ERP levels by sector for all sites.
    8. Coverage prediction plots generated.
    9. *C/I* prediction plots generated.
    10. Differences noted between analog and digital reuse patterns and groups.
    11. Cochannel reusers identified by channel, SAT, and DVCC.
    12. Adjacent channel cell sites identified by channel, SAT, and DVCC.
    13. SAT assignments checked for cochannel and adjacent channel interference.
    14. Link budget balance checked.
- Control channel assignments
    1. Reason for change.
    2. Coverage prediction plots generated.

3. Cochannel *C/I* plots generated.
4. Proposed ERP levels by sector.
5. Cocontrol channel reusers identified by channel and DCC.
6. Adjacent control channel reusers identified by channel.
7. DCC assignments checked for dual originations.
8. 333/334 potential conflict checked.

- Digital control channel assignments
  1. Reason for change.
  2. Coverage prediction plots generated.
  3. Digital control channel assignments matched to preferred channel list.
  4. Cochannel *C/I* plots generated.
  5. Proposed ERP levels by sector.
  6. Codigital control channel reusers identified by channel and DCC.
  7. Adjacent digital control channel reusers identified by channel.
  8. DCC assignments checked for dual originations.
- Frequency design reviews
  1. RF design engineer
  2. Performance engineer
  3. Engineering managers
  4. Adjacent markets (if required)

Q. *What is the GSM retune checklist?*

A. The following checklist can be used for a regional or systemwide retune for a GSM system:

- Channel assignments
  1. Reason for change.
  2. Number of radio channels predicted for all sites for analog.
  3. Number of time slots and therefore radios predicted for all sites.

4. Spectrum allocations restrictions.
5. Guardband defined.
6. New sites expected to be added.
7. Proposed ERP levels by sector for all sites.
8. Coverage prediction plots generated.
9. *C/I* prediction plots generated.
10. Cochannel reusers identified by channel and DCC.
11. Adjacent channel cell sites identified by channel and DCC.
12. Link budget balance checked.

- Frequency design reviews
    1. RF design engineer
    2. Performance engineer
    3. Engineering managers
    4. Adjacent markets (if required)

Q. *What is the iDEN retune checklist?*

A. The following checklist can be used for a regional or systemwide retune for an iDEN system:

- Radio channel assignments
    1. Reason for change.
    2. Number of time slots predicted for all sites for 3:1 interconnect traffic.
    3. Number of time slots predicted for all sites for 6:1 interconnect traffic.
    4. Number of time slots predicted for dispatch traffic.
    5. Number of time slots predicted for DCCH traffic.
    6. Total number of radios, BRs, defined from time slots required.
    7. Spectrum allocation available for the market.
    8. Guardband defined.
    9. New sites expected to be added.
    10. Proposed ERP levels by sector for all sites.
    11. Coverage prediction plots generated.

12. *C/I* prediction plots generated.
13. Cochannel reusers identified by channel and DVCC.
14. Adjacent channel cell sites identified by channel and DVCC.
15. Link budget balance checked.

- Frequency design reviews
  1. RF design engineer
  2. Performance engineer
  3. Engineering managers
  4. Adjacent markets (if required)

Q. *What is the PCS CDMA retune checklist?*

A. The following checklist can be used for a regional or systemwide retune for a CDMA system. This retune is based on the PN codes rather than the radio frequency allocations. However, there is a check for microwave clearance that is included for reference:

- CDMA carrier assignments
  1. Reason for change.
  2. Number of traffic channels predicted for all sites for analog.
  3. Spectrum allocation restrictions.
  4. CDMA channels available defined.
  5. Guardband defined.
  6. New sites expected to be added.
  7. Proposed pilot power level distribution defined.
  8. Coverage prediction plots generated.
  9. Pilot pollution problems identified.
  10. PN codes defined.
  11. PN reusers identified, including shift.
  12. Link budget balance checked.
- Frequency design reviews
  1. RF design engineer
  2. Performance engineer

3. Engineering managers
4. Adjacent markets (if required)

Q. *What is a cellular CDMA retune checklist?*

A. The following checklist can be used for a regional or systemwide retune for a cellular CDMA system. The checklist assumes a dual-mode situation for cellular:

- Voice channel assignments
  1. Reason for change.
  2. Number of radio channels predicted for all sites for analog.
  3. Number of traffic channels predicted for CDMA carrier.
  4. Digital and analog spectrum allocations.
  5. Guardband defined.
  6. New sites expected to be added.
  7. Proposed ERP levels by sector for all sites.
  8. Coverage prediction plots generated.
  9. *C/I* prediction plots generated.
  10. Cochannel reusers identified by channel and SAT.
  11. Adjacent channel cell sites identified by channel and SAT.
  12. SAT assignments checked for cochannel and adjacent channel interference.
  13. Proposed pilot power level distribution defined.
  14. Pilot pollution problems identified.
  15. PN codes defined.
  16. PN reusers identified, including shift.
  17. Hard handover sites identified.
  18. Link budget balance checked.
- Control channel assignments
  1. Reason for change.
  2. Coverage prediction plots generated.

3. Cochannel C/I plots generated.
4. Proposed ERP levels by sector.
5. Cocontrol channel reusers identified by channel and DCC.
6. Adjacent control channel reusers identified by channel.
7. DCC assignments checked for dual originations.
8. 333/334 potential conflict checked.

- Frequency design reviews
  1. RF design engineer
  2. Performance engineer
  3. Engineering managers
  4. Adjacent markets (if required)

Q. *What is the post-retune process?*

A. The post-retune process that is recommended is listed below. Obviously, more detail could be added, but the steps listed provide the foundation upon which to build:

1. Identification of the key objectives and desired results prior to the retune taking place.
2. Statistical analysis for the two weeks prior to the effort, using the same time frames and reference points (very difficult).
3. Full cooperation of operations, implementation, customer service, MIS, and of course engineering for staffing levels.
4. Staffing during changes.
5. Post-retune statistical analysis.
6. Identification of the most problematic areas in the network. Initial statistical analysis is performed at this time during or right after the configuration is checked for the network. The problem sites need to be identified by following the key metrics, as listed before:
   - Lost calls
   - Attempt failures

- Blocks
- BER/FER
- Channel failures
- Usage/RF loss
- Customer complaints
- Field reports from the drive test team(s)

7. The initial drive test data are then analyzed for the key potential problem.
8. Drive test data are then analyzed for the general runs for the rest of the retune area involved, again focusing on any problems that occurred, phase 2.
9. The third phase of testing analysis involves the followup tests and postchange corrections needed to the network, and if required, additional tests are then performed.
10. Over the ensuing two weeks, a daily statistics and action report is generated showing the level of changes and activities associated with the effort.

Q. *What are the types of drive testing performed on a wireless system?*

A. There are several types of drive tests that are performed on a wireless system such as the following:

1. Pre-site qualification
2. SQT
3. Performance testing
4. Prechange and postchange testing
5. Competition evaluations
6. Post-cell turnon
7. Post-retune efforts

Q. *What types of performance tests are commonly used?*

A. There are several types of performance tests that can be conducted with drive tests, and the following is a listing of the major items.

1. Interference testing (cochannel and adjacent channel)
2. Coverage problem identification
3. Customer complaint validation
4. Cell site parameter adjustments
5. Cell site design problems
6. Software change testing
7. Postchange testing
8. Miscellaneous

Q. *What site activation method is most commonly used?*

A. The method or philosophy of site turnon, or activation, varies from wireless company to wireless company. Some of the philosophies are driven by engineering, while others are driven by financial objectives. One philosophy is that when a site is finished being constructed, it should be activated into the network. The second philosophy is that the site's depreciation should be minimized or maximized depending on the accounting method employed by the company. The third philosophy is that when a site is not activated until the implementation, the plan put forth by engineering dictates the timing of the activation. A fourth philosophy or activation involves a combination of the second and third philosophies.

Q. *What issues need to be checked for a new cell site prior to activation?*

A. The items listed below need to be checked for a new cell site prior to its activation to ensure that there are no new problems introduced into the network as a result of the physical configuration of the site. The objective of a new site investigation is to validate that the site is built to the design specifications put forth by engineering. Some of the key areas to validate involve the following:

Antenna system orientation

Antenna system integrity

Radio power settings

Cell site parameters

Hardware configuration

Grounding system

Q. *What is a new cell site performance checklist?*

A. A new cell site performance checklist is a detailed list of issues that need to be verified by RF engineering to ensure that the potential new site is ready for integration into the network. A basic list of items to have checked is shown below:

- Antenna system
  - Installation completed.
  - Installation orientation and mounting verified.
  - Feed-line measurements made and recorded.
  - Return loss measurements made and acceptable.
  - Feed lines grounded and waterproofed.
- FAA compliance
  - Lighting and marking completed (if required).
  - Alarming system installed and operational.
- Receive and transmit filter system
  - Bandpass filter performance validated.
  - Notch filter performance validated (if required).
  - Transmit filter performance validated.
- EMF power budget provided
- Cell site parameters
  - Frequency assignments validated.
  - Handoff topology lists checked.
  - Cell site parameters checked.
  - Cell site software load for all devices validated.
- Spectrum check
  - Sweep of transmit and receive spectrum for potential problems.
  - Collocated transmitter identified.

Q. *What needs to be checked when visiting an existing cell site in a wireless network?*

A. When conducting visits to existing cell sites, it is recommended that the following checklist of items be used to help expedite the inspection:

1. Review statistics for the cell site in question and the surrounding cells.
2. Review frequency plan and that of the surrounding cells.
3. Review handoff topology and the site's neighbors.
4. Review expected problems and possible remedial actions.
5. Review cell site's hardware configuration.
6. Review maintenance issues for the site over the preceding month.
7. Secure site access for the location.
8. Obtain maps of the area and directions to the site.
9. Obtain test equipment needed for the investigation.

Q. *What needs to be done to conduct an intermodulation test?*

A. There are numerous steps that need to be taken prior to and during a site investigation carried out for the identification and resolution of a potential intermodulation problem. The following steps should be completed at the very least, but they can be expanded upon:

Previsit Work

1. Talk to the cell site technician, and have him or her go over the nature of the problem and all the steps already taken to correct the problem.
2. Examine the site-specific records for the location to see if a previous problem was investigated and if there were any changes made recently to the site.
3. Determine if there are any colocated transmitters at the facility, and conduct an intermodulation report looking for hits in the site's own band, or in another band, based on the nature of the problem.

4. Collect statistical information on the site to try to determine any problem patterns.
5. Review maintenance logs for the site.
6. Formulate a hypothesis for the cause of the problem, and generate a test plan to follow.
7. Secure the necessary test equipment and operations support for the site investigation.
8. Allocate sufficient time to troubleshoot the problem.

Site Work

1. Devise an initial test plan.
2. Isolate the problem by determining if the problem is internal or external to the cell site.
3. Verify that all connectors are secure and tight.
4. Monitor the spectrum for potential intermodulation products determined from the report.
5. When intermodualtion products appear, determine common elements that caused the situation.

Q. *What is downtilting?*

A. *Downtilting* is the altering of the antenna inclination of a cell site to alter the site's coverage. The reasons to alter the concatenation of the cell site's antenna system can vary. Sometimes antennas are downtilted to reduce interference, to improve in-building penetration, to improve coverage, or to limit the coverage area. The concatenation of the antenna system can have a major impact on the actual performance of the cell site itself.

Q. *What material needs to be reviewed for a change in an antenna design?*

A. The items that need to be reviewed at the antenna change design review meeting are the following:

1. The objective for the antenna alteration

2. Defined coverage area for the cell site and its sectors
3. The preliminary tilt angles desired based on items 1 and 2 above
4. The pretest and posttest plans
5. Review pass/fail criteria

Q. *What is EMF compliance?*

A. *Electromagnetic force* (EMF) radiation must be monitored for compliance with federal regulations. The FCC standard C95.1 is the basic reference that establishes the allowable electromagnetic field strength. Cellular technology currently utilizes the C95.1-1982 specification while PCS utilizes C95.1-1992.

The C95.1-1992 specification has two basic sets of criteria that must be followed in determining if a communication facility meets the regulations. The two basic criteria pertain to a controlled and uncontrolled environment. The specification defines controlled and uncontrolled environments.

Q. *What is a controlled environment?*

A. A *controlled environment* is one in which the workers for the communication company, usually the RF technicians, are monitored for EMF exposure.

Q. *What is an uncontrolled environment?*

A. An *uncontrolled environment* is one in which everyone—that is, the general public—is monitored for EMF exposure. The primary difference between controlled and uncontrolled environments is the difference in power levels to which a person may be exposed (see Table 12.1). As a general rule, the uncontrolled environment exposure limit is about one-fifth that of the controlled environment. The C95.1-1982 standard, however, does not differentiate between a controlled and an uncontrolled environment. Therefore, many cellular operators are following C95.1-1992 already, if for no other reason than for public relations.

Q. *What is a switch CPU loading study?*

TABLE 12.1 POWER DENSITY (mW/cm²) CHART

	Equation	Frequency, MHz 880	900	1800	1900	2000
C95.1-1982	*f*/300	2.933	3.0	6.0	6.333	6.666
C95.1-1992						
Controlled	*f*/300	2.933	3.0	6.0	6.333	6.666
Uncontrolled	*f*/1500	0.586	0.6	1.2	1.266	1.333

Notes: (1) *f* (megahertz) is the particular frequency of concern.
(2) Power-density levels are the maximum permissible exposure levels.
(3) Values are taken from MPE tables for controlled and uncontrolled environments (C95.1).

A. A *switch CPU loading study* is an evaluation of all the CPUs that are part of a network's switching equipment that contain processors whether the node architecture is of a distributed or hierarchical design. These processors are usually divided between the main or central processor and the subordinate or regional processors it controls.

The processor loads of a switch with a hierarchy-based design structure should take into account all levels of processors and their specific function in the delivery of a mobile telephone call. It should be determined which processors have the highest traffic levels and thus, which are the most susceptible to reaching an upper threshold and causing problems as the traffic in the system increases. A processor load study for a switch architecture based on a central processor will be mainly concerned with measuring this processor's traffic load rather than any of the secondary processors it controls. However, the secondary processor loads should be reviewed as well, just not as frequently.

Q. *What is switch call processing efficiency?*

A. *Switch call processing efficiency* is a value that equals the call volume to the CPU loading ratio. This metric can be used to monitor the call processing efficiency of the switch as well as to indicate if a possible problem exists in the system. For example, a handoff border may need optimizing, or a registration interval may need adjusting.

Q. *What is the switch/node total erlangs and call volume?*

A. The *switch/node total erlangs and call volume* is the total erlangs and calls carried during the system busy hour. This number is necessary to determine the percent of the system that the switch is carrying and to determine the nodes' call processing efficiency. This metric is also used to trend the loading of the nodes in the system. This is important in determining where the next critical loading limit will be encountered in the system and when to begin plans to relieve this load.

Q. *What is a switching node performance report?*

A. A *switching node performance report* is a summary of the major metrics used to measure the performance of a network node:

1. Measure the load on critical node processors.
2. Measure the processing efficiency of the node.
3. Track and monitor the outages of the node.
4. Measure the load on the switch service circuits.
5. Measure and report on the total erlangs and calls processed by the node during the system busy hour.
6. Monitor the node alarms.
7. Track and monitor the node's memory utilization and capacity.
8. Check the node timing sources for a stradum 2 level accuracy.

Q. *What is included in a call delivery report?*

A. Table 12.2 is an example of a call delivery report that should be available for any operator independent of the technology platform used.

TABLE 12.2 SYSTEM X CALL DELIVERY STATISTICS REPORT

System Call Mix	
Percent total system calls (M-L) = 80.0%	
Percent total system calls (L-M) = 15.0%	
Percent total system calls (M-M) = 5.0%	
System Calls by Category	
Category	**Total system calls, %**
Completed calls	78.00
Unacknowledged mobile units	09.45
Invalid mobile units	00.05
Invalid ESN	00.10
RF channel unavailable	00.23
Land trunk unavailable	00.06
IMT unavailable%	00.05

Q. *What steps are needed to ensure the integrity of the call routing tables used?*

A. The following is a summary of the steps used in measuring and maintaining the routing performance of the system translation tables:

1. Obtain a copy of the translations tables for reference.
2. Obtain statistics on the utilization of the system's recorded announcements.
3. Obtain statistics on the network call delivery.
4. Monitor the utilization of the recorded announcements and the call delivery category percentages and note any unusual changes in the values.

5. Correlate large changes in the data to possible errors in the routing tables or the current network design.

6. Devise formal procedures to track and correct any errors that may develop in the routing tables. Also, keep a current copy of the tables in a secure place at all times, and limit access to the tables to only experienced authorized personnel.

Q. *What needs to be included in the network performance report?*

A. The following is a list of general system performance and design data that can be used as a measure of a system's operation and capacity. These data will also be helpful in specific performance and growth studies of the switch.

1. Cell site data
   - Number of cell sites currently in service on a switch-by-switch basis
   - Number of cell sites projected to be cut in service by the end of the year
   - Number of macrocell sites currently in service on a switch-by-switch basis
   - Number of projected macrocell sites to be cut in service by the end of the year

2. Subscriber data
   - Number of subscribers currently assigned in the switch
   - Number of subscribers projected to be assigned by the end of the year
   - Number of subscribers assigned with the voice mail feature
   - Number of subscribers assigned with other network features (traffic information service, etc.)

3. System and switch traffic data
   - Average system busy hour usage (erlangs) (10 high day average per month)
   - Estimated network maximum traffic level
   - Average switch busy hour usage on a switch-by-switch basis

- Estimated switch maximum traffic level
- The usage per subscriber (erlangs)
- Percentage of subscribers registered and active in the system during the busy hour
- Number of registration attempts on the system during the system busy hour
- The registration interval assignments on a switch-by-switch basis
- Number of call attempts in the system during the system busy hour
- Number of calls completed in the system during the system busy hour
- Number of blocked calls in the system during the system busy hour
- Number of dropped calls in the system during the system busy hour
- Number of intrasystem handoffs during the system busy hour
- Number of intersystem handoffs during the system busy hour
- Average call holding time (seconds)
- Average call setup time (seconds)
- Switch processor load on a switch-by-switch basis
- Switch efficiency percentages on a switch-by-switch basis
- Total number of voice channels in service on a switch-by-switch basis (both RF and land)

Q. *What are the troubleshooting steps involved with call testing?*

A. The following list of steps should be taken when testing calls in a wireless system for their integrity and investigating problems reported involving call delivery or feature problems:

1. Develop a testing lab complete with test cells and terminals assigned to every switch in the network for use in troubleshooting and completing network hardware and software acceptance call testing.

2. Develop a test call log sheet.
3. Obtain the customer trouble tickets from the customer service department for determining the type of problem calls, the number of the subscriber, the features assigned to that customer, and the quantity and magnitude of the customer complaints.
4. Obtain a mobile phone and an unassigned number in the switch for call testing. Choose a mobile phone that is easy to program and a test number that is in the same 100s block that the customer's number is in if possible.
5. Set up the subscriber profile of the test mobile unit in the switch database. Specify the type of service the mobile unit is to have and the features needed for testing.
6. Notify the other departments in the company of this call testing by using standardized forms and formal agreed-upon procedures.
7. Program the mobile phone for operating on the designated network switch. Check the phone for proper operation.
8. Obtain for reference a listing of all the recordings on the network switches and a list of the mobile unit codes in the system and the nodes that are actually loaded and/or stored. Obtain a list of the service categories used in the system.
9. Begin call testing by trying to mimic the type of mobile unit call in question. Record the findings on the test call log sheets.

Q. *What methodology should be used for system performance troubleshooting?*

A. The methodology that should be used for system performance troubleshooting is best defined by using the seven steps listed below. These steps will apply to any technology or situation:

1. Identify objective.
2. Remove variables.
3. Isolate the system components.

4. Make a test plan.
5. Communicate the test results.
6. Take action to correct any problems.
7. Conduct a postanalysis.

Q. *What courses does an RF engineer need to take to be able to work on wireless systems?*

A. The following is a preliminary list of courses that an RF engineer needs to take to work on wireless systems. Some of the courses should be taken at regular intervals as refresher courses.

Antenna theory
Basic cellular communications
Call processing algorithms
CDMA
CDPD
Cell site grounding
Cell site installation
Cell site maintenance
Cellular call processing
Digital radio design
Disaster recovery
EMF compliance
Frequency planning
GSM
iDEN
Interconnect
Interview techniques
Microcell design
Microwave system design
Network architecture
Network design
PCS
Performance troubleshooting
Presentation techniques
Project management
Rerad design
RF design
Statistics theory
Switch architecture
TDMA
Traffic theory
WPBX

Q. *What courses should a technical manager for a wireless system take?*

A. The following is a preliminary list of courses that a technical manager for a wireless system should take. Some of the courses should be taken at regular intervals as refresher courses.

Basic cellular communications	Network design
Budget training	Network fundamentals
CDMA	Operations and maintenance
Cellular call process	PCS
Digital radio design	Presentation techniques
Disaster recovery	Project management
Fraud management	Real estate acquisition
General management	RF design
GSM	SS7
iDEN	Statistics theory
Interconnect	Switch architecture
Interview techniques	Tariffs
IS41	TDMA
MFJ/descent decree	Traffic theory

Q. *What courses should a network engineer for a wireless system take?*

A. The following is a preliminary list of courses that a network engineer for a wireless system should take. Some of the courses should be taken at regular intervals as refresher courses.

AIN	Basic cellular communications
ATM	Call processing algorithms

- CDPD
- Cellular call processing
- DACS
- Data and voice transport
- Disaster recovery
- Equipment grounding
- Fiber optics
- Inter- and intra-LATA
- Interconnect
- Interview techniques
- IS-41
- iSDN
- LAN/WAN topology
- MFJ/descent decree
- Network architecture
- Network design
- Network fundamentals
- Network maintenance
- Network services
- Numbering plans
- PCM and ADPCM
- Performance troubleshooting
- Presentation techniques
- Project management
- Real estate acquisition
- RF design
- SONET
- SS7
- Statistics theory
- Switch architecture
- Switch maintenance
- Traffic theory
- Translations cell site
- Translations switch
- Tariffs
- Voice mail
- WPBX

Q. *What are some of the advantages of using a centralized approach for a technical organization?*

A. The centralized approach has the advantage of potentially achieving economies of scale. The economies of scale are achieved through the elimination of redundant functions in each of the markets or areas. An example of a centralized function is new technology research for the company, which does not have to be accomplished with every department or division within a company. The idea of having one group lead the effort will ensure uniformity, accountability, and the probability that the direction picked is coordinated among the various organizations in the company.

Q. *What are some of the disadvantages of using a centralized approach for technical organizations?*

A. The centralized approach has the disadvantage of being defocused on the market requirements. The defocusing of market requirements can come about through not having any local knowledge of the technical configuration for the network. An example of defocusing can occur over a simple matter of switch port assignments or roads a cell site actually covers.

Q. *What are some advantages of using a decentralized approach for a technical organization?*

A. The decentralized approach has the advantage of being more market sensitive and flexible than the centralized approach. An example of the adaptability to the market environment would involve the continuous configuration of the network based on handoff and call origination traffic. The idea of the decentralized approach is that the decisions that will affect the market are brought as close to the customer as possible.

Q. *How should the technical departments be organized?*

A. The technical organization structure utilized by a company should be driven by functional requirements, not by personalities. Personalities, however, often define the organization's structure in that the organization is arranged by who is in the group rather than who should be doing the work.

Q. *Should the technical departments be centralized or distributed?*

A. The technical organization's role for a wireless system should be both centralized and distributed. A blended approach works most efficiently for a company's technical structure.

Q. *What is the disadvantage of using a decentralized approach for technical organizations?*

A. The disadvantage of using the decentralized approach is the amount of redundant work that is performed. The decentralized approach lends itself to localized procedures that foster inefficiencies and a lack of knowledge transport. The lack of knowledge transport often leads to the problem or situation being repeated in another market when some simple communication could convey how it could possibly be avoided. The decentralized approach also does not lend itself to any engineering practice procedures, which are essential in the rapidly changing world of wireless communications.

Q. *What is the difference between lost calls for the system and lost call segments?*

A. The difference between lost calls for the system and lost call segments is best represented by a simple example below using the same system performance statistics.

$$\text{Originations and terminations (O\&Ts)} = 10{,}000$$

$$\text{Handoffs (HOs)} = 10{,}000$$

$$\text{Lost calls (LCs)} = 200$$

$$\text{Overall system lost calls} = \text{LCs/O\&Ts} = 2 \text{ percent}$$

$$\text{Lost call segments} = \text{LCs/(O\&Ts} + \text{HOs)} = 1 \text{ percent}$$

The lost calls for the overall system show what the subscribers experience, while the lost call segments indicate the performance relative to the cell site as a measure of facility usage assuming that the more handoffs that occur, the better the system performance must be.

Q. *What is the structure of a centralized technical organization?*

A. The centralized organization normally involves a span of control that can range from five to possibly seven or eight, as shown in Fig. 12.1.

Q. *What is the structure of a decentralized technical organization?*

A. The decentralized organization structure shown in Fig. 12.2 can be varied as required. Specifically, the variations would pertain to whether the company has a centralized budget department encompassing all departments. The decentralized organization normally has a span of control of five, or six if the administration support is included. Exceeding a span of control beyond seven or eight for any organization becomes very difficult to manage.

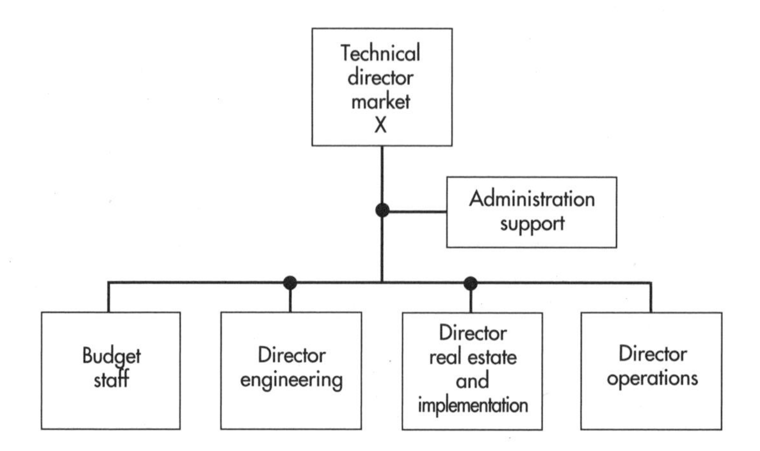

Figure 12.1 Centralized technical organization by market.

Q. *What is a downtilt procedure?*

A. A downtilt procedure is used to assure that there are no negative effects upon the system when making any antenna changes. The steps listed below are general in nature and apply to any technology platform used.

1. Identification of the problem
2. Operations check
3. Physical observations
4. Drive testing
5. Define cell's coverage area
6. Design review
7. Test plan
8. Closure

Q. *What are performance goals?*

A. *Performance goals* are goals that are used to improve the RF or switched network to improve a customer service or the company's financial situation. Performance goals need to match the overall goals for the wireless company, and they are usually defined for a period of one year with either quarterly, six-month, or yearly targets.

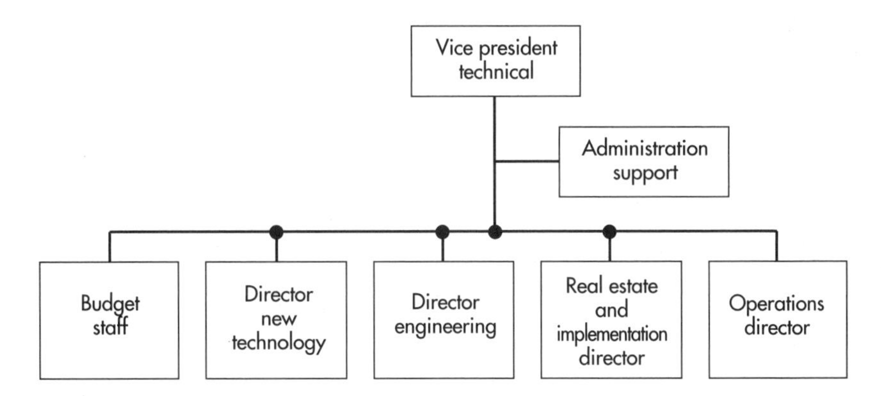

Figure 12.2 Decentralized technical organization chart.

Q. *What are performance objectives?*

A. *Performance objectives* can be either performance goals or a subset of the performance goals.

Q. *Who should receive engineering reports?*

A. A procedure for distributing engineering reports should be well thought out. The following basic issues for report generation and dissemination need to be addressed for every engineering report:

1. What is the purpose of the report?
2. Who will generate the report on an ongoing basis?
3. Who will act on the information that is in the report?
4. Who will receive the report?
5. Is the report needed?
6. What format should the report be in—that is, electronic or paper?
7. How will the report be processed?

Q. *What are network performance reports?*

A. *Network performance reports* are the various types of reports used to monitor and improve the performance of the wireless network from a fixed network aspect. A multitude of network performance reports need to be generated, disseminated, analyzed, and acted upon in a wireless system. The reports generated should match the organization's goals and objectives plus any performance criteria that are deemed necessary to operate from:

Bouncing busy hour traffic report

Telephone number inventory report

Facility usage and traffic report

Facilities interconnect report

Switch and node metric report

System traffic forecast report

Network configuration report

Q. *What types of reports are needed in the technical organization?*

A. Table 12.3 lists the types of reports needed for the technical organization for any wireless system.

Q. *What is the telephone number inventory report?*

A. The telephone number inventory report, which is vital for monitoring usage by mobile telephones, is a record of the actual numbers assigned to a system subscriber base at a given time. The report is used for predicting, projecting, and trending the directory

TABLE 12.3 REPORTS NEEDED WITHIN THE TECHNICAL ORGANIZATION

Department/report	Internal/external	Frequency
Director engineering		
Key performance	External	Monthly
Facilities	Internal	Monthly
Growth plan	External	Quarterly
Software loads	Internal	Monthly
Project plan	Internal	Weekly
Cell site progress		
System performance	Internal	Weekly
Equipment status	Internal	Weekly

TABLE 12.3 REPORTS NEEDED WITHIN THE TECHNICAL ORGANIZATION (*CONTINUED*)

Department/report	Internal/external	Frequency
Director engineering (*Continued*)		
Network briefing	External	Monthly
Directory inventory	External	Monthly
Network systems		
Individual node performance report	Internal	Weekly
Network link performance report	Internal	Weekly
Network routing report	Internal	Weekly
Network software report	Internal	Biweekly
Facilities management		
Systems facility interconnect plan	Internal	Quarterly
Data network		
Systems facility interconnect plan	Internal	Quarterly
Network link performance report	Internal	Weekly
Software engineering		
Software configuration	Internal	Weekly
FOA	External	Biweekly
Translations status	Internal	Weekly
RF engineering		
Cell site status	External	Weekly
FAA compliance	External	Weekly
Frequency management	Internal	Weekly
RF design	Internal	Weekly
Digital radio (CDMA/TDMA)	Internal	Weekly
CDPD	Internal	Weekly
System performance		
Performance report	External	Weekly
Equipment engineering		
Equipment status	External	Weekly

inventory growth. By knowing the telephone number growth patterns, engineering can order additional mobile unit codes for future use from the LEC.

Q. *How often is the telephone number inventory report generated?*

A. This report needs to be generated on a monthly basis and should indicate as a minimum the complete breakdown of all the directory numbers used in the network.

Q. *How should the telephone number inventory report be structured?*

A. The telephone number inventory report should be structured as shown in Fig. 12.3.

Q. *What are handover, or handoff, drops?*

A. *Handover,* or *handoff, drops* are calls that are lost as a result of handoffs. Depending on the technology platform and the vendor for the infrastructure used, the inclusion of the handover drops may or may not be integral to the lost call peg counts. However, handover drops are an important performance metric. Handover drops can occur with all the wireless technologies.

Q. *How are signal quality estimates used?*

A. Signal quality estimates (SQEs) are used extensively in an iDEN system. SQE is very similar to *C/I* for most of its range from, say, 10 to 25, but after 25 the relationship with *C/I* tends to deviate. A design goal tends to be 22 SQE, which is similar to 22 dB *C/I*.

Q. *What is a handoff candidate?*

A. A *handoff candidate* is a cell site that is being considered or is in the process of being the target for a subscriber to hand off from one cell to another.

Q. *What is directed retry?*

A. *Directed retry* is a parameter setting that is used to redirect subscribers from one cell to another when accessing the network. The use of directed retry is meant solely for a short-term relief that is needed for severe blocking situations that can and do

Figure 12.3 Telephone number inventory report.

Telephone Number Inventory Report

System name:

Week ending:

Date:

NPA	NXX	XXXX	Available	Central office	Resident switch	Route name	Tandem	Tested	No. released	No. active	Utilization rate	Comments
914	365	7000-7999	8/15/95	ORB	WDB1/003	BBEN9	ZERK	1000	1000	500	50%	
201	968	2000-2999	8/26/95	PARM	ERU07/011	CPG2/4	NWK	1000	500	200	20%	

arise. Specifically, directed retry on origination will direct a subscriber from cell A to cell B when there are no channels for use at cell A. However, a call that has originated on a best-serving cell site cell A which then switches to call B will most likely degrade because the frequency plan probably was not designed for that situation.

Q. *Can directed retry be used to resolve blocking problems?*

A. Directed retry can be used effectively to reduce blocking at a cell site on a short-term basis; however, using this device will be at the expense of the quality of service and will increase lost calls for the network. Directed retry to resolve blocking problems is a tool that should only be used on a short-term basis and then removed once the channels are installed at the correct cell site to resolve the blocking problem.

Q. *Should directed retry be used in every cell site?*

A. Directed retry should not be used in every cell site because it can create interference and call degradation.

Q. *Where is SQE used?*

A. SQE is used in an iDEN system for many call processing functions for both interconnect and dispatch. A poor SQE will prevent handoffs and originations on a system.

Q. *What is a mobile assisted handoff?*

A. *Mobile assisted handoff* (MAHO) *technology* is used by IS-136, GSM, and iDEN systems, to mention a few. An MAHO is a process by which the subscriber unit measures the RSSI received from the downlink signals, or beacons, of all the neighbor cell sites that are on the neighbor list and compares the RSSI value to the current serving signal. If a neighbor or neighbor cell site presents a better downlink RSSI, then a handover is initiated.

Q. *What is coding, or processing, gain?*

A. *Coding,* or *processing, gain* is a measure of improvement in signal reception, expressed as *S/N*, that occurs due to the coding sequence used. Coding gain is normally associated with spread-spectrum systems, or CDMA systems. (See Fig. 12.4.)

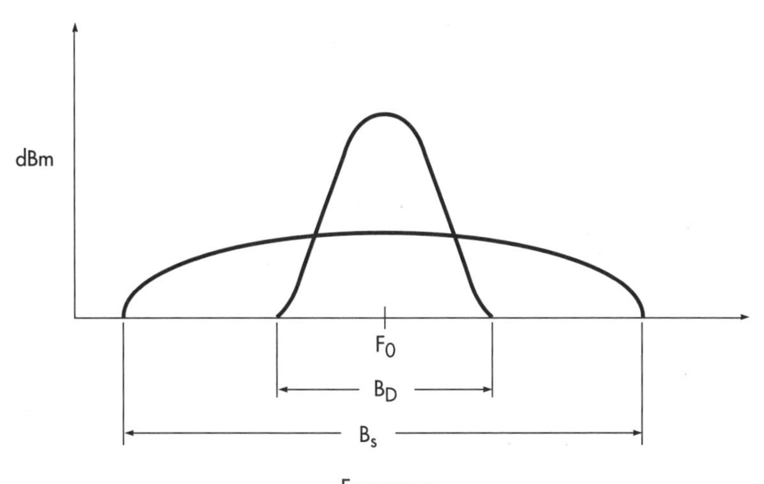

Figure 12.4 Coding, or processing, gain. B_D = bandwidth of initial signal, B_S = bandwidth of initial signal spread, processing gain = $G_p = B_S/B_D$.

Q. *What are glares?*

A. The term *glares* has two different meanings depending on whether it is used in the context of landlines or radio environments. For a landline, *glares* is a situation in which both ends of a DS-1 are attempting to seize the same time slot. The landline conflict is resolved by assigning control to one end of the DS-1.

In an RF environment, a *glare* is a situation in which a mobile unit attempts to set up on two separate cell sites at the same time, usually on different switches. Glares are a direct result of the reuse of the control channel having the same DCC. RF glares are usually resolved through better management of the frequency plan.

Q. *What is double access?*

A. *Double access* is a situation in which a wireless system mobile unit attempts to set up its call on two separate cell sites at the same time. The term *double access* is used in place of the terms *glares* or *RF glares*. Double-access issues are usually resolved through better management of the frequency plan.

Q. *What are ping-pong handoffs?*

A. *Ping-pong handoffs* are handoffs that occur repeatedly between cell sites. More specifically, a ping-pong handoff would be the situation in which a mobile unit hands off a call from cell A to cell B and almost immediately hands the call back to cell A. This situation can occur numerous times, and it is usually a result of either an incorrect parameter setting or a maintenance problem at the target cell site.

Q. *What should be included in a bouncing busy hour report?*

A. Figure 12.5 illustrates the basic format for a bouncing busy hour report.

Q. *What is hysteresis?*

A. *Hysteresis* is a parameter value that is normally associated with a neighbor list in a wireless system. Hysteresis is used to prevent ping-pong handoffs and also traffic steering. Most systems have an ability to add hysteresis in the neighbor list selection process. The usual default value that is used, regardless of vendor or technology platform, is 4 dB. With a hysteresis of 4 dB, a handoff will not occur from cell A to cell B unless cell B is detecting the mobile unit better than cell A by 4 dB. This principle could also apply to the downlink RSSI as well.

Figure 12.5 Cell site bouncing busy hour report.

Cell Site Bouncing Busy Hour Report

System name:

Date:

Cell	Time	BBH traffic erlangs	Usage	Radios assigned	Radios OOS	Blocking	Radios needed
1A	1600	5.3					
1B	1700						
1C	1700						
2A	1700						
2B	1800						
2C	1700						

Q. *What is a neighbor list?*

A. A *neighbor list* is a list of potential handoff candidates that is used to determine potential cells that a subscriber unit can hand off to. For instance, cell A could have cell B, cell C, and cell E on its neighbor list, and therefore a subscriber unit could hand off from cell A to either cells B, C, or E. However, the subscriber could not hand off to cell D since it is not on the neighbor list. The selection of the neighbor list candidate is determined by coverage area as well as system performance in functions such a frequency reuse, or *C/I*.

Q. *What is a handoff, or candidate, list?*

A. A *handoff,* or *candidate, list* is a list of neighbor cell sites that are potential handoff candidates.

Q. *What is a neighbor candidate?*

A. A *neighbor candidate* is a cell site that is being considered as a possible target for a handoff. For instance, if cell B is on the neighbor list for cell A, then during the course of a call when a handoff request process is begun, cell B is considered as a potential neighbor candidate.

Q. *Does CDMA technology have frequency reuse capability?*

A. CDMA utilizes frequency reuse in that it uses the same channel in every cell site but utilizes a PN code to differentiate cell sites, or pilots, from each other. CDMA technology has a reuse factor of 1—that is, $N = 1$.

Q. *What is call dragging?*

A. *Call dragging* is a situation in which the interconnect call, or cellular call, is carried beyond the coverage area of a cell site, which causes call degradation to occur. An example of call dragging would be when a subscriber is on cell A and its coverage area, or design, is meant to cover from 14th Street to 20th but the subscriber does not hand off or hands off at, say, 22nd Street. Call dragging is normally a result of an incorrect parameter setting or neighbor list problems and can be easily corrected in most cases.

Q. *What is E_b/N_o?*

A. E_b/N_o is a power ratio, and it is similar to *C/I*. E_b/N_o is a value that is used as a performance and design metric from CDMA systems. The usual value for E_b/N_o in CDMA systems is about 7.5 E_b/N_o, which can be plugged into a link budget as 7.5 dB of power over the effective noise floor for reliable communication.

Q. *Does N = 4 work?*

A. Yes, *N* = 4 works as a frequency reuse pattern and provides the highest erlangs per square kilometer of any of the cellular and PCS reuse patterns.

Q. *What is interference?*

A. *Interference,* or *radio interference,* is the situation in which either a cochannel, same channel, or an adjacent channel degrades call quality. The amount of interference allowed or designed for is determined by the *C/I* criteria. However, from a customer's point of view, interference could be any form of call degradation ranging from engineering's view of interference to a poor subscriber unit that is distorting the call quality.

Q. *What is the frequency reuse factor?*

A. The *frequency reuse factor* is a measure of spectral efficiency and indicates how often a radio frequency is reused in a network:

$$\text{Frequency reuse factor} = \frac{\text{total channels used}}{\text{radio frequencies available}}$$

Therefore, if there were 200 frequencies and 1000 channels used in a network, the frequency reuse factor would be 5.

Q. *What is dispatch dragging?*

A. *Dispatch dragging* is a situation in an iDEN system in which a subscriber is utilizing the dispatch mode and does not get handed off to the proper cell during the dispatch process. Dispatch dragging often results in a reselection process for iDEN.

Q. *What is incorrect orientation?*

A. *Incorrect orientation* is the situation in which the antenna or antennas of a sector are misaligned and pointing in the wrong direction. The incorrect orientation occurs when the antennas are oriented at 10 degrees true north when they should have been oriented at 0 degrees true north. The incorrect orientation will negatively impact performance for the cell site.

Q. *What is orientation?*

A. *Orientation* is the direction or orientation to which the antennas for a sector or cell site are directed. The orientation is a reference to true north (TN) and is an azimuth reference only. Usually a three-sector cell site will have each of its sectors oriented 120 degrees from each other. Therefore, sector 1 would have an orientation of 0 degrees TN while sector 2 would be 120 degrees TN, and sector 3 would be 240 degrees TN.

Q. *What is intermodulation?*

A. *Intermodulation* (IM) is another name for *intermod* (IMD) and also *intermodulation distortion.* Intermodulation is the mixing of two or more signals to produce a third or fourth frequency that is undesired. All cell sites produce intermodulation since there is more than one channel at a site. However, the fact that there is intermodulation products produced does not mean there is a problem.

Various intermodulation products are shown below for reference. The values used are simplistic so as to facilitate the examples. In each of the examples, $A = 880$ MHz, $B = 45$ MHz, $C = 931$ MHz, and D is the intermodulation product. The examples listed below do not represent all the perturbations possible.

Second order: $A + B = D$, or 925 MHz

$A - B = D$, or 835 MHz

Third order: $A + 2 = D$, or 970 MHz

$A - 2B = D$, or 790 MHz

$A + B + C = D$, or 1856 MHz

$A - B + C = D$, or 1766 MHz

Fifth order: $2A - 2B - C = D$, or 739 MHz

Q. *What are quarterly drive tests?*

A. *Quarterly drive tests* is another name for *competitive drive tests*—that is, tests that attempt to simulate the call quality for the system relative to the subscriber's point of view. Also during the quarterly drive test, the competitors' system or systems may also be tested for comparative reasons. The quarterly drive tests utilize the same roads and calling patterns from one quarter to the next for the purposes of recording improvements or degradations to the network that are either systemwide or localized.

The testing involves driving defined routes and placing calls of a fixed duration and recording the amount of times the tester is able to gain access to the system, the call quality using an MOS test, and any problems encountered.

Quarterly drive tests are often referred to as *benchmarking tests* and are used not only to evaluate the performance of the network, to compare the system against the competition, but also to compare the performance of the network to other sister companies.

Q. *What is the received signal strength?*

A. The *received signal strength* (RSSI) is the power value in decibels, milliwatts, or decibels, microvolts, that is received at the receiver at either the cell site or the subscriber unit. The RSSI value is typically given in terms of decibels, milliwatts, for cellular and PCS systems.

$$\text{RSSI, dBm} = \text{transmitted power, dBm} - \text{path loss} + \text{system gains, dB}$$

Therefore, if the transmitted power, dBm = +10 dBm, then

$$\text{Path loss, dB} = 100 \text{ dB}$$

$$\text{System gains, dB} = 10 \text{ dB}$$

$$\text{RSSI, dBm} = +10 \text{ dBm} - 100 \text{ dB} + 10 \text{ dB} = -80 \text{ dBm}$$

The −80 dBm value is the RSSI value.

Q. *What is the system call completion ratio?*

A. The *system call completion ratio* is the ability of a system—RF and switching—to deliver a call successfully and to have it answered as compared to all the attempts that are made. The system call completion ratio, however, does not indicate what is a billable call versus what is not a billable call but rather is a systemwide view of the network's ability to process a call and assign all the facilities correctly for the purposes of processing wireless telephony.

Q. *What is call completion?*

A. *Call completion* has several meanings depending on which system node is being referenced. For instance, there is call completion, RF call completion, switch call completion, and billing call completion. *Call completion* refers to the ability to complete the facility's assignment successfully for the purposes of handling wireless telephony calls.

Q. *What is RF call completion?*

A. An *RF call completion ratio* expresses the ability of the system to successfully assign and also connect, without problems, a radio circuit for use by subscribers when they wish to place or receive a wireless call. The RF completion value should be around 95 percent or greater, but that does not mean the call was successful in terms of being answered, passed to the PSTN, or billed. *RF call completion* refers specifically to a situation in which a subscriber unit needed a radio facility and it was given one successfully.

Q. *What is switch call completion?*

A. A *switch call completion ratio* expresses the ability of the MSC to successfully deliver a call to not only the radio environment but also to the PSTN. An unsuccessful call would be a case in which the party receiving the call—whether it is the PSTN or the wireless network—does not answer the call or is blocked due to facilities not being available. The switch call completion ratio is therefore different from the RF call completion ratio.

Q. *What is billing call completion?*

A. The *billing call completion ratio* expresses the number of calls that are successfully completed can be billed, in comparison to all the other attempts to complete a call, which also could mean calls that are not billed.

Q. *What is audio quality?*

A. *Audio quality* is the perceived quality of a wireless telephony call. The audio quality is a subjective measurement, and depending on who is referencing the situation, it can either mean that the call quality is good or bad. Audio quality should be referenced to the MOS test scores to assure some level of consistency.

Q. *What is usage?*

A. *Usage* is the amount of traffic handled by a wireless system, the radio environment, or the switching environment. Usage is either expressed in centa calls per second (CCS) or erlangs.

Q. *What is call delivery?*

A. *Call delivery* is the method used for processing a telephony call in a wireless system. Specifically, call delivery includes the ability to route incoming and outgoing calls for the wireless system and the process with which they are treated.

Q. *What is the facilities usage report format?*

A. Figure 12.6 is an example of a facility usage report.

Q. *What is the facilities interconnect report for data?*

A. The *facilities interconnect report for data* is intended to display the current configuration and performance of the network data links for detecting and resolving data problems and to plan for the dimensioning of these facilities as the network traffic grows. These are not the switch-to-cell site links but rather the switch-to-switch and/or switch-to-network node links responsible for call processing and database inquiries.

Q. *What should the facilities interconnect report for data include?*

A. The facilities interconnect report for data should include the following items as a minimum:

Figure 12.6 Facility usage report.

Facilities Usage Report

System name:

Month:

Date:

Trunk no.	**Usge**	**Band**	**% band usage**	**% variation previous month**	**Design %**
NJ004-05	8545	1	25	4	25
		2	35	1	35
		3	30	5	25
		4	10	6	15

Basic link configuration data

- Defined network point codes
- Number of defined link sets in the network
- Number of links defined in each link set

Performance data

- Link traffic load (erlangs): System busy hour data only
- Link set traffic load (erlangs): System busy hour data only
- System traffic load (erlangs): System busy hour data only
- Link active and inactivity times: Total peak service hours (0700–2000)
- Link set change over count: Total peak service hours (0700–2000)
- Link retransmission, percent: Total peak service hours (0700–2000)

Q. *How often should the facilities interconnect report be generated?*

A. The facilities interconnect report should be generated on a weekly basis.

Q. *What does the facilities interconnect report look like?*

A. Figure 12.7 is an example of a facilities interconnect report.

Q. *What is a switch and node metrics report?*

A. A *switch and node metrics report* is used to assist in the monitoring and dimensioning of the switches in the network. It should be generated once a week for each switch and node in the system.

Q. *What should the switch and node metrics report include?*

A. The switch and node metrics report should include, as a minimum, the data fields shown in Table 12.4.

Q. *How often should the facilities usage and traffic report be generated?*

A. The report should be issued on a monthly basis.

Q. *What is a facility usage and traffic report?*

A. A *facility usage and traffic report* tracks the various interconnect facility usage levels in the network. The facility usage report should be used to determine the best locations for the point-of-presence locations in the network. The selection of the POP locations, if done properly, will minimize the network infrastructure cost for delivering calls.

The report needs to also be used for verifying that the interconnect bills received for operating the network are valid. The interconnect bills should be reconciled against the facility usage report to ensure that there is nothing out of the ordinary being reported or billed. The reconciliation of the facility usage bill is normally the responsibility of a revenue assurance department.

Q. *What is the format for the switch and node metrics report?*

A. The format for the switch and node metrics report is shown in Fig. 12.8.

Q. *What is a daily exception report?*

A. A *daily exception report* is a report that identifies any changes or problems within a wireless system over the last 24 hours. The

Figure 12.7 Facilities interconnect report.

Facilities Interconnect Report

System name: System *X*

Date:

Network point code assignments:

Node name: 255 - 1 - 1

Node name: 255 - 1 - 2

Network link definitions:

Link set: 255 - 1 - 1	Links: SLC - 1
	SLC - 2
Link set: 255 - 1 - 2	Links: SLC - 1

Link traffic data:

Link set: 255 - 1 - 1	0.38 erlang	SLC - 1	0.20 erlang
		SLC - 2	0.18 erlang
Link set: 255 - 1 - 2	0.15 erlang	SLC - 1	0.15 erlang

System data link traffic load: 0.43 erlang

Link service data:

Link set: 225 - 1 - 1	Links: SLC - 1	No outages	
	SLC - 2	0900–1000 h, 02:00 min	
	Link change over count:	2	
	Link routing error count:	0	
	Link retransmission:	SLC - 1	0.15%
		SLC - 2	0.01%
Link set: 225 - 1 - 2	Links: SLC - 1	No outages	
	Link change over count:	0	
	Link routing error count:	0	
	Link retransmission:	SLC - 1	0.01%

TABLE 12.4 TYPICAL SWITCH AND NODE METRICS REPORT

Performance data	Recommended sample times
Central processor unit load-utilization value	System busy hour, %
Secondary processor (SP)/ load utilization value	System busy hour, %
Port capacity assigned	Weekly average
Port capacity available	Actual
Subscriber capacity assigned	Weekly average
Subscriber capacity available	Weekly average
Memory capacity assigned	Weekly average
Memory capacity available	Weekly average
Switch and node I/O capacity assigned	Weekly assignments
Switch and node I/O capacity available	Actual
Service circuits load/utilization values (for senders, receivers, tone generators, etc.)	System busy hour, %
Switch and node outages (no. and duration of occurrences)	Daily recordings

daily exception report usually has system performance metrics included in it to not only communicate the health and well-being of the network but also to prevent the generation of a second report that would contain only that information. The daily exception report is one of the most valuable tools available for troubleshooting on a daily basis.

Q. *What is the format of the daily exception report?*

A. The daily exception report structure is shown in Fig. 12.9.

Figure 12.8 Switch and node metrics report.

Switch and Node Metrics Report

System name:	System *X*
Date:	
Report week:	
Node:	Switch 1
CPU load/processor occupancy:	47%
Secondary processor 1 load:	29%
Secondary processor 2 load:	20%
Secondary processor 3 load:	23%
Node port capacity:	985/1200 matrix ports
Node subscriber capacity:	45,000/70,000 subscriber records
Node memory capacity:	10 M/35 M
Node I/O capacity:	12/21 I/O ports
Service circuits loading:	
Sender circuits:	30%
Receiver circuits (MF/DTMF):	43%
Conference circuits:	23%
Node outage data:	No outage for this report period
Start time of outage:	
End time of outage:	
Duration of outage:	
Reason for outage:	

Figure 12.9 Daily exception report.

Daily Exception Report

Date: 10/15/98

Previous day's busy hour data:

Total calls completed:	100,000
Usage:	75,000
Lost calls:	1.9%
BER/FER:	1.25%
Attempt failures:	1.85%
RF blocking:	1.5%

Cell site outage:	1
T-1 outage:	2
Radios OOS:	50

Planned outages (description):

Unplanned outages (description):

Q. *What is a customer service, or customer care, report?*

A. A *customer service report,* or *customer care report,* is one of the key metrics for receiving information about the quality of the network. The customer care report identifies the type and frequency of various technical trouble tickets. Included with the type of technical trouble tickets should also be a reference to the volume as well as the number of trouble tickets closures, pendings, and those still open.

Q. *How often should the customer care report be generated?*

A. The customer care report should be generated biweekly; however, if the volume of trouble tickets that arrive to the technical community is excessive, a weekly report might be better.

Q. *What does a customer care report look like?*

A. Figure 12.10 is an example of a customer care report.

Q. *What is a project status report?*

A. A *project status report* is used to track all the major projects currently under way or proposed by engineering for the coming year. The report should match the organization and company goals and objectives. This report needs to be generated and issued on a biweekly basis to the managers in engineering, operations, and implementation. The report should have a rolling one-year projection.

Q. *What does a project status report look like?*

A. Figure 12.11 is an example of a project status report.

Q. *What material should be included in a project plan?*

A. The material that should be included in a project plan is listed below. It is important to note that every project that is undertaken should have a project plan associated with it.

Project Plan Report Format

1.0 Executive summary (one page)

- 1.1 Objective
- 1.2 Expected time
- 1.3 Manpower and infrastructure requirements
- 1.4 Projected cost
- 1.5 Positive network impact
- 1.6 Negative network impact

2.0 Project description

- 2.1 Project leader
- 2.2 Project team
- 2.3 Project milestones
- 2.4 Related documents

Figure 12.10 Customer care report.

Customer Care Report

System name:

Date:

	Region 1	Region 2	Region 3	System
Network complaints	11	12	17	40
No. lost calls	3	2	5	10
No. interference complaints	5	9	5	19
No. did not get onto system	3	1	7	11
No. trouble reports issued	4	6	5	15
No. trouble reports closed	3	6	5	14
No. outstanding trouble reports	1	0	0	1

3.0 Design criteria
- 3.1 Basic design criteria
- 3.2 Project review dates
- 3.3 Method of procedure
- 3.4 Hardware changes
- 3.5 Software changes
- 3.6 Pass-fail criteria
- 3.7 Preimplementation test plan
- 3.8 Postimplementation test plan

4.0 Resources
- 4.1 Personnel projections
  - ■ Weekly
  - ■ Monthly
- 4.2 Infrastrucutre requirements

Figure 12.11 Project status report.

Project Status Report

System name:

Date:

Dept	Project name	Priority	Originator	Lead dept/person	Due date	Project plan	Capital funding	Comments
RF Eng	Alpha	1	Marketing	RF/Smith	8/15/95	7/16/95	7/17/95	On target
Netwrk	Tree	2	Engineering	Netwk/Gervelis	8/15/95	7/20/95	7/24/95	On target

4.3 Interdepartment resources required

4.4 External department impact

- Marketing
- Customer care
- MIS
- Operations
- Implementation
- Vendors

5.0 Budget

5.1 Total project cost

5.2 Capital budget impact

5.3 Expense budget impact

5.4 Comparison of project costs to budget, planned, and actual

Q. *What is a network configuration report?*

A. A *network configuration report* is a collection of diagrams showing the network configuration as it exists. The report consists of the following material that is made up of a combination of diagrams and tables containing all the pertinent information.

Voice interconnect diagrams

- Mobile unit system to PSTN LEC central office and IXC tandems
- Internal system voice facilities for call delivery and handoffs

Data network diagram

- SS7 data links to external networks
- SS7 data links for internal system

System switch-to-cell site assignments

Auxiliary system interconnection diagrams

- Voice mail
- Validation systems

Q. *What does a network configuration report look like?*

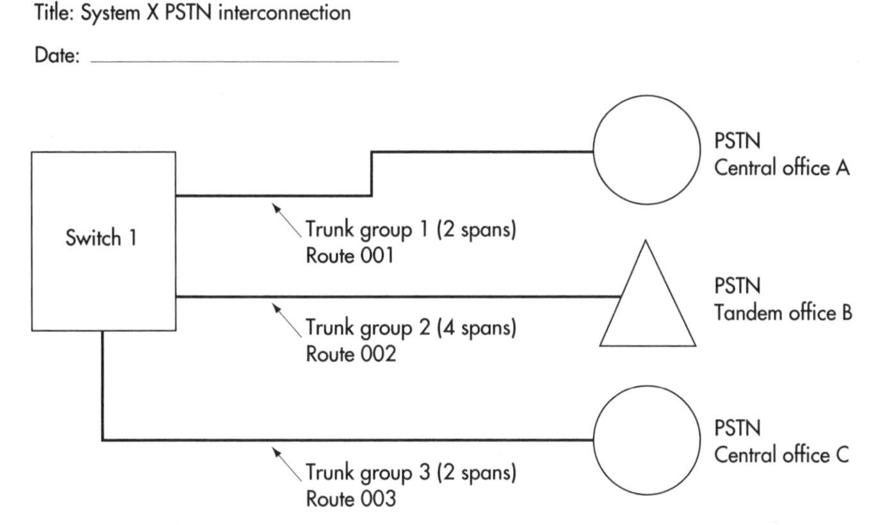

Figure 12.12 Network configuration report.

A. A network configuration report is prepared as shown in Fig. 12.12.

Q. *How often should a network configuration report be generated?*

A. A network configuration report should be generated after every major network change and as a minimum once every three months.

Q. *What is an exception report, or morning report?*

A. An *exception report,* or *morning report,* identifies either a network element, node, cell site, or performance metric that is not operating within acceptable operating parameters. An exception report could be generated for all cell sites that have no usage on them or have lost calls greater than a certain percentage. An exception report can also log only the worst-performing cells, sectors, or nodes in a wireless system.

Q. *What is a system software report?*

A. A *system software report* identifies the current software configuration for the network. The report should be only one page in length and issued on a biweekly or monthly basis.

Q. *What does a system software report look like?*

A. A system software report is shown in Fig. 12.13.

Figure 12.13 System software report.

System Software Report

System name

Date:

	Current	Tested	Next load and expected date
Switch CPU:			
Switch matrix:			
Switch database:			
Voice mail:			
DACS:			
Cell site:			

Q. *What is the key critical parameter for setting an access threshold?*

A. The key critical parameter used in a wireless system for setting an access threshold value should be revenue. Unless there is a compelling case, clearly identified, then the value established for the access threshold should be no higher than 5 dB above the noise floor of the individual cell.

Q. *What is an upper-management report?*

A. The *upper-management report* is a series of quick and concise briefings pertaining to the configuration and performance of the wireless system. The exact information content that is included in an upper-management report will vary depending on whether the upper management is in technology or marketing departments.

Q. *What does an upper-management report look like?*

A. The format of an upper-management report will vary considerably. However, Fig. 12.14 can be used as an upper-management report for a technical group.

Figure 12.14 Upper-management report.

Management Report

System name: ______________________

Date: ____________________

	Present	Prior month	Variance
System measurements			
Total Mobiles			
Calls Handled			
Total System Erlangs			
Avg BBH Erlangs			
Erlang/Sub			
RF			
% Lost Calls			
% RF Blocks			
% Attempt Failures			
% Radios OOS			
Network			
CPU Load/Utilization Rates			
Subscriber Capacity			
Switch/Node I/I Capacity			
Memory Capacity			
Configuration			
Switches			
Cell Sites			
WPBX			
RF Radios			
Microwave T-1s			
PSTN Trunks			
Total Trunks			
DACs			
POPs			

Q. *What are some of the issues that need to be addressed in designing a network for a wireless system?*

A. The issues that need to be addressed in designing a network for a wireless system are listed below. The time durations that accompany each of the steps are not included because they will change with the size of the system as well as with the time-to-market requirements. The plan below can be used for both a new system and an existing system that is trying to expand by introducing a new MSC to the network.

Wireless Fixed Network Design

Network design criteria.

- Network design objectives.
- Network system capacity.
- Network criteria design review.

MSC design.

Specify network equipment for MSC.

Specify network equipment.

Network equipment design review.

MSC power requirements.

- Review MSC power budget.
- Specify power distribution frames.
- Specify inverter power.
- Review power requirements per bellcore.
- Power budget report.
- Power budget design review.
- Order power equipment.
- Order battery equipment.
- Power and battery integration test plan.
- Power integration design review.

MSC network equipment placement.

- Specify network equipment layout.

- Network equipment layout report.
- Network equipment layout design review.

Test cell located at MSC.

- Equipment layout.
- Cell site parameters.
- Switch translations.
- Design review.
- Order RF and network equipment.
- Install cell site.
- Integrate cell site.

Toll room design.

- Define toll room requirements.
- Specify toll room equipment layout.
- Toll room equipment layout report.
- Toll equipment layout design review.
- Order toll room equipment

NOC design.

- Define NOC criteria.
- Review NOC design against bellcore standards.
- NOC design review report.
- Room layout.
- Interconnect design.
- Alarm display and/or reporting platform.
- Terminals.
- Dark MSC monitoring.
- Report generation.
- NOC design review.
- NOC equipment ordering.
- NOC implementation plan.
- NOC integration plan.
- NOC integration.

Acceptance test procedure.

- Develop ATP plan for switches.

- Develop ATP plan for auxiliary equipment.
- Develop ATP plan for power plant.
- Develop ATP for HLR.
- Develop ATP for network echo cancellers (if required).
- Develop ATP for STP/SP.
- Develop ATP for voice mail.
- Develop ATP for billing system.
- Develop ATP for protocol conversion (if required).
- Develop ATP for IS-41.
- Develop ATP for auxiliary systems.

Develop data network plan.

- Data network design criteria.
- Data network design.
- Data network design report.
- Data network design review.
- Test data network.

Define network software.

- System software criteria.
- Software design.
- Software design report.
- Design review.
- Switch database.
- Design criteria.
- Switch functionality.
- Service provider requirements.
- Switch provider requirements.

Billing system input.

- Customer service subsystem.
- Design criteria report.
- Design and implementation.
- Develop initial generic tape.
- Generate tape.

- Test generic system.
- Design parameter tape.
- Generate tape.
- Test major features.
- Design user translations.
- Generate tape.
- Test feature administration.
- Test routing.
- Test trunk provisioning.
- Configuration report.

HLR database.

- Design criteria.
- Design and implementation.
- Configuration report.
- Design review.

SCP database.

- Design criteria.
- Design and implementation.
- Configuration report.
- Design review.

STP database.

- Design criteria.
- Design and implementation.
- Configuration report.
- Design review.

Voice mail database.

- Design criteria.
- Design and implementation.
- Configuration report.
- Design review.

Short-messaging node database.

- Design criteria.

- Design and implementation.
- Configuration report.
- Design review.

Authentication database.

- Design criteria.
- Design and implementation.
- Configuration report.
- Design review.

BSC database.

- Design criteria.
- Design and implementation.
- Configuration report.
- Design review.

Cell site database.

- Design criteria.
- Design and implementation.
- Configuration report.
- Design review.

Define system dialing plan.

- Design criteria.
- Design.
- Design report.
- Design review.

Define system routing plan.

- Design criteria.
- Determine IXC.
- RFI issued.
- RFI received and reviewed.
- IXC(s) selected.
- Secure NXXs.
- Routing plan design.
- Design report.
- Design review.

Interconnect plan.

- Design criteria.
- Carrier meetings.
- Interconnect provider decision.
- Design report.
- Hub decision (if applicable).
- Design review.
- Generate report.
- Order facilities.
- Acceptance test plan.
- Integration and testing.

Telephone code management.

- Design criteria.
- Design report.
- Design review.
- Order numbers.
- Track mobile codes.
- Acceptance test plan.
- Integration and testing.

STP/SP design.

- Design criteria.
- Design.
- Design report.
- Design review.
- STP provider agreement.
- Integration plan.
- STP integration.

Call delivery testing.

- Design criteria.
- Design.
- Design report.
- Design review.
- Test call delivery scenarios.

- Billing system validation.
- Posttest report.

Implement billing system.

- Design criteria.
- Design.
- Design report.
- Design review.
- Billing system validation.
- Billing system integration.
- Alternative interconnect.
- Design criteria.
- Design.
- Design review.

Implementation plan.

- FCC/PUC filings.
- Integration.
- System integration.
    1. Design criteria.
    2. RF integration plan.
    3. Network integration plan.
    4. Software integration plan.
    5. Integration design review.
    6. System integration.

Postcut integration testing.

Q. *What are some of the issues to be addressed in designing an in-building or tunnel system for a wireless system?*

A. The issues that need to be addressed in an in-building or tunnel system design for a wireless system are listed below. The time durations that accompany each of the steps are not included because they depend directly upon the size of the system as well as the time-to-market requirements.

Project Kickoff Meeting

- Antenna mount installation
- Equipment installation
- UPS system installation
- Equipment rack installation
- Equipment installation
- Microcell installation
- Cable installation to microcell
- Radiax cable and/or distributed antenna system installation
- Antenna installation

RF Engineering Project

- Design kickoff meeting
- Establishment of responsibility centers
- Performance criteria
- Review system designs
- Microcell system review
- Macrocell system review
- Coverage requirements
- Handover
- Cable design
- Performance criteria
- Tunnel operator system design review
- System requirements
    1. Design criteria defined
    2. Link budget
    3. Intermodulation
    4. Noise levels
    5. Filter requirements
    6. Microcell system review
    7. Radiax system review

8. RF plumbing design
9. Path analysis
10. Antenna system
11. Antenna selection
12. Lightning protection
13. Tx combining method
14. Frequency plan
15. Translations
16. Parameter adjustments
17. Design review

Establishment of Equipment List and Ordering Acceptance Test Plans

- Criteria establishment
- Test plan generated
- ATP issued
- Tests run
  1. Facilities
  2. Fiber testing
  3. Power acceptance
  4. Drive testing
  5. RF engineering
  6. Operations
  7. Testing support

ATP Signoff

Q. *What are some of the issues that need to be addressed in designing an RF network for a wireless system?*

A. The design issues for an RF network for a wireless system are listed below. The time durations that accompany each of the steps are not included because they depend directly upon the size of the system as well as the time-to-market requirements. The plan below can be used for both a new and an existing system that is being expanded by introducing a new technology platform to the network.

Overall (phase A) system design

Kickoff meeting

System design guidelines

RF design requirements

RF design criteria

- Antenna design criteria
- Cell site design criteria
- Link budget

Preliminary RF design

Preliminary RF design review

Final RF design

Final cell plan

- Antenna configuration
- Cell site count and configuration
- Cell site configuration design
- Cell site count defined by region

Design review

Frequency plan

Traffic analysis completed

Design review

RF design review

Search areas released

OFS relocation plan (if applicable)

Bibliography

Chapter 1

Smith, Clint, *Practical Cellular and PCS Design,* McGraw-Hill, New York, 1998.

Smith, Clint, and Curt Gervelis, *Cellular System Design and Optimization,* McGraw-Hill, New York, 1996.

Chapter 2

The ARRL 1986 Handbook, 63d ed, American Radio Relay League, Newington, Conn., 1986.

Code of Federal Regulations, CFR 47 Parts 1, 17, 22, 24, and 90.

DeRose, James F., *The Wireless Data Handbook,* Quantum Publishing, Inc., Mendocino, Calif., 1994.

Dixon, Robert C., *Spread Spectrum Systems,* 2d ed, Wiley, New York, 1984.

Engineering and Operations in the Bell System, 2d ed, AT&T Bell Laboratories, Murray Hill, N.J., 1983.

Jakes, W. C., *Microwave Mobile Communications,* IEEE Press, New York, 1974.

Lee, W. C. Y., *Mobile Cellular Telecommunications Systems,* 2d ed, McGraw-Hill, New York, 1996.

MacDonald, "The Cellular Concept," *Bell Systems Technical Journal,* vol. 58, no. 1, 1979.

Mouly, M., and M-B Pautet, *The GSM System for Mobile Communications,* self-published, 1992.

Rappaport, Theodore S., *Wireless Communications,* Prentice Hall, 1996.

Smith, Clint, *Practical Cellular and PCS Design,* McGraw-Hill, New York, 1998.

Smith, Clint, and Curt Gervelis, *Cellular System Design and Optimization,* McGraw-Hill, New York, 1996.

CHAPTER 3

Lee, W. C. Y., *Mobile Cellular Telecommunications Systems,* 2d ed, McGraw-Hill, New York, 1996.

MacDonald, "The Cellular Concept," *Bell Systems Technical Journal,* vol. 58, no. 1, 1979.

Miller, Nathan, *Desktop Encyclopedia of Telecommunications,* McGraw-Hill, New York, 1998.

Newton, Harry, *Newton's Telcom Dictionary,* 14th ed, Flatiron Publishing, 1998.

Rappaport, Theodore S., *Wireless Communications,* Prentice Hall, 1996.

Smith, Clint, *Practical Cellular and PCS Design,* McGraw-Hill, New York, 1998.

Smith, Clint, and Curt Gervelis, *Cellular System Design and Optimization,* McGraw-Hill, New York, 1996.

Winch, Robert, *Telecommunication Transmission Systems,* 2d ed, McGraw-Hill, New York, 1998.

CHAPTER 4

The ARRL 1986 Handbook, 63d ed, American Radio Relay League, Newington, Conn., 1986.

Carlson, A. B., *Communications Systems,* 2d ed, McGraw-Hill, New York, 1975.

Dixon, Robert C., *Spread Spectrum Systems,* 2d ed, Wiley, New York, 1984.

Fink, Donald, and Donald Christiansen, *Electronics Engineers Handbook,* 3d ed, McGraw-Hill, New York, 1989.

Jakes, W. C., *Microwave Mobile Communications,* IEEE Press, New York, 1974.

Johnson, R. C., and H. Jasik, *Antenna Engineering Handbook,* 2d ed, McGraw-Hill, New York, 1984.

Kaufman, M., and A. H. Seidman, *Handbook of Electronics Calculations,* 2d ed, McGraw-Hill, New York, 1988.

Lee, W. C. Y., *Mobile Cellular Telecommunications Systems,* 2d ed, McGraw-Hill, New York, 1996.
Rappaport, Theodore S., *Wireless Communications,* Prentice Hall, 1996.
Reference Data for Engineers, 7th ed, Sams, 1985.
Sklonik, M. I., *Introduction to Radar Systems,* 2d ed, McGraw-Hill, New York, 1980.
Smith, Clint, *Practical Cellular and PCS Design,* McGraw-Hill, New York, 1998.
Smith, Clint, and Curt Gervelis, *Cellular System Design and Optimization,* McGraw-Hill, New York, 1996.
Steele, Raymond, *Mobile Radio Communications,* IEEE, New York, 1992.
Winch, Robert, *Telecommunication Transmission Systems,* 2d ed, McGraw-Hill, New York, 1998.

Chapter 6

Code of Federal Regulations, CFR 47 Parts 1, 17, 22, 24, and 90.
Smith, Clint, *Practical Cellular and PCS Design,* McGraw-Hill, New York, 1998.
Smith, Clint, and Curt Gervelis, *Cellular System Design and Optimization,* McGraw-Hill, New York, 1996.

Chapter 7

The ARRL 1986 Handbook, 63d ed, American Radio Relay League, Newington, Conn., 1986.
Carlson, A. B., *Communications Systems,* 2d ed, McGraw-Hill, New York, 1975.
Carr, J. J., *Practical Antenna Handbook,* TAB Books–McGraw-Hill, Blue Ridge Summit, Pa., 1989.
DeRose, James F., *The Wireless Data Handbook,* Quantum Publishing, Mendocino, Calif., 1994.
Dixon, Robert C., *Spread Spectrum Systems,* 2d ed, Wiley, New York, 1984.
Engineering and Operations in the Bell System, 2d ed, AT&T Bell Laboratories, Murray Hill, N. J., 1983.
Fink, Donald, and Donald Christiansen, *Electronics Engineers Handbook,* 3d ed, McGraw-Hill, New York, 1989.
Jakes, W. C., *Microwave Mobile Communications,* IEEE Press, New York, 1974.
Johnson, R. C., and H. Jasik, *Antenna Engineering Handbook,* 2d ed, McGraw-Hill, New York, 1984.

Kaufman, M., and A. H. Seidman, *Handbook of Electronics Calculations,* 2d ed, McGraw-Hill, New York, 1988.
Lathi, B. P., *Modern Digital and Analog Communication Systems,* CBS College Publishing, New York, 1983.
Lee, W. C. Y., *Mobile Cellular Telecommunications Systems,* 2d ed, McGraw-Hill, New York, 1996.
Rappaport, Theodore S., *Wireless Communications,* Prentice Hall, 1996.
Reference Data for Engineers, 7th ed, Sams, 1985.
Schwartz, M., W. R. Bennett, and S. Stein, *Communication Systems and Techniques,* IEEE, New York, 1996.
Sklonik, M. I., *Introduction to Radar Systems,* 2d ed, McGraw-Hill, New York, 1980.
Smith, Clint, *Practical Cellular and PCS Design,* McGraw-Hill, New York, 1998.
Smith, Clint, and Curt Gervelis, *Cellular System Design and Optimization,* McGraw-Hill, New York, 1996.
Steele, Raymond, *Mobile Radio Communications,* IEEE, New York, 1992.
Webb, W. T., and L. Hanzo, *Modern Amplitude Modulations,* IEEE, New York, 1994.
Winch, Robert, *Telecommunication Transmission Systems,* 2d ed, McGraw-Hill, New York, 1998.

Chapter 8

The ARRL 1986 Handbook, 63d ed, American Radio Relay League, Newington, Conn., 1986.
Carlson, A. B., *Communications Systems,* 2d ed, McGraw-Hill, New York, 1975.
Jakes, W. C., *Microwave Mobile Communications,* IEEE Press, New York, 1974.
Kaufman, M., and A. H. Seidman, *Handbook of Electronics Calculations,* 2d ed, McGraw-Hill, New York, 1988.
Lee, W. C. Y., *Mobile Cellular Telecommunications Systems,* 2d ed, McGraw-Hill, New York, 1996.
Reference Data for Engineers, 7th ed, Sams, 1985.
Sklonik, M. I., *Introduction to Radar Systems,* 2d ed, McGraw-Hill, New York, 1980.
Smith, Clint, *Practical Cellular and PCS Design,* McGraw-Hill, New York, 1998.
Smith, Clint, and Curt Gervelis, *Cellular System Design and Optimization,* McGraw-Hill, New York, 1996.

Chapter 9

The ARRL 1986 Handbook, 63d ed, American Radio Relay League, Newington, Conn., 1986.

Carr, J. J., *Practical Antenna Handbook,* TAB Books–McGraw-Hill, Blue Ridge Summit, Pa., 1989.

Jakes, W. C., *Microwave Mobile Communications,* IEEE Press, New York, 1974.

Johnson, R. C., and H. Jasik, *Antenna Engineering Handbook,* 2d ed, McGraw-Hill, New York, 1984.

Kaufman, M., and A. H. Seidman, *Handbook of Electronics Calculations,* 2d ed, McGraw-Hill, New York, 1988.

Lee, W. C. Y., *Mobile Cellular Telecommunications Systems,* 2d ed, McGraw-Hill, New York, 1996.

Reference Data for Engineers, 7th ed, Sams, 1985.

Smith, Clint, *Practical Cellular and PCS Design,* McGraw-Hill, New York, 1998.

Smith, Clint, and Curt Gervelis, *Cellular System Design and Optimization,* McGraw-Hill, New York, 1996.

Chapter 10

Engineering and Operations in the Bell System, 2d ed, AT&T Bell Laboratories, Murray Hill, N.J., 1983.

Lee, W. C. Y., *Mobile Cellular Telecommunications Systems,* 2d ed, McGraw-Hill, New York, 1996.

MacDonald, "The Cellular Concept," *Bell Systems Technical Journal,* vol. 58, no. 1, 1979.

Mouly, M., and M-B Pautet, *The GSM System for Mobile Communications,* self-published, 1992.

Smith, Clint, *Practical Cellular and PCS Design,* McGraw-Hill, New York, 1998.

Smith, Clint, and Curt Gervelis, *Cellular System Design and Optimization,* McGraw-Hill, New York, 1996.

Steele, Raymond, *Mobile Radio Communications,* IEEE, New York, 1992.

Chapter 11

Engineering and Operations in the Bell System, 2d ed, AT&T Bell Laboratories, Murray Hill, N.J., 1983.

Lee, W. C. Y., *Mobile Cellular Telecommunications Systems,* 2d ed, McGraw-Hill, New York, 1996.

MacDonald, "The Cellular Concept," *Bell Systems Technical Journal,* vol. 58, no. 1, 1979.

Miller, Nathan, *Desktop Encyclopedia of Telecommunications,* McGraw-Hill, New York, 1998.

Mouly, M., and M-B Pautet, *The GSM System for Mobile Communications,* self-published, 1992.

Newton, Harry, *Newton's Telcom Dictionary,* 14th ed, Flatiron Publishing, 1998.

Smith, Clint, *Practical Cellular and PCS Design,* McGraw-Hill, New York, 1998.

Smith, Clint, and Curt Gervelis, *Cellular System Design and Optimization,* McGraw-Hill, New York, 1996.

Winch, Robert, *Telecommunication Transmission Systems,* 2d ed, McGraw-Hill, New York, 1998.

Chapter 12

Engineering and Operations in the Bell System, 2d ed, AT&T Bell Laboratories, Murray Hill, N.J., 1983.

Lee, W. C. Y., *Mobile Cellular Telecommunications Systems,* 2d ed, McGraw-Hill, New York, 1996.

MacDonald, "The Cellular Concept," *Bell Systems Technical Journal,* vol. 58, no. 1, 1979.

Mouly, M., and M-B Pautet, *The GSM System for Mobile Communications,* self-published, 1992.

Smith, Clint, *Practical Cellular and PCS Design,* McGraw-Hill, New York, 1998.

Smith, Clint, and Curt Gervelis, *Cellular System Design and Optimization,* McGraw-Hill, New York, 1996.

Index

B

C

N

S

U

About the Author

Clint Smith, P.E., is Vice President of Operations for o2Wireless Solutions, and former Vice President of Technical Services for Communication Consulting Services, Inc., and former Director of Engineering at NYNEX Mobile Services. He is also the author of *Practical Cellular and PCS Design* and *LMDS*, and the co-author of *Cellular System Design and Optimization*, all published by McGraw-Hill.